CISM COURSES AND LECTURES

The series presents lecture notes, monographs, edited works and proceedings in the field of Mechanics, Engineering, Computer Science and Applied Mathematics.
Purpose of the series is to make known in the international scientific and technical community results obtained in some of the activities organized by CISM, the International Centre for Mechanical Sciences.

INTERNATIONAL CENTRE FOR MECHANICAL SCIENCES

COURSES AND LECTURES - No. 439

MANIPULATION AND CONTROL OF JETS IN CROSSFLOW

EDITED BY

ANN R. KARAGOZIAN
UNIVERSITY OF CALIFORNIA

LUCA CORTELEZZI
McGILL UNIVERSITY, MONTREAL

ALFREDO SOLDATI
UNIVERSITY OF UDINE

Springer-Verlag Wien GmbH

This volume contains 163 illustrations

Originally published by Springer-Verlag Wien New York in 2003
SPIN 10918869

In order to make this volume available as economically and as rapidly as possible the authors' typescripts have been reproduced in their original forms. This method unfortunately has its typographical limitations but it is hoped that they in no way distract the reader.

ISBN 978-3-211-00753-2 ISBN 978-3-7091-2792-6 (eBook)
DOI 10.1007/978-3-7091-2792-6

PREFACE

The Centre International des Sciences Mecaniques (CISM) in Udine, Italy brought together in June, 2001 a group of experts with specialized interests in the modeling, experimental interrogation, manipulation, and control of jets, with special focus on the jet in crossflow or transverse jet. The result was a week-long Advanced School on the Manipulation and Control of Transverse Jets. There was an extensive series of lectures by five academic specialists on subjects relevant to jets and their control (Professors Luca Cortelezzi, Ari Glezer, Ann Karagozian, Petros Koumoutsakos, and M. Godfrey Mungal). as well as a series of workshop presentations by a wide variety of academic as well as industry speakers from around the globe.

The outstanding lectures and discussions which took place in this Advanced School are captured in the series of papers contained within the present text. The text is organized into a series of chapters providing the state-of-the-art in terms of: 1) a fundamental background on and characteristics of transverse jet and related jet systems, 2) a fundamental background on modeling and dynamical systems control, with special focus on jet flows, 3) descriptions of multiphase and/or reactive jet in crossflow studies, and 4) descriptions of the actively controlled jets in crossflow and related jet systems. It is anticipated that this text will be a useful guide for researchers in fluid mechanics and/or in dynamical systems control. Recent technological advances in both fields have made the control of complex fluid flows a more practical reality, one that is ripe for further exploitation.

The co-organizers wish to thank CISM for its support of this Advanced School, with special thanks to Ms. Paola Agnola for her assistance in the organization of the meeting and of this text. Professor Carlo Tasso also provided outstanding assistance during the Advanced School and in the production of this text. Finally, the co-organizers wish to thank the retired Secretary General of CISM, Professor Giovanni Bianchi, for his immeasurable support, encouragement, and strong role in making this Advanced School and text possible.

Ann R. Karagozian, Luca Cortelezzi, and Alfredo Soldati

CONTENTS

Part I

Fundamental Non-Reactive Jets in Crossflow and Other Jet Systems

Background on and Applications of Jets in Crossflow

Ann R. Karagozian[1] *

Department of Mechanical and Aerospace Engineering, University of California, Los Angeles

Abstract. The jet in crossflow or transverse jet has been studied extensively because of its relevance to a wide variety of flows in technological systems, including fuel or dilution air injection in gas turbine engines, thrust vector control for high speed airbreathing and rocket vehicles, and exhaust plumes from power plants. These widespread applications have led over the past 50+ years to experimental, theoretical, and numerical examinations of this fundamental flowfield. This summary provides a background on these studies and applications, setting the stage for the specific topics described later in this text which are relevant to transverse jet modeling, manipulation, and control.

1 Canonical Jet Flows

Jets are used in widespread engineering systems and as such constitute a canonical set of flowfields. The most common goal of jet flows is to promote mixing of the injectant fluid with the fluid into which the jet is introduced. The impulse created by a jet additionally allows introduction of a localized force into the flowfield, hence thrust or force vector control may also be achieved through the introduction of jets. While the focus of the present text lies, for the most part, on the jet in crossflow, characteristics of other jet flows are relevant to the understanding of transverse jets, and hence their general features are provided here.

The "free" round jet or two-dimensional slot issuing into quiescent surroundings constitutes a fundamental flowfield occuring in a variety of technological contexts. Classical studies of the laminar 2D jet or slot allow extraction of a similarity solution (Schlichting (1933)), in which the jet's velocity distribution takes a $sech^2$ form, and in which the rate of mass flow across any perpendicular jet slice scales as $x^{1/3}$, where x is the downstream distance. As a consequence, the jet width increases as $x^{2/3}$. Similarity solutions are also possible for the laminar round jet (Schlichting (1933)). Even relatively low speed free jets exhibit clear instability mechanisms from laminar to turbulent flow via vortex rollup (Huerre & Monkewitz (1990)). Early analyses of the turbulent slot (Gortler (1942)) attempted scaling approaches similar to that of the laminar slot, yet with the reasoning that the width of the jet increases linearly with x, so that the $sech^2$ form for the 2D jet's mean velocity profile is similarly recovered.

Turbulent free jets historically have been characterized as being composed of a number of distinct regions: the potential core, a mixing region adjacent to the potential core, a transitional region, and fully developed turbulent flow, as indicated in Figure 1. The mixing region is one in which the transfer of mass and momentum between the surroundings and the jet is accelerated. This nearfield mass and momentum exchange process, often characterized here (and in the

* Work by this author on transverse jets has been supported over the years by NASA Dryden and Glenn Research Centers, by the CEC Energy Innovations Small Grants program, and by the University of California Energy Institute.

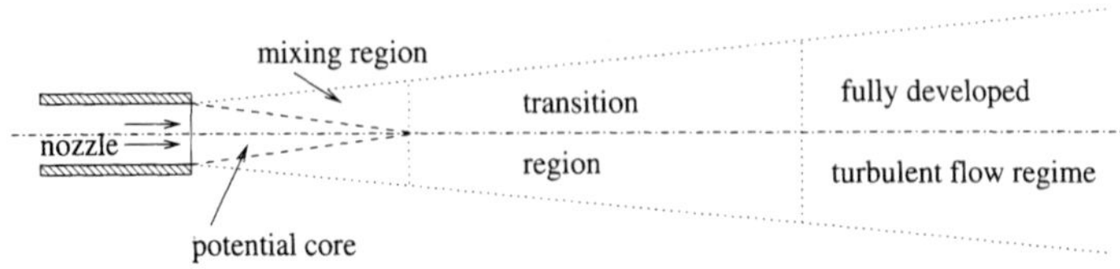

Figure 1. Regions of the free jet, as described in Beer and Chigier (1972).

farfield) by the "entrainment" of surrounding fluid into the jet, occurs within about five jet diameters downstream of the jet exit plane. The transition region between the potential core and fully developed turbulent jet flow proceeds roughly ten jet diameters beyond the core. Ricou & Spalding (1961) quantified jet entrainment by examining essentially free (yet broadly confined) turbulent jets in porous chambers. Among their findings are that the jet mass flux increases linearly downstream, while the mass entrainment, that is, the ratio of local mass flux to that at the jet exit, becomes independent of jet Reynolds number for large Reynolds number, where Reynolds number Re is based on the jet orifice diameter.

These early ideas concerning jet entrainment, spread, and characteristic regimes have been extended by a number of researchers to similar jet systems, that is, to jets in non-quiescent surroundings such as coflow and crossflow, as will be discussed below and in subsequent chapters of this text. Modern experimental diagnostic techniques, including particle image velocimetry (PIV) as well as planar laser-induced fluorescence (PLIF) imaging, have been used to quantify such jet parameters as entrainment, in addition to scalar field mixedness, turbulent kinetic energy, scalar dissipation rates, and Reynolds' stresses (Liepmann (1991), Kailasanath, et al. (1993)). Modern computational methods have been used extensively to simulate jet flows, including direct numerical simulations (DNS) of relatively low Re jets, enabling resolution of length scales for the flowfield down to the Kolmogorov scale (Brancher, et al. (1994), Freund, et al. (2000)). Similar approaches, in particular, large eddy simulations (LES), as well have begun to be applied to the flowfield associated with the jet in crossflow (Yuan, et al. (1999), Salvetti (2003)).

2 The Jet in Crossflow (JICF)

Similar to the fundamental flowfields above, the jet in crossflow is a canonical flowfield used in a variety of technological applications for its ability to mix rapidly with crossflow and to introduce a controlled jet force into flowfield. A schematic diagram of the generic JICF is shown in Figure 2, and a photo of the typical "instantaneous" structure of the JICF, obtained via smoke visualization, is shown in Figure 3. While the jet may issue at a steady rate into the crossflow, the unsteady nature of the evolution of flow structures is readily evident. The periodic nearfield rollup of vortical structures, indicated in this smoke visualization with a short time exposure, is thought to contribute to the formation of vortical structures which dominate the ensemble-averaged flowfield, principally the counter-rotating vortex pair (CVP) structure observed to dominate the transverse jet's cross-section in the mean, as shown in Figure 2. Extensive references on the exploration of transverse jet flows and their applications may be found in Margason (1993).

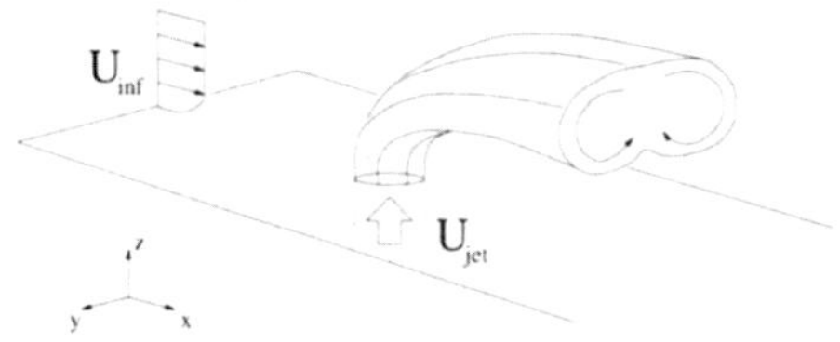

Figure 2. Global flowfield associated with a jet in crossflow, in which a counter-rotating vortex pair (CVP) is formed.

Figure 3. Smoke visualization of the jet in crossflow at an average jet-to-crossflow velocity ratio of R $\equiv \bar{\mathrm{U}}_{\mathrm{j}}/\mathrm{U}_\infty = 2.58$, where $\bar{\mathrm{U}}_{\mathrm{j}} = 3.1$ m/s, $\mathrm{U}_\infty = 1.2$ m/s, and the jet Reynolds number $Re \equiv \bar{U}_j d/\nu$ is of the order 1500. The exposure time for the photograph is 1/2000 sec, yielding an "instantaneous" image (from M'Closkey, et al. (2002)).

2.1 Applications of Jets in Crossflow

Applications of the JICF are widespread in engineering systems. Jets in crossflow arise commonly in air-breathing engine systems (dilution air jets, fuel/air mixers, film cooling systems, scramjet fuel injectors, V/STOL aircraft) as well as in rocket engine systems (thrust vector control) and effluent stacks and plumes. The superior mixing properties associated with the transverse jet as compared, say, with the jet in quiescent surroundings (Steward (1970)), make the flowfield especially attractive for implementation when rapid injectant mixing, with or without the presence of a chemical reaction, is desired.

Dilution air injection in air-breathing gas turbine engines has motivated extensive fundamental as well as applied studies on transverse jets over the past half-century. Dilution jets are principally introduced downstream of the primary and/or secondary combustion zones within a gas turbine engine as a means of cooling high temperature combustion products before they enter the turbine section of the engine (see Figure 4). Arrays of transversely injected jets are typically used; the behavior of such arrays is discussed in a review article by Holdeman (1993). For gas turbine engine applications, peak temperatures entering the turbine section, $T_{4,peak}$, as compared with the mean turbine entrance temperature $T_{4,mean}$ and the combustor entrance temperature T_3, are scaled through a parameter known as the "temperature pattern factor" PF:

$$PF \equiv \frac{T_{4,peak} - T_3}{T_{4,mean} - T_3} \tag{1}$$

Reductions in PF, often possible only through enhanced and/or controlled transverse dilution air mixing, allow lower peak temperatures to be present, yet under these circumstances, often higher average turbine inlet temperatures can be tolerated. Thus the overall performance of the engine in terms of thrust and specific impulse can be enhanced (Hill and Petersen (1992), Kerrebrock (1977)) through controlled dilution jet injection.

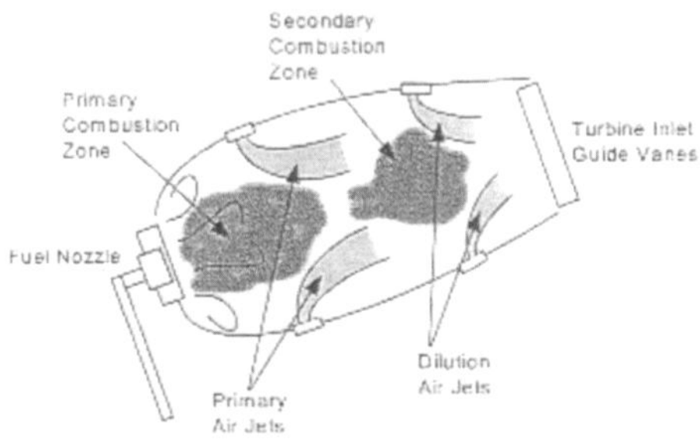

Figure 4. Schematic diagram of primary air and dilution air jets injected transversely into the combustor section of a gas turbine engine. Drawing courtesy of Dr. Jeff Cohen, United Technologies Research Center.

Transversely injected air jets are also used in the primary zone of gas turbine combustors as a means of controlling the air-fuel mixture ratio and hence the emissions of nitrogen oxides (NO_x), also indicated in Figure 4. Tunable air-fuel mixing can allow simultaneous control of NO_x and CO emissions. This can constitute quite a benefit in terms of emissions, since NO_x emissions are often minimized in a combustor design at conditions for which the conversion of CO to CO_2 is inhibited, thereby increasing CO emissions (Bowman (1992)). Active transverse air jet mixing control in fact could be used to optimize fuel-air mixing at high power to achieve lower NO_x, but then could be used to de-tune such mixing at low power to achieve low CO emissions. Passive or active film cooling air injection is similarly used within high pressure turbine blades; flows created by such injection are often that of jets in crossflow (Hill and Petersen (1992)).

Jets in crossflow are also found within high speed air breathing engines, such as ramjets or scramjets, as a means of fuel injection into high speed air entering the combustion chamber (Kerrebrock (1977), Northam & Anderson (1986), Abbitt, et al. (1993), Wang, et al. (1995), Karagozian, et al. (1996)). Rapid fuel penetration, mixing with crossflow, and ignition and sustainment of combustion processes are highly desirable, since high air speeds entering the combustion chamber (supersonic in the case of a scramjet) require very rapid completion of the reaction process (Curran (2001)). Transverse fuel jet injection within a scramjet combustion configuration has been proposed to take place behind a rearward-facing step, allowing the flow to slow sufficiently in the immediate upstream region of the transverse fuel jet so that adequate fuel jet penetration, and associated mixing with air, can take place (Abbitt, et al. (1993), Wang, et al. (1995), Karagozian, et al. (1996)). Sustained ignition and burning of the fuel can be assisted by the presence of the rearward-facing step, which can act simultaneously as a flameholder. In the absence of a rearward-facing step, the reactive fuel jet in crossflow nevertheless allows mixing, ignition, and sustained combustion, even in supersonic crossflow (McMillin, et al. (1994)).

Transverse jets, both liquid and gaseous, are used for thrust vector control, principally for rocket engines (Humble, et al. (1995)). Here an array of transverse jets can be injected into the rocket engine's nozzle, whereby the flow in the nozzle is deflected due to the jet impuse, either with or without a reaction. This deflection allows the thrust of the vehicle to be altered in direction and, to a limited extent, in magnitude, through control of the jets.

Finally, jets in crossflow are also found during takeoff, hover, and transition to wing-borne flight in vertical/short take-off and landing (V/STOL) aircraft such as the Harrier "jump jets" originating in the 1960s. Control of the lift and thrust vectors is complicated by losses experienced by the vehicle during such transition, in addition to ground effects. Temporal control of the velocities and direction of the (transverse) jet exhaust allows for minimization of undesirable losses.

2.2 General Non-Reactive Transverse Jet Characteristics

Operating conditions for the jet in crossflow often are characterized in terms of a variety of parameters. These include the mean jet-to-crossflow momentum flux ratio, $\mathrm{J} \equiv \rho_{\mathrm{jet}}\bar{\mathrm{U}}_{\mathrm{j}}^2/\rho_\infty \mathrm{U}_\infty^2$, and the mean scaled jet-to-crossflow velocity ratio, $\mathrm{R} \equiv \mathrm{J}^{1/2}$, noting that R reduces to the mean jet-to-crossflow velocity ratio, $\bar{U}_j/U_\infty$, for equal density flows ($\rho_{jet} = \rho_\infty$). Additional relevant parameters include the jet Reynolds number $\mathrm{Re_j} \equiv \bar{\mathrm{U}}_\mathrm{j}\mathrm{D}/\nu_{\mathrm{jet}}$ and the crossflow Reynolds number $\mathrm{Re}_\infty \equiv \mathrm{U}_\infty \mathrm{D}/\nu_\infty$, where D is the jet nozzle diameter. These parameters, especially R or J, are used to scale characteristic features of the JICF. Trajectories of the local jet velocity maxima, the local temperature maxima or minima, the local concentration maxima, the centroid of the CVP, etc., are often characterized in terms of R or J (Kamotani & Greber (1972), Kelso, et al. (1996), Smith & Mungal (1998)). For example, characteristic incompressible transverse jet velocity centerline trajectories are often scaled using the formula

$$\frac{z}{RD} = A\left(\frac{x}{RD}\right)^m \tag{2}$$

where x and z are the downstream and transverse spatial coordinates of the jet, defined in Figure 2, and A and m are constants. Typical experimental values of m are of the order 0.28 (Margason (1993)) and A is of the order 2.0 (Pratte & Baines (1967)). Downstream decay of parameters such as centerline velocity and concentration of jet fluid are similarly characterized in terms of R or J. In the case of both the centerline velocity and the centerline concentration, as for the free jet, there is no significant decay in the region of the potential core, but beyond about 5 jet diameters, there appears to be an exponential decay which is dependent on R (Fearn & Weston (1974), Smith & Mungal (1998)). Characteristic length scales for the JICF are also often defined in terms of R, for example, RD or $\mathrm{R}^2\mathrm{D}$. It has long been known that the decay of velocity and jet concentration for the JICF is more rapid than for the free jet, indicating the potential of this flowfield for rapid mixing enhancement. These scaling issues are discussed more fully in Smith & Mungal (1998) and, in this volume, in Mungal & Smith (2003).

2.3 General Reactive Flow Characteristics

The reacting jet in crossflow has been studied in the context of both gaseous (fuel-air) reactions and liquid (acid-base) reactions. Both types of experiments allow examinations of jet/flame trajectories as well as the dependence of flame length on flow conditions. For a liquid reaction, in which a dilute solution of base is injected into a crossflow containing dilute acid, with a pH indicator or fluorescent dye mixed with the base, the "end" of the visualized jet indicates principally where the reaction process ends (Broadwell & Breidenthal (1984)). For a gas phase reaction, the luminosity of sooting fuel jets historically has been used to characterize jet flame shape and length, as has the condition where the fuel has been diluted to its lean flammability limit (Brzustowski (1977)). Recent experimental studies employ optical diagnostics such as OH PLIF (Hasselbrink & Mungal (2001), Mungal & Hasselbrink (2003)) to more accurately resolve, for a diffusion flame, the location of the flame end.

Fundamental studies of gas phase reactive jets in crossflow (e.g., Brzustowski (1977)) suggest that the jet/flame trajectory follows that of the non-reactive, "cold" jet rather closely (Pratte & Baines (1967)). These studies indicate that, as one increases the crossflow velocity from zero, the flame length actually decreases then increases, suggesting critical values of R or J for which mixing and reaction processes are maximized. This critical mixing condition occurs for a variety of fuels, and is also seen in corresponding liquid phase (acid-base) reactive transverse jet experiments (Broadwell & Breidenthal (1984)). Scaling arguments relative to entrainment of crossflow by the vorticity associated with the reactive jet, leading to flame burnout, have been proposed by Broadwell & Breidenthal (1984) and apply well for R values below about 10. These arguments, coupled with a nearfield vorticity-generation model, are used by Karagozian (1986a) and Karagozian & Nguyen (1986) to predict flame lengths for various fuels as a function of R or J. Specific effects of reaction-induced heat release on the transverse jet are discussed in Mungal & Hasselbrink (2003).

2.4 Vortex Systems Associated with the Transverse Jet

Fundamental characteristics of the JICF are dominated by a complex, inter-related set of vortex systems, some of which are shown schematically in Figure 5. These include the counter-rotating vortex pair (CVP), observed to dominate the jet cross-section, especially in the farfield (Kamotani & Greber (1972), Fearn & Weston (1974)) but with evidence of its nearfield initiation (Kelso, et al. (1996), Smith & Mungal (1998), Cortelezzi & Karagozian (2001)). In addition, there are jet shear layer vortices which are formed as jet vorticity rolls up in the nearfield (Fric & Roshko (1994)), horseshoe vortices (Kelso & Smits (1995)), which form in the plane of the injection wall and surround the upstream portion of the jet, and upright wake vortices ((Fric & Roshko (1994)), which are formed within the wall boundary layer and are shed beyond the JICF, allowing fluid to be drawn from the boundary layer into the jet itself. Mixing enhancement by the JICF is often associated with the development and sustenance of the CVP structures.

The CVP and its Origins. A variety of interpretations of the counter-rotating vortex pair and its origins have evolved over the years. These include suggestions that: 1) CVP formation is "similar to (vortices) in the wake of a cylindrical bluff body" (Beer and Chigier (1972)), 2) the "axial vortex pair in the flow arises from the jet momentum normal to the free stream, the momentum flux being equivalent to a normal force, i.e., to a lift" (Broadwell & Breidenthal (1984)), and 3) the "crossflow shears the jet fluid along the lateral edges downstream to form a kidney shaped cross-section...this shearing folds the downstream face over itself to form a vortex pair which

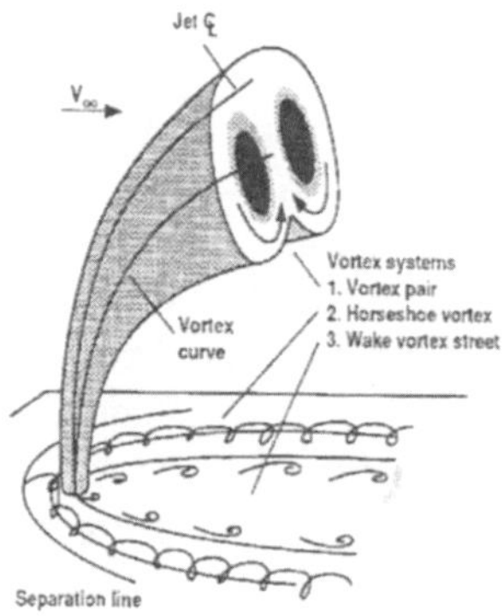

Figure 5. Schematic diagram of the vortex systems associated with the jet in crossflow, from Margason (1993).

dominates the (farfield) flow" (Margason (1993)). In general, however, CVP formation is widely viewed to be formed by the vortex sheet or thin shear layer emanating from the jet nozzle (Kelso, et al. (1996), Fric & Roshko (1994)).

It is generally accepted that "the shear layer of the jet folds and rolls up very near to the pipe exit, leading to or contributing to the formation of the CVP" (Kelso, et al. (1996)), although there are questions pertaining to the nature of vortex rollup near the jet exit (Lim, et al. (2001)). Tilting and folding of the vortical structures are seen to contribute downstream components of vorticity which form, on an average basis, counter-rotating vortical structures. These notions are consistent with the results of numerical simulations of the transverse jet by Cortelezzi & Karagozian (2001); a summary of this work is provided in Cortelezzi & Karagozian (2003). Transient, three-dimensional vortex simulations reveal that periodic vortices rolling up in the nearfield of the jet are deflected by the crossflow in a manner similar to that seen in the experiments of Lim, et al. (1998). Moreover, when the vorticity field is averaged over a single rollup period in the computations, the vorticity isosurfaces of the largest magnitude coincide with the downstream tilted "CVP structures" identified in the experiments of Lim, et al. (1998). More details on the computations and comparisons with experiments are provided in Cortelezzi & Karagozian (2003). The origins and development of the CVP are important because control of vorticity generation and evolution is a means of controlling transverse jet mixing and, potentially, reaction processes.

Horseshoe Vortex Systems. Horseshoe vortex systems result from the interaction between the laminar (wall) boundary layer and the round transverse jet. Horseshoe vortex system are found to be steady, oscillating, or coalescing (Kelso & Smits (1995), Krothapalli, et al. (1990)). Frequencies of oscillation have been found to be correlated with periodic motions of upright wake vortices (Fric & Roshko (1994), Krothapalli, et al. (1990)). Notional views on the development of the interactions of the horseshoe vortex system with periodic motions of upright wake vortices are suggested by Kelso, et al. (1996). Critical node and saddle points present in this topology suggest the development of and linkages among wall-attached vortical structures and horseshoe vortices.

Upright Wake Vortices. Clear visualization of upright wake structures associated with the jet in crossflow is provided via the smoke wire technique in air by Fric & Roshko (1994). These researchers find that vortices in the wake result from convection, turning, stretching, and diffusion of vorticity in the injection wall boundary layer. Boundary layer fluid outside the horseshoe vortices is seen to "wrap around the jet and then separates on its lee side". The frequency of boundary layer separation events can be found to be close to frequency of these upright vortices. Such motions have been captured in high resolution numerical simulations of the transverse jet flowfield (Rudman (1996)). Kelso, et al. (1996) suggest that wake vortices can be a link between wall vortices and the CVP, while Kelso & Smits (1995) suggest that wake motions tend to lock onto the horseshoe vortex frequency or a subharmonic thereof.

Implications of Vortical Systems for Mixing Enhancement. The ability of the jet in crossflow to enhance mixing appears to be intimately connected with the vortex systems present. This is especially true for the CVP, in a quasi-steady mode of operation. Connection among characteristic frequencies of various unsteady vortex systems (shear layer vortices, wake vortices, horseshoe vortices) are yet not well understood nor characterized, but this connection could have a significant effect on the ability to control JICF mixing processes. The capability of the JICF to significantly amplify penetration, mixing, and spread may well depend on the ability to properly actuate and sense important vorticity generating flow mechanisms, hence the importance of such systems.

2.5 UCLA Research on Jets in Crossflow

Our interest in the JICF developed in the mid-1980s, for applications relevant to dilution air injection and mixing in gas turbine engines. Locally 2D vortex modeling of the incompressible JICF, with reaction (Karagozian (1986a)) and without reaction (Karagozian (1986b)) was followed by studies on multiple jets in crossflow (Karagozian, et al. (1986)), and effects of heat release on the reactive jet in crossflow (Karagozian & Nguyen (1986)). Applications of the JICF to scramjet fuel injection schemes led not only to numerical simulations of the liquid jet in compressible crossflow (Heister, et al. (1989), Nguyen & Karagozian (1992)) and gaseous jet in subsonic and supersonic crossflow (Heister & Karagozian (1990a), Heister & Karagozian (1990b)) but also to in-flight iodine PLIF imaging experiments on F-104G aircraft at NASA Dryden Flight Research Center (Wang, et al. (1995), Karagozian, et al. (1996)). More recent work on the JICF has involved transient, three-dimensional numerical simulations (Cortelezzi & Karagozian (2001)) and experimental studies on actively controlled low speed jets in crossflow (M'Closkey, et al. (2002)); some details on these studies are provided in this volume in Cortelezzi & Karagozian (2003) and M'Closkey, et al. (2003).

3 Concluding Remarks

The unique characteristics of the jet in crossflow, especially with respect to its variety of complex vortical structures, make the flowfield one whose interrogation, analysis, and control can have widespread benefits. It is hoped that, in succeeding chapters of this text, specific mechanisms for the experimental and computational interrogation, analysis, and control will be clear and will inspire future researchers to study this important canonical flowfield via contemporary diagnostic methods.

References

Abbitt, J. D. III, Segal, C., McDaniel, J. C., Krauss, R. H., and Whitehurst, R. B. (1993). Experimental Supersonic Hydrogen Combustion Employing Staged Injection Behind a Rearward-Facing Step. In *J. Propul. Power*, 9:472–478.

Beer, J. and Chigier, N. (1972). *Combustion Aerodynamics*, Kreiger Publishing Company.

Bowman, C. T. (1992). Control of Combustion-Generated Nitrogen Oxide Emissions: Technology Driven by Regulation. In *24th Symposium (Intl.) on Combustion*, pp. 859–878.

Brancher, P., Chomaz, J. M., and Huerre, P. (1994). Direct numerical simulations of round jets: Vortex induction and side jets. In *Physics of Fluids*, 6(5), pp. 1768–1774.

Broadwell, J. E. and Breidenthal, R. E. (1984). Structure and mixing of a transverse jet in incompressible flow. In *J. Fluid Mech.*, 148:405–412.

Brzustowski, T. A. (1977). In *Turbulent Combustion: Progress in Astronautics and Aeronautics*, 58:407–430.

Cortelezzi, L. and Karagozian, A. R. (2001). On the Formation of the Counter-Rotating Vortex Pair in Transverse Jets. In *J. Fluid Mech.*, 446: 347-373.

Cortelezzi, L. and Karagozian, A. R. (2003). Three-dimensional vortex modeling of unforced transverse jets. In *Manipulation and Control of Transverse Jets*, Springer-Wein, New York.

Curran, E. T. (2002). Scramjet engines: The first forty years. In *J. Propul. Power*, 17(6): 1138–1148.

Fearn, R. and Weston, R. (1974). Vorticity associated with a jet in a cross flow. In *AIAA J.*, 12:1666–1671.

Freund, J.B., Lele, S.K., and Moin, P. (2000). Numerical simulation of a Mach 1.92 turbulent jet and its sound field. In *AIAA J.*, 38(11), pp. 2023–2031.

Fric, T. F. and Roshko, A. (1994). Vortical structure in the wake of a transverse jet. In *J. Fluid Mech.*, 279:1–47.

Gortler, H. (1942). In *Z. Angew.Math. Mech*, 22:244–254.

Hasselbrink, E. F., Jr. and Mungal, M. G. (2001). Transverse Jets and Jet Flames. Part 2. Velocity and OH Field Imaging. In *J. Fluid Mech.*, 443:27-68.

Heister, S. D., Nguyen, T. T., and Karagozian, A. R. (1989). Modeling of Liquid Jets Injected Transversely into a Supersonic Crossflow. In *AIAA J.* 27:1727-1734.

Heister, S. D. and Karagozian, A. R. (1990a). Vortex Modeling of Gaseous Jets in a Compressible Cross Flow. In J. Prop. Power, 6:85-92.

Heister, S. D. and Karagozian, A. R. (1990b). Gaseous Jet in Supersonic Crossflow. In *AIAA J.*, 28: 819-827.

Hill, P. G. and Peterson, C. R. (1992). *Mechanics and Thermodynamics of Propulsion*, 2nd Ed., Addison-Wesley Publishing Co.

Holdeman, J. D. (1993). Mixing of Multiple Jets with a Confined Subsonic Crossflow. In *Prog. Energy Comb. Sci.*, 19(1):31–70.

Huerre, P. and Monkewitz, P. A. (1990). Local and Global Instabilities in Spatially Developing Flows, In *Ann. Rev. Fluid Mech.*, 22:473–537.

Humble, R. W., Henry, G. N., and Larson, W. J. (1995). *Space Propulsion Analysis and Design*, Revised edition, Mc-Graw-Hill Publishing Co.

Kailasanath, P., Sreenivasan, K. R., and Saylor, J. R. (1993). Conditional Scalar Dissipation Rates in Turbulent Wakes, Jets, and Boundary Layers. In *Phys. Fluids A*, 5(12): 3207–3215.

Kamotani, Y., and Greber, I. (1972) Experiments on a turbulent jet in a cross flow In *AIAA J.*, 10:1425–1429.

Karagozian, A. R. (1986a). The flame structure and vorticity generated by a chemically reacting transverse jet. In *AIAA J.*, 24:1502-1507.

Karagozian, A. R. (1986b). An analytical model for the vorticity associated with a transverse jet. In *AIAA J.*, 24:429–436.

Karagozian, A. R. and Nguyen, T. T. (1986). Effects of heat release and flame distortion in the transverse fuel jet. In *21st Symp. (Intl.) on Combustion*, 1271-1279.

Karagozian, A. R., Nguyen, T. T., and Kim, C. N. (1986). Vortex modeling of single and multiple dilution jet mixing in a crossflow. In *J. Propul. Power*, 2(4):354-360.

Karagozian, A. R., Wang, K. C., Le, A.-T., and Smith, O. I. (1996). Transverse Gas Jet Injection Behind a Rearward-Facing Step In *J. Propul. Power*, 12:1129–1136.

Kelso, R. M. and Smits, A. J. (1995). Horseshoe vortex systems resulting from the interaction between a laminar boundary layer and a transverse jet. In *Phys. Fluids*, 7:153–158.

Kelso, R. M., Lim, T. T., and Perry, A. E. (1996). An experimental study of round jets in cross-flow. In *J. Fluid Mech.*, 306:111–144.

Kerrebrock, J. L., (1977). *Aircraft Engines and Gas Turbines*, The MIT Press.

Krothapalli, A., Lourenco, L., and Buchlin, J. M. (1990). Separated flow upstream of a jet in a crossflow. In *AIAA J.* 28:414–420.

Liepmann, D. (1991). Streamwise Vorticity and Entrainment in the Near Field of a Round Jet. In *Phys. Fluids A*, 3(5):1179–1185.

Lim, T. T., Kelso, R. M., and Perry, A. E. (1998). A visual study of vortex rings fired transversely into a crossflow. In *3th Australian Fluid Mechanics Conference*, Monash University, Melbourne, Australia.

Lim, T.T., New, T.H., and Luo, S.C. (2001). On the development of large-scale structures of a jet normal to a cross flow. In *Phys. Fluids*, 13(3):770-775.

Margason, R. J. (1993). Computational and Experimental Assessment of Jets in Cross Flow. In *AGARD-CP-534*, 1:1–141.

M'Closkey, R. T., King, J. M., Cortelezzi, L., and Karagozian, A. R. (2002). The actively controlled jet in crossflow. In *J. Fluid Mech.*, Vol. 452, pp. 325–335.

M'Closkey, R. T., King, J. M., Cortelezzi, L., and Karagozian, A. R., (2003). The actively controlled jet in crossflow. In *Manipulation and Control of Transverse Jets*, Springer-Wein, New York.

McMillin, B. K., Seitzman, J. M., and Hanson, R. K. (1994). Comparison of NO and OH Planar Fluorescence Temperature Measurements in Scramjet Model Flowfields. In *AIAA J.*, 32(10):1945–1952.

Mungal, M. G. and Hasselbrink, E. F., Jr. (2003). Jets in Crossflow – Effects of Heat Release. In *Manipulation and Control of Transverse Jets*, Springer-Wein, New York.

Mungal, M. G. and Smith, S. H. (2003). Jets in Crossflow - Scalar Mixing via PLIF. In *Manipulation and Control of Transverse Jets*, Springer-Wein, New York.

Nguyen, T. T. and Karagozian, A. R. (1992). A Liquid Fuel Jet in Subsonic Crossflow. In *J. Propul. Power*, 8:21-29.

Northam, G. B. and Anderson, G. Y. (1986). Supersonic Combustion Ramjet Research at Langley. In AIAA Paper No.86-0159.

Pratte, B. D. and Baines, W. D. (1967). Profiles of the round turbulent jet in a cross flow In *J. Hydronaut. Div. ASCE*, 92:53–64.

Ricou, F.P. and Spalding, D.B. (1961). In *J.Fluid Mech.*, 11:21 - 32.

Rudman, M. (1996). Simulation of the Near Field of a Jet in a Cross Flow. In *Experimental Thermal and Fluid Science*, 12:134–141.

Salvetti, M. V. (2003). Large Eddy Simulation of Jets: State of the Art and Open Problems. In *Manipulation and Control of Transverse Jets*, Springer-Wein, New York.

Schlichting, H. (1933). In *Z. Angew.Math. Mech*, 13:260–263.

Smith, S. H. and Mungal, M. G. (1998). Mixing, structure and scaling of the jet in crossflow. In *J. Fluid Mech.*, 357:83–122.

Steward, F. R. (1970). In *Comb. Sci. and Tech.*, 2:203.

Wang, K. C., Smith, O. I., and Karagozian, A. R. (1995). In-Flight Imaging of Transverse Gas Jets Injected into Subsonic and Supersonic Crossflows. In *AIAA J.*, 33(12):2259–2263.

Yuan, L. L., Street, R. L., and Ferziger, J. H. (1999). Large-eddy simulations of a round jet in crossflow. In *J. Fluid Mech.*, 379: 71–104.

Jets in Crossflow – Scalar Mixing via PLIF

M. G. Mungal and S. H. Smith

Mechanical Engineering Department, Stanford University, Stanford, CA 94305-3032, USA

Abstract. The scalar concentration field of the round jet in a uniform crossflow is studied for a range of jet to crossflow velocity ratios, r, with emphasis on $r = 10$ and 20, using planar laser-induced fluorescence (PLIF) of acetone. For a top-hat jet exit velocity profile, the emerging presence of jet fluid in the wake structures is seen for for $r > 10$. The near-field region is characterized by a centerline concentration decay of $s^{-1.3}$, an improvement over the mixing rate s^{-1} of the free jet. The far-field is marked by a branching away from the $s^{-1.3}$ decay and approaching a decay of $s^{-2/3}$ a rate derived from modelling efforts. When normalized by r^2d, the branch points occur at a uniform jet position, $x/r^2d \sim 0.2$. The view of the branch points as points of transition in the flow is reinforced by the probability density function (pdf) results along the upper edge of the jet.

1 Introduction

The jet in crossflow consists of four known vortical structures: the horseshoe vortices, the jet shear layer, the wake structures and the counter rotating vortex pair (Fric and Roshko, 1994). The horseshoe vortices form upstream of the jet exit, wrapping around the exiting jet column. The jet shear layer consist of the ring vortices shed from the jet boundary layer. The wake structures form downstream of the jet column, and persist and convect far downstream of the exit nozzle. The jet column forms the counter-rotating vortex pair (CVP), the dominant vortical structure of the transverse jet after the jet has turned in the crossflow direction.

The resulting flowfield depends primarily upon the ratio of the jet momentum to the crossflow momentum. It is customary to define the effective velocity ratio, r, as the square root of the momentum ratio, J, or $r = J^{1/2} = (\rho_j U_j^2/\rho_{cf} U_{cf}^2)^{1/2}$. In the case of equal density flows $r = U_j/U_{cf}$. Three length scales are used to normalize jet behavior: (1) the jet exit diameter, d, (2) the rd length scale is used by Pratte and Baines (1967) to collapse the centerline trajectories of jets of different r, using the correlation, $y/rd = A(x/rd)^m$ with $A = 2.05$ and $m = 0.28$ for $r = 5$ to 35. Jet trajectory is still most often correlated by a functional form based on jet diameter, $y/d = Ar^n(x/d)^m$, see Margason (1993) for examples. Broadwell and Breidenthal (1984) also considered the jet exit as a point source of momentum flux (thereby analyzing a region far downstream of the jet nozzle) and concluded that the only global length scale is rd. (3) A third length scale found in the transverse jet is given by Keffer and Baines (1963), and is equal to $Jd = r^2d$. They normalize the jet trajectory of $r = 2$ to $r = 10$ jets at distances up to 8 jet diameters from the jet nozzle along the jet centerline. Their $r = 6$, 8, 10 cases collapse on the same curve using r^2d.

Maximum centerline concentration decay results have been presented by Patrick (1967), Broadwell and Breidenthal (1984) and Hasselbrink and Mungal (1996) and Yuan (1997) . Patrick (1967) identifies a maximum centerline concentration decay proportional to $s^{-1.2}$ (s is the centerline coordinate following the maximum local concentration). The Broadwell and Breidenthal (1984) model predicts a far-field decay proportional to $x^{-2/3}$, a result matched by Hasselbrink and Mungal (1996), using a simple algebraic expressions based on an integral approach to conservation of mass and momentum. The analysis of Hasselbrink and Mungal (1996), and to a lesser extent that of Broadwell and Breidenthal (1984), relies upon the jet obtaining self-similarity through a range of velocity ratio r. Temperature profiles in the $z = 0$ plane have been shown to collapse to a Gaussian error curve using the functional form $(T - T_o)/(T_{max} - T_o)$ based on distances y normalized by the half-width at half maximum (Kamotani and Greber, 1972, Sherif and Pletcher, 1991, Papaspyros et al., 1997). Velocity profiles have also been shown to collapse using $(U - U_o)/(U_{max} - U_o)$ (Keffer and Baines 1963). Neither Kamotani and Greber (1972) nor Keffer and Baines (1963) conclude the transverse jet is a self-similar flow. Keffer and Baines (1963) and Patrick (1967) state the transverse jet does not achieve self-similarity, while Chassaing et al. (1974) do refer to the far-field region as the zone of velocity profile similarity.

The possibility of self-similarity is in direct contrast to a growing body of evidence that the transverse jet shows evolving characteristics which depend upon r. Kelso and Smits (1995) and Kelso et al. (1996) define different vortical interactions at the nozzle depending upon velocity ratio (and Reynolds number). Similarly, Fric and Roshko (1994) focus on the $r = 4$ jet because the bursting of the boundary layer fluid into the wake structures is most pronounced there. At the opposite extreme Andreopoulos and Rodi (1984) presents the $r = 0.5$ jet as a flow that, downstream of the jet exit, reentrains into the boundary layer with no memory of the jet. This is certainly not the case in the range from $r = 5$ to $r = 25$ studied here.

The purpose of this work is to present the experimental results from an extensive study of the scalar mixing and concentration field of the jet in crossflow using planar laser induced fluorescence (PLIF) of acetone which is seeded into the jet. The instantaneous images obtained show the instantaneous vortical structure of the jet, and statistical quantities derived from these images reveal insights into the jet mixing downstream of the jet exit. The results concern both the vortex interaction region of the jet and the jet far-field. They are unified by a continuing analysis and presentation of the global length scales affecting the flowfield. Full details can be found in Smith and Mungal (1998).

2 Experimental Methods

The flow facility test section is 54 cm by 54 cm and 94 cm in length. The jet is placed flush with the tunnel wall and discharges from a plenum with a rapid contraction leading to a top-hat jet exit velocity profile. In this experiment, nozzle diameter, d, is changed with, r, to keep the rd product a constant. The nozzle diameters available (and their design velocity ratio) are 20 mm ($r = 2.5$, not studied), 10 mm ($r = 5$), 5 mm ($r = 10$), 3.3 mm ($r = 15$), 2.5 mm ($r = 20$) and 2.0 mm ($r = 25$). Keeping rd constant holds Re_j constant as well, but the different jet diameters used changes the relative sizes of the jet boundary layer, crossflow boundary layer and jet exit diameter.

In the current experiment, acetone PLIF (Lozano et al., 1992) is used exclusively to measure the concentration of the jet fluid. Acetone absorbs in the uv (225-320 nm; 278 nm peak) and emits in the visible (350-550 nm, 435 nm peak), allowing for the use of non-gated, non-intensified CCD arrays, giving high dynamic range (upwards of 200) and high signal to noise (upwards of 200) images. A XeCl excimer laser provides a 400 mJ, 308 nm pulse which is formed into a sheet and focused to a waist in the test section. Sheet thickness is 0.6 mm for a field of view that is typically about 20 cm. The jet flow is seeded with acetone vapor (10% by volume) by bubbling the jet fluid through a pair of heated acetone containing baths and then discharged into a crossflow of air. A 512 by 512 CCD array images the fluorescence. The camera is non-intensified, thermoelectrically cooled and has 16 bit pixel resolution. The jet potential core is normalized to a concentration of 100% after careful corrections for background level, laser sheet non-uniformities and laser absorption (Smith, 1996).

3 Results

3.1 Jet trajectory

The instantaneous images and the ensemble average acquired for each data set are used to obtain trajectory, structure and scaling data. The jet trajectory presented here is defined to be the locus of points in the $z = 0$ plane at which the maximum concentration value occurs in a line perpendicular to the jet direction. This trajectory is based upon local maximum concentration, and not upon velocity measurements. Trajectory data are displayed in Fig. 1 for $r = 5$ to $r = 25$. The x and y coordinates in the trajectory plot are normalized by the length scale rd. The rd-scaled trajectories, Fig. 1, do not collapse completely, as the data of Pratte and Baines (1967) which extends to $100rd$ and is acquired using photographs of smoke issuing from a pipe that extends into the crossflow. The figure also suggests that the $r = 5$ case belongs to a different family and is likely influenced by its image system in the wall as is seen in side-view images (not shown). The best fit of the trajectory data in Fig. 1 gives values of $A = 1.5$ and $m = 0.27$ in the Pratte and Baines formulation described above.

3.2 Jet asymmetry

The ensemble-averaged end views, Fig. 2, reveal the jet imaged in this experiment is not symmetric about the $z = 0$ plane. Symmetry in the jet in crossflow has seen limited previous discussion and may have been implicitly assumed by previous researchers. The results from this study suggest that the jet in crossflow is non-symmetric in the scalar field, to within the experimental accuracy of setting up the experiment. The worst case of asymmetry found is the $r = 20$ jet at $x = 4rd$; in that case the peak concentration in the $z > 0$ region is twice that of the $z < 0$ region. This factor of two difference is not caused by any obvious laboratory defect. In fact the images or $r = 20$ at $x = 1rd$ and $2rd$ are from the same tunnel conditions. In this case, the asymmetry does not develop until after $x = 1rd$. Again, it is emphasized that this asymmetry exists in the scalar field but it is presently unknown if the same effect is present in the velocity field.

3.3 Instantaneous jet structure

Instantaneous realizations of the jet concentration field are very different than the corresponding ensemble-averaged fields. Figure 3 shows a typical instantaneous side-view image through the $z = 0$ plane of an $r = 20$ jet. Concentration profiles through the jet are shown at five locations; the ensemble-averaged profiles are included on each instantaneous profile. While the average is characterized by a smooth Gaussian shape, the instantaneous profiles are marked by sharp vertical rises in concentration resulting in small plateaus of high concentration fluid. The Gaussian-shaped average is a result of the motion of the high concentration regions.

In side-view images, the primary jet path is visualized as are parts of the known vortical structures. The jet shear layer is shown, but the vortex interaction region lacks the resolution necessary for a clear presentation. The horseshoe vortex system is never seen because the jet boundary layer is not seeded in this experiment. The side-view image, Fig. 3, shows freestream fluid penetrating deeply into the jet. Penetrations from the upper edge occur past the jet centerline. These freestream penetrations, especially from the upper edge, and the sharp rises in the instantaneous jet concentration field result in a clearly defined edge between the presence of jet fluid and its absence. A fixed probe placed along the upper edge of the jet would find fluctuating periods between 0% (no jet fluid) and a finite value (jet fluid present). The lower edge is characterized by weak concentrations of fluid in the wake structures, but images not shown here suggest that the fluid penetration into the wake structures occurs progressively as r increases from 10 to 15 and is fully established by $r = 20$ (for a top-hat jet exit profile).

End-view images clearly show the CVP. Figure 4 displays an $r = 20$ jet at $x = 2rd$. In this image, the CVP is tilted to the right in an asymmetric fashion. The peak concentration is in the left CVP lobe (see the horizontal profile). The experience of seeing multiple images (Smith 1996) tells us that the CVP is most often asymmetric, with one lobe typically maintaining a 5% to 15% greater concentration than the other; that value can rise to as high as 50% in individual images. In the $r = 20$ jet the lower-concentration lobe always lies below the higher-concentration lobe, and the higher-concentration lobe may be larger in size as well. The dominant lobe can be either the right or left vortex of the CVP. This figure shows that the jet fluid that enters the wake consists of thin fingers of fluid that can be as high as 0.5% concentration. These fingers remain primarily along the centerline of the jet as can be seen in Figs. 3, 4; they connect to the jet body, but not to the tunnel floor. The connection to the jet body can occur to either CVP lobe, or to the jet center, but it is always connected to the weaker lobe; connection to the right CVP lobe is shown in figure 4. Plan views (not shown) generally show the dominant CVP lobe switching from side to side. In addition, both axisymmetric and sinusoidal modes are revealed in the concentration field.

3.4 Centerline concentration decay

Figure 5 displays the centerline concentration decay along the centerline coordinate s for $r = 5$ to $r = 25$. The $r = 15$, $r = 20$ and $r = 25$ cases connect data taken using different jet diameters. The connection occurs at $s \approx 1rd$, and is necessary because near the jet nozzle, the relative size of the smaller nozzles and the relative thickness of the laser sheet causes the maximum concentration values to be underpredicted. Figure 5 is shown as a log-log plot, so an exponential decay appears as a straight line. A decay of $s^{-2/3}$ is the far-field rate predicted in modelling efforts by Broadwell and Breidenthal (1984) and Hasselbrink and Mungal (1996) (both predictions are in x, but $s \sim x$ in the far-field). Decay rates of $s^{-1.3}$ (best available near-field fit) and s^{-1} (free jet decay rate) are also shown.

For each jet, the centerline concentration remains 100% through the potential core, then decays initially at $s^{-1.3}$, an improvement over the decay rate of a free jet, s^{-1}. During the shared -1.3 decay, the $r = 10$ to $r = 25$ jets nearly lie upon the same line when s is normalized with rd. At s/rd locations specific to each value of r, the centerline decay branches away from the -1.3 decay rate.

These branch points are labeled by the marker "x". Note the $r = 25$ jet is not deemed to have reached a branch point yet, and the different path of the $r = 5$ jet is consistent with it belonging to a different regime of transverse jets. The branch points cause the jet centerline concentration to decay at a rate slower than the free jet. Much interest in the transverse jet results from its enhanced mixing properties, making the regions before and just after the branch points important areas of study.

The maximum centerline concentration decay plotted against s/r^2d is shown in Fig. 6, and this scaling aligns the branch points at $s/r^2d \approx 0.3$, which corresponds to $x/r^2d \approx 0.2$. The position of the branch points at $x/r^2d \approx 0.2$ appears to be a good use of normalizing by r^2d and provides a convenient definition of the jet far-field, $x/r^2d \approx 0.2$. Based upon the lack of self-similarity seen in the end-views, the region $x/r^2d \approx 0.2$ is labeled the near-field. The near-field label is due in part to the belief that effects from the d-scaled vortex interaction region still persist at $x/r^2d < 0.2$.

The ensemble-averaged end-view data (not shown) suggests self-similarity may occur beginning at the branch points. The possibility is raised by the close appearance between the $r = 10$ jet at $x = 2rd$, and $r = 20$ at $x = 4rd$. Additional data are necessary to determine if self-similarity occurs beginning from the branch point positions. Because of the changes in the transverse jet which occur at the branch points, the branch points are viewed as a transition region in the flow. This view is supported by examining the probability density function (pdf) of concentration (not shown), which reveals a non-marching character before the branch points, and a tilted character after the branch points. The change in character is consistent with the deeper penetrations along the upper edge of the $r = 20$ jet than the $r = 10$ jet. The $r = 10$ jet has already passed its branch point and is in a tilted-pdf mixing region, characterized by a more diffusive nature, and showing a decreasing preferred concentration at positions farther from the jet centerline. The $r = 20$ jet is not yet at its branch point and is in a non-marching mixing region, characterized by an invariant preferred concentration.

4 Summary and Conclusions

The jet in crossflow is a complex flow that shows features consistent with three length scales d, rd and r^2d. The vortex interaction region shows boundary layer separation and vortex roll-up events that scale on jet diameter. The d-scaling allows for wake structure effects that are dependent on r; these include the jet fluid in the wake structures which form from $r = 10$ to $r = 15$ and persist at higher r. The second pertinent r-dependent effect is the delayed formation of the CVP in higher velocity ratio jets. It is the r-dependent effect that cause us to label the region following the vortex interaction region as the near-field of the transverse jet.

The trajectory and physical dimensions of the transverse jet scale with rd in both the near- and far-field. (The scaling of the physical dimensions of the jet does transition initially in the near-field from d-scaling to rd-scaling.) The maximum centerline concentration has a near-field decay of $s^{-1.3}$, an improvement over the free jet, and a far-field decay that is less than the free jet. When normalized by rd, the centerline concentration decay lines separate at their branch points, recovering a $1/r$ maximum concentration scaling. The branch points occur at a uniform position of $x/r^2d \approx 0.2$. The r^2d-scaling provides a convenient boundary at which to separate the near- and far-field. The enhanced mixing properties of the transverse jet in comparison to the free jet are seen to be a near-field effect. It should be noted that the near-field is the region of CVP development, and the far-field is the region of a fully developed CVP. Thus the CVP itself is not viewed as a good mixer since the far-field region has reduced mixing compared to the free jet. Rather, it is the structural formation of the CVP that corresponds to the enhanced mixing region, the near-field.

References

Andreopoulos, J. and Rodi, W. (1984). Experimental investigation of jets in a crossflow. *J. Fluid Mech.* 138:93-127.

Broadwell, J.E. and Breidenthal, R.E. (1984). Structure and mixing of a transverse jet in incompressible flow. *J. Fluid Mech.* 148:405-412.

Chassaing, P., George, J., Claria, A. and Sananes, F. (1974). Physical characteristics of subsonic jets in a cross-stream. *J. Fluid Mech.* 62:41-64.

Fric, T.F. and Roshko, A. (1994). Vortical structure in the wake of a transverse jet. *J. Fluid Mech.* 279:1-47.

Hasselbrink, E.F. and Mungal, M.G. (1996). An analysis of the time-averaged far-field properties of the tranverse jet. AIAA Paper 96-0201, *AIAA Thirty-fourth Aerospace Sciences Meeting and Exhibit*, Reno, NV.

Kamotani, Y. and Greber, I. (1972). Experiments on a turbulent jet in a cross flow. *AIAA J.* 10:1425-1429.

Keffer, J. F. and Baines, W. D. (1963). The round turbulent jet in a cross-wind. *J. Fluid Mech.* 15:481-496.

Kelso, R.M., Lim, T.T. and Perry, A.E. (1996). An experimental study of round jets in cross-flow. *J. Fluid Mech.* 306:11-144.

Kelso, R.M. and Smits, A.J. (1995). Horseshoe vortex systems resulting from the interaction between a laminar boundary layer and a transverse jet. *Phys. Fluids A* 7:153-158.

Lozano, A., Yip, B. and Hanson, R. K. (1992). Acetone: a tracer for concentration measurements in gaseous flows by planar laser-induced fluorescence. *Exps. Fluids* 13:369-376.

Margason, R.J. (1993). Fifty years of jet in cross flow research. *AGARD-CP 534*, Paper 1.

Papaspyros, J. N. E., Kastrinakis, E. G. and Nychas, S. G. (1997). Coherent contribution to turbulent mixing of a jet in cross flow. *Applied Scientific Research*, 57:291-307.

Patrick, M. A. (1967) Experimental investigation of the mixing and penetration of a round turbulent jet injected perpendicularly into a transverse stream. *J. Hydronaut. Div. ASCE* 45:16-31.

Pratte, B.D. and Baines, W.D. (1967). Profiles of the round turbulent jet in a cross flow. *J. Hydronaut. Div. ASCE* 92:53-64.

Sherif, S.A. and Pletcher, R.H. (1991). Jet-wake thermal characteristics of heated turbulent jets in crossflow. *J. Thermophysics* 5:181-191.

Smith, S.H. (1996). The Scalar Concentration Field of the Axisymmetric Jet in Crossflow. Ph.D. thesis, Stanford University.

Smith, S.H. and Mungal, M.G. (1998). Mixing, structure and scaling of the jet in cross flow. *J. Fluid Mech.* 357:83-122.

Sykes, W., Lewellen, W. and Parker, S. (1986). On the vorticity dynamics of a turbulent jet in a crossflow. *J. Fluid Mech.* 168:393-413.

Yuan, L.L. (1997). *Large eddy simulations of a jet in crossflow.* Ph.D. Thesis, Stanford University.

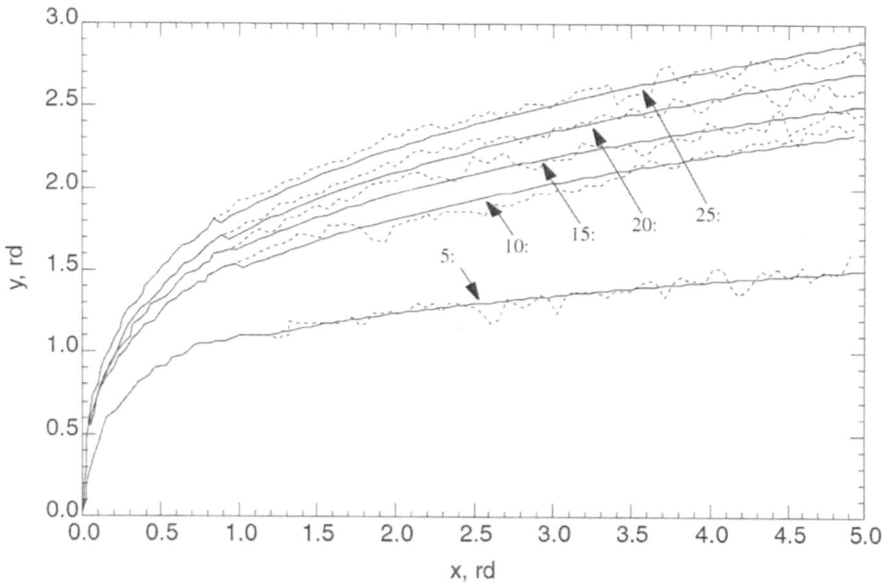

Fig. 1 Centerline trajectories normalized by *rd*

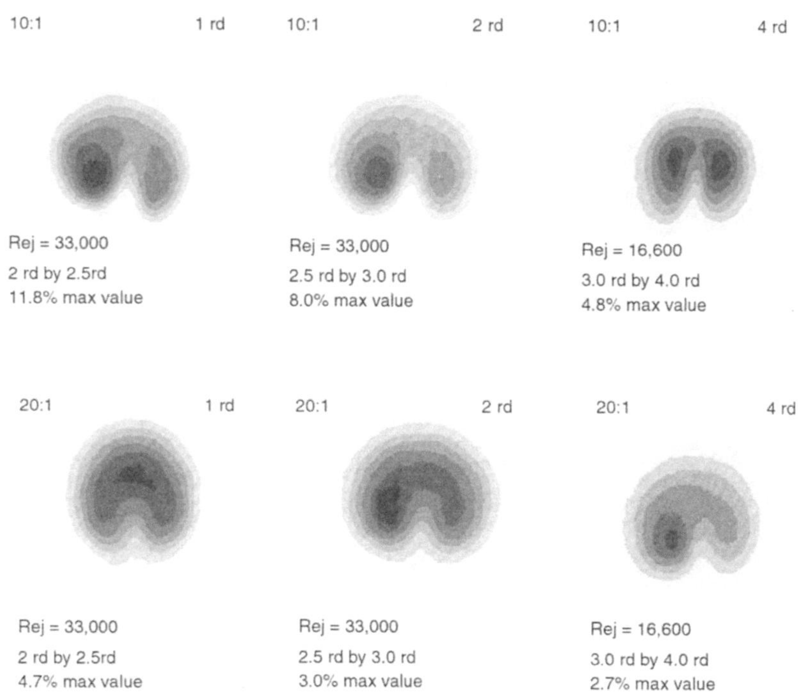

Fig. 2 Ensemble-averaged end-view concentration fields for $r = 10$, 20 showing asymmetry

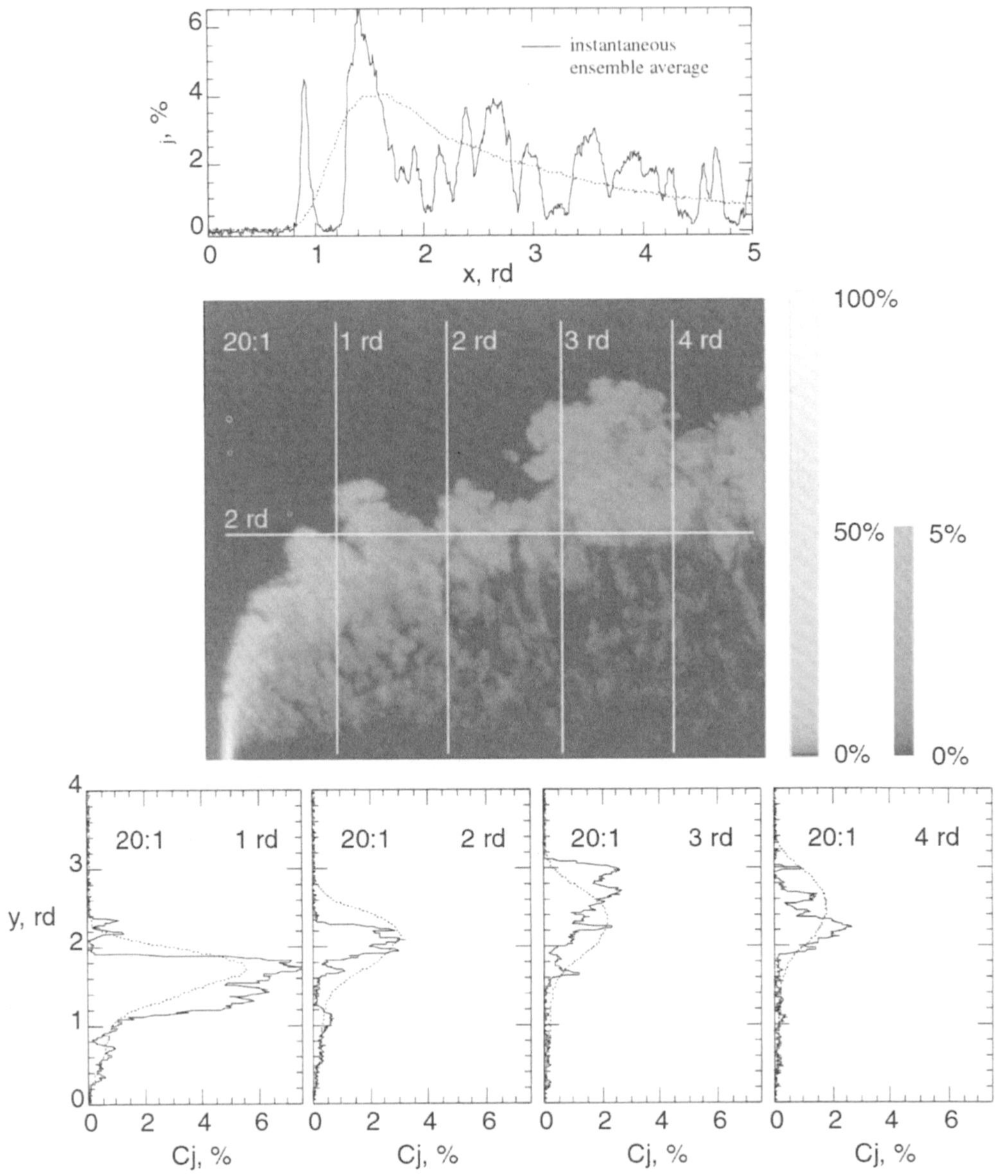

Fig. 3 Instantaneous jet concentration field, side view, $r = 20$, $Re_j = 16{,}600$

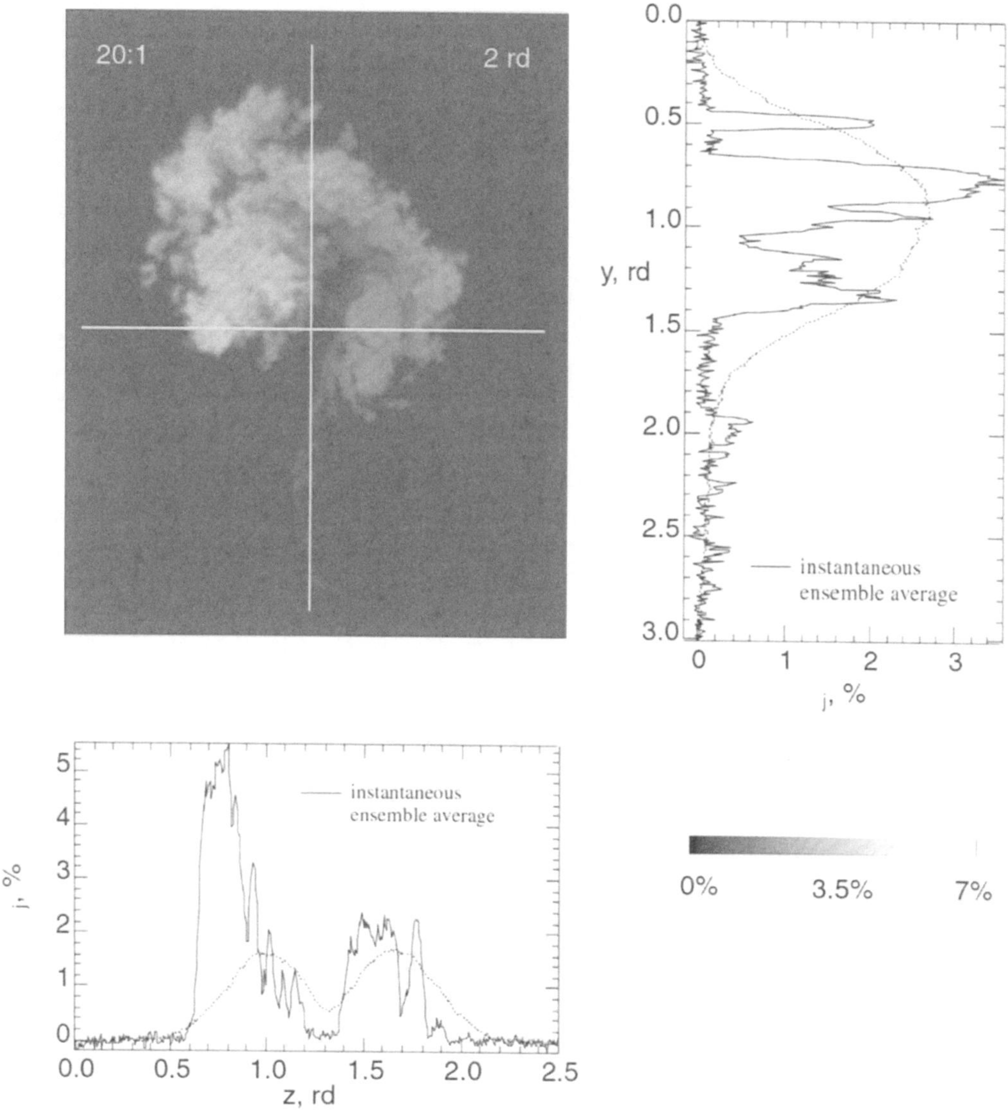

Fig. 4 Instantaneous jet concentration field, end view, $r = 20$, $x = 2rd$, $Re_j = 33{,}000$

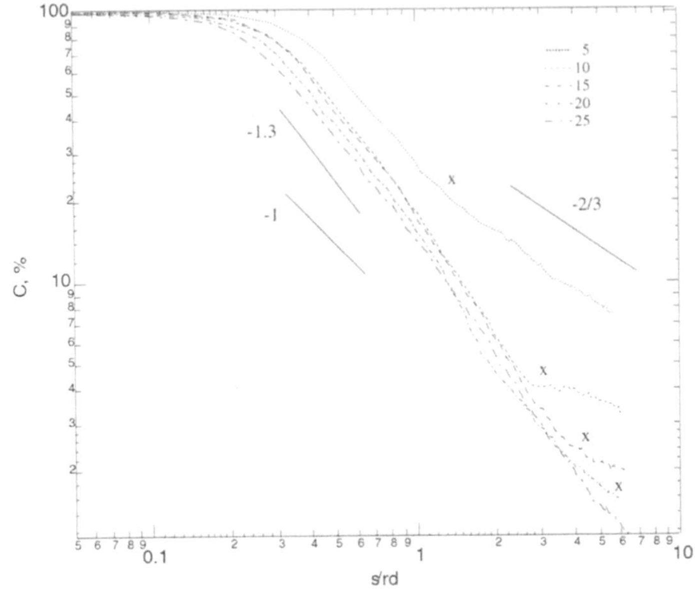

Fig. 5 Maximum centerline concentration decay plotted vs downstream distance, *s*, normalized by *rd*

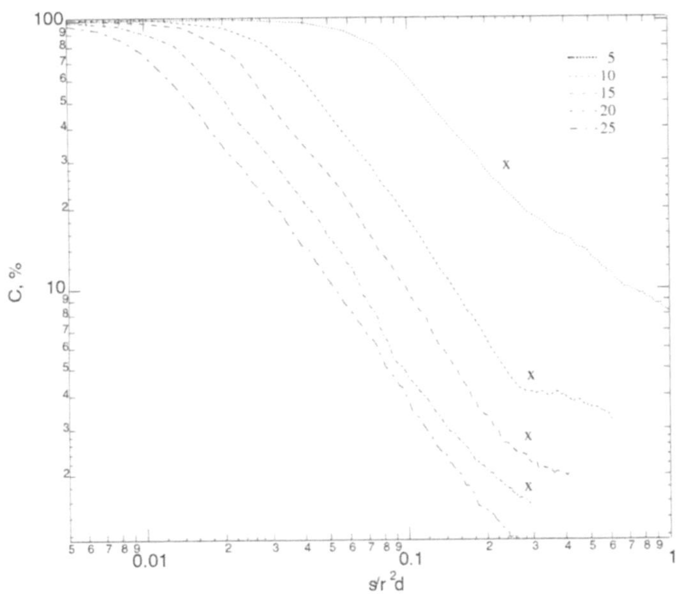

Fig. 6 Maximum centerline concentration decay plotted vs downstream distance, *s*, normalized by r^2d

Jets Injected Normally into Compressible Crossflow

Ann R. Karagozian, K.-S. Charles Wang, and Owen I. Smith *

Department of Mechanical and Aerospace Engineering, University of California, Los Angeles

Abstract. An overview of in-flight experiments and modeling relevant to the compressible transverse jet, in subsonic and transonic crossflow, is described here. Modeling efforts employed a low order, locally two-dimensional representation of the transverse jet, relying on the presence and dynamics of the counter-rotating vortex pair that is observed to dominate the jet cross-section. The experiments involved injection of iodine-seeded nitrogen into compressible air flow, with trajectory visualization via planar laser-induced fluorescence (PLIF) imaging of the iodine. The experiments were conducted in a flight test fixture situated under the fuselage of an F-104G aircraft, allowing examinations of crossflows in the high subsonic and transonic flow regimes. Both uniform crossflows and the crossflow created by compressible flow over a rearward-facing step were considered in these studies. Further details on these studies may be found in separate papers (Heister & Karagozian (1990a), Heister & Karagozian (1990b), Wang, et al. (1995), Karagozian, et al. (1996)).

1 Introduction

Gas phase jets injected into subsonic, transonic, and/or supersonic crossflows have relevance to a variety of flowfields present in high speed combustion engines, including dilution or primary air injection in high subsonic gas turbine engines, fuel injection in scramjet engine systems, and thrust vector control systems for rocket engine exhaust. Rapid fuel penetration, mixing with crossflow, and ignition and sustainment of combustion processes are highly desirable, as high air speeds entering the combustion chamber (supersonic in the case of a scramjet) require very rapid completion of the reaction process (Curran (2001)). Transverse fuel jet injection within a scramjet combustion configuration has been proposed to take place behind a rearward-facing step, allowing the flow to slow sufficiently in the immediate upstream region of the transverse fuel jet. Potentially, this could allow adequate fuel jet penetration and associated mixing with air to take place (Abbitt, et al. (1993), Wang, et al. (1995), Karagozian, et al. (1996)). Sustained ignition and burning of the fuel can be assisted by the presence of the rearward-facing step, which can act simultaneously as a flameholder, yet the step creates greater flow complexity in the vicinity of the fuel jet exit. In the absence of a rearward-facing step, the reactive fuel jet in crossflow nevertheless allows mixing, ignition, and sustained combustion, even in supersonic crossflow (McMillin,, et al. (1994)).

As a consequence of these applications, examinations of jets injected normally into compressible crossflow, especially supersonic crossflow, have been ongoing since the 1960s, particularly in connection with scramjet related flows. Experiments by Spaid & Zukoski (1966) demonstrated that high speed transverse jet penetration scales with jet-to-freestream stagnation pressure ratio.

---

* Work by these authors on compressible transverse jets has been supported by NASA Dryden Flight Research Center. Mr. Wang is now a Member of the Technical Staff of the Aerospace Corporation.

Schetz, Billig, and co-workers (e.g., as described in Orth, et al. (1969), Schetz & Billig (1987)) performed extensive experimental and modeling efforts on compressible jets in crossflow, specifically examining the effects of underexpanded vs. perfectly expanded jets, the effects of injection angle, and effects of multiple jets. More recent studies have used planar laser-induced fluorescence (PLIF) imaging of seeded species to visualize and quantify jet penetration and mixing (McDaniel & Graves (1988), Hollo, et al. (1994), McMillin,, et al. (1994)). Computational studies on compressible jets in compressible crossflow have largely involved RANS simulations of the flowfield with mixing length, one-, or two-equation turbulence models (e.g., Uenishi, et al. (1989), Rogers & Weidner (1983)). These studies enable a great deal of information pertaining to jet structure, spread, mixing, and reaction to be extracted.

The present paper describes studies dealing with experimental in-flight examinations (conducted at NASA Dryden Flight Research Center) and numerical simulations of gas phase jets in high speed crossflows, with and without the presence of an upstream rearward-facing step. Details on these studies may be found in separate papers (Heister & Karagozian (1990a), Heister & Karagozian (1990b), Wang, et al. (1995), Karagozian, et al. (1996)).

2 Gas Jets Injected into Uniform Compressible Crossflow

2.1 In-Flight Experiments

In cooperation with the NASA Dryden Flight Research Center at Edwards AFB, CA, our group utilized the F-104G research aircraft (numbered 826) as a "flying wind tunnel" platform for the transverse jet experiments. During the 1960's, the Lockheed F-104G was modified to carry a low aspect ratio fin on the underside of the fuselage for the purpose of conducting panel flutter tests in a flight environment. This "flight test fixture" (FTF) evolved into a highly versatile facility for fluid mechanics research. The capabilities of the FTF include a large flight Mach number envelope (0.4 to 2.0), including the region through Mach 1.0 with no adverse effects, and a larger Reynolds number and dynamic pressure envelope than is available in most wind tunnels. The present flight experiments illuminate the fundamental structure of the transverse jet while removing the effects of wall confinement which are present in wind tunnel experiments.

A schematic diagram of the FTF utilized in the present experiments is shown in Figure 1. A photograph of the actual flight test fixture beneath the F-104G is shown in Figure 2. As illustrated in Figure 1, in the present experiment the gas jet was injected normally into the crossflow from the side of the FTF, turning immediately into the freestream. The nozzle exit was machined to an internal diameter of 7.2 mm. Iodine-seeded nitrogen was used as the injectant since previous work has suggested that jet penetration and spreading are insensitive to injectant molecular weight (Torrence (1971)), hence the injectant here could represent air or fuel. As noted in Figure 1, the FTF created an oblique shock and expansion fan upstream of the jet injection location, but the characterization of the FTF through extensive prior flight testing by NASA enabled accurate upstream conditions to be known. For supersonic crossflow conditions, a bow shock is known to form upstream of the transverse jet, as indicated. This often creates a situation in which the pressure surrounding the jet exit, p_b, could differ from the static pressure within the jet at the exit plane, p_j, hence forming an underexpanded or overexpanded jet. In fact, much of the existing wind tunnel data for compressible transverse jets are for highly underexpanded jets (Spaid & Zukoski (1966)).

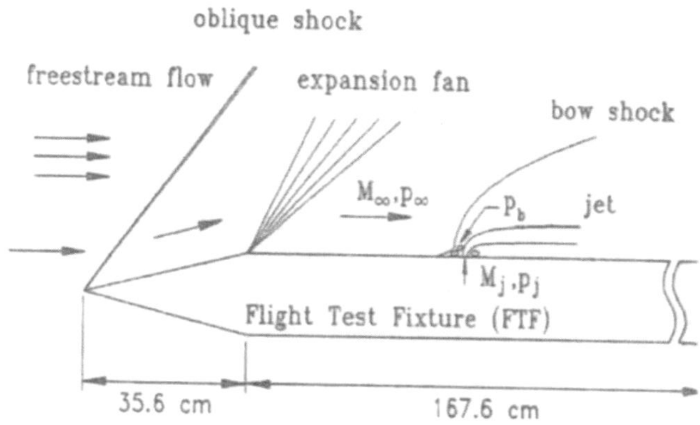

Figure 1. Schematic diagram of flight test fixture and jet injection system.

Figure 2. Photograph of the F-104G aircraft in flight, with the flight test fixture attached to the fuselage.

The trajectory and spreading of the jet were determined by illuminating a streamwise planar cross-section of the nitrogen jet with a sheet of laser light. Planar laser-induced fluorescence (PLIF) of the iodine seed allowed jet images to be visualized. The 2nd harmonic of a Nd:YAG laser at 532.25 nm was used to pump several lines in the R and P branches of the $B \leftarrow X$ transition for I_2. A Q-switched (pulsed) diode pumped Nd:YAG laser (Adlas DPY 321 QD) was used, operating in the broadband mode with visible green/yellow light. Flowfield images were captured by a high-speed gated intensified video camera synchronized to the laser pulse rate and equipped with appropriate optical filters to improve the signal-to-noise ratio. Details on the laser diagnostics may be found in Wang, et al. (1995).

Transverse jet behavior has been found to be strongly correlated with the jet-to-freestream momentum flux ratio (Heister & Karagozian (1990a), Heister & Karagozian (1990b), McDaniel & Graves (1988)). For an ideal gas, the momentum flux ratio is equivalent to the jet-to-freestream dynamic pressure ratio and is expressed as:

$$J \equiv \frac{\gamma_j p_j {M_j}^2}{\gamma_\infty p_\infty {M_\infty}^2} = \frac{\rho_j {U_j}^2}{\rho_\infty {U_\infty}^2} \quad (1)$$

where the subscript j corresponds to jet exit conditions and ∞ corresponds to freestream conditions; if there is a bow shock, ∞ corresponds to conditions ahead of the bow shock. The ratios of specific heats for jet and crossflow were approximately the same in the present experiments. The jet Mach number was fixed near $M_j = 1$ since modeling of the gas jet in supersonic crossflow (Heister & Karagozian (1990b)) indicates that jet penetration can be maximized when the injection Mach number is slightly greater than unity. The freestream Mach number in these experiments was varied from approximately $M_\infty = 0.6$ to $M_\infty = 1.2$.

The experimental apparatus was comprised of three subsystems: one containing the laser and its associated optics, one containing the intensified camera and its filters and mirrors responsible for imaging the fluorescent jet, and a jet formation subsystem which included all components involved in the storage, pressure regulation, heating, and flow control of the nitrogen gas and iodine seed. Details on these subsystems, in addition to photographs of the FTF and its internal components, are provided in Wang, et al. (1995). Upon reaching a desired flight test condition, the pilot fired the experiment, then stabilized the flight Mach number, altitude, and angle-of-attack as well as possible for the duration of the experiment. The entire experiment was operated solely by the pilot in flight. Images of the illuminated jet were viewed in real-time during the flights as well as recorded on VHS and Beta tape formats for post-flight review.

2.2 2D Vortex Modeling

An analytical/numerical model was developed to describe the behavior of gaseous jets injected transversely into compressible crossflow. Details of the model are in Heister & Karagozian (1990a) and Heister & Karagozian (1990b). The cross-section of the jet is modeled as an inviscid, compressible counter-rotating vortex pair, consistent with experimental observations of the transverse jet cross-section (McDaniel & Graves (1988), Kamotani & Greber (1972)). A schematic diagram of the important features for this compressible transverse jet, and the model assumptions, is shown in Figure 3. The flowfield associated with the compressible vortex pair is computed numerically (Heister, et al. (1990)), and forces $\pm F_y$ acting to separate the vortices in "flow time", as the jet evolves spatially downstream, are determined as a function of flow conditions. Additional forces $\pm F_{yp}$ acting to separate vortices are present if the transverse jet is underexpanded, and these too are computed from the compressible vortex pair solution.

The local orientation of the jet, ϕ_v, is determined from that for vortex pair circulation Γ, which is again dependent on flow conditions. Consistent with the analysis done by Karagozian (1986b), the circulation is assumed to arise from the jet's impulse as well as the viscous forces acting at the jet periphery; these contributions to circulation lie perpendicular to one another. Finally, force balances in the plane of the jet cross-section (section A-A in Figure 3) and mass and momentum balances along the jet trajectory are performed to produce governing differential equations written in terms of the vortex separation distance (h), the local average density of the jet within the cross-section (with the allowance of cross-stream entrainment), the local axial velocity of the jet, and the jet orientation angle ϕ_v. These equations are integrated numerically using

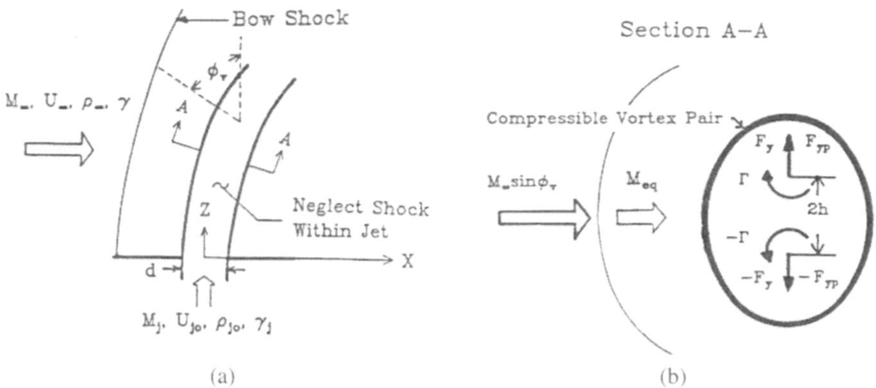

Figure 3. Schematic diagram of (a) transverse jet parameters and (b) jet cross-section.

Huen's method (Johnson & Riess (1982)), and predictions of the actual vortex pair trajectory can be made.

2.3 Results

Four individual flight tests (take-offs and landings) were successfully performed at NASA Dryden in January of 1994. Two of the flights, numbered 1410 and 1411, were for gas jets injected into uniform crossflow, the cases under examination here. The other two flights, 1412 and 1415, contain test data for a jet injected downstream of a rearward-facing step, a case reported in Section 3 and in detail in Karagozian, et al. (1996). For most flights, several supersonic crossflow test points and a number of subsonic points were achieved. The various jet and flight conditions run for these experiments are summarized in Wang, et al. (1995). From flight test data, jet conditions were evaluated using embedded pressure transducers within the nozzle, while freestream conditions were determined from a pitot probe situated in the noseboom of the F-104G.

Iodine concentration contours were produced by averaging iodine PLIF images over approximately 135 frames, representing the latter 75% of the duration of jet injection, and then analyzing the images. The nozzle temperature varied over the initial 25% of the jet injection period, hence these images were not included in the averaging. The latter 75% of the jet injection period produced very stable transverse jets.

Typical iodine contours for subsonic and supersonic crossflow conditions are shown in Figure 4. Highest concentrations of iodine were apparent in the near field "potential core" region of the jet, which disappeared within a few jet diameters downstream of injection. Contours appearing below the injection wall resulted from scattered light off the wall and should be ignored. The spread of the jet downstream was evident through the spatial growth of the regions of lower iodine concentrations. As would be expected, the degree of jet turning was increased with higher crossflow Mach number. From images such as those in Figure 4 it was possible to determine the locations of maxima in iodine concentration. The loci of concentration maxima roughly corresponded to the trajectory of the vortex pair associated with the jet cross-section, since most of the jet fluid became concentrated within a few diameters into the vortical structures. It should

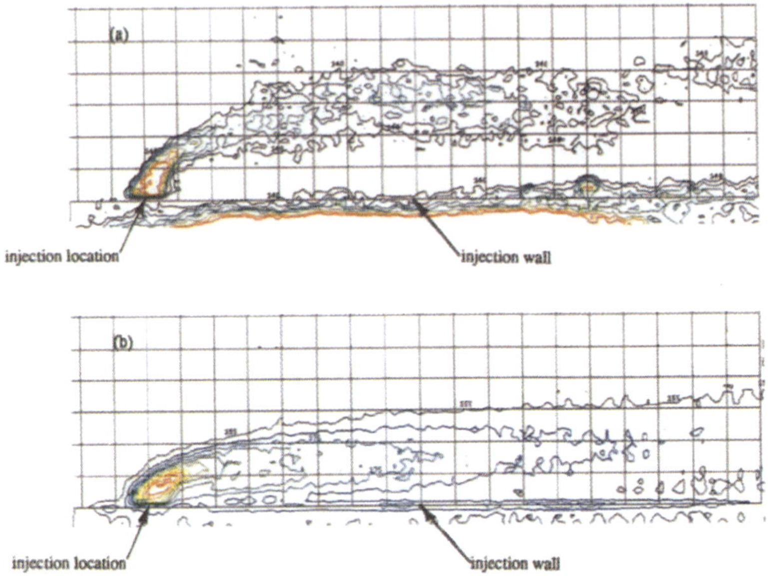

Figure 4. Iodine concentration contours for the uniform jet in crossflow, with a) crossflow Mach number 0.61, jet Mach number 1.0, and momentum flux ratio 2.75, b) crossflow Mach number 0.975, jet Mach number 0.955, and momentum flux ratio 1.04. High concentrations of iodine appear in red and yellow; lower concentrations appear in green and blue. Each grid marking shown represents one jet diameter D.

be noted that these maxima in iodine concentrations did not correspond exactly to the loci of velocity maxima, however, but somewhat lagged this actual jet "centerline", as does the vortex pair trajectory (Kamotani & Greber (1972)).

Measured loci of concentration maxima thus may be compared with vortex pair trajectories predicted for compressible transverse jets by models described in Section 2.2. These comparisons are shown in Figure 5, where experimental data are shown by the symbols and where the mass balance modification introduced in Karagozian, et al. (1996) was used for the model results shown by the lines in Figure 5.

For these gas jets injected into uniform subsonic, transonic, and supersonic crossflow we see a reasonable correspondence between the model and experimental data. It should also be noted that, for all of the subsonic crossflow comparisons shown, the jet-to-crossflow momentum flux ratio, J (called R in Figure 5), exceeded unity. Under these circumstances, earlier experiments (McDaniel & Graves (1988)) indicate that the jet momentum is probably high enough for a vortex pair to form and be dominant in the jet farfield; hence the model is appropriate for these cases. One also finds interesting trends in the experimental data. In both subsonic and supersonic crossflow cases, the jets often experience a rather sudden increase in penetration near six jet diameters downstream of injection. Prior wind tunnel studies (McDaniel & Graves (1988), Orth, et al. (1969)) also indicate slight increases in penetration downstream of the injection point; this had previously been thought to be associated with the influence of the opposite wall of the

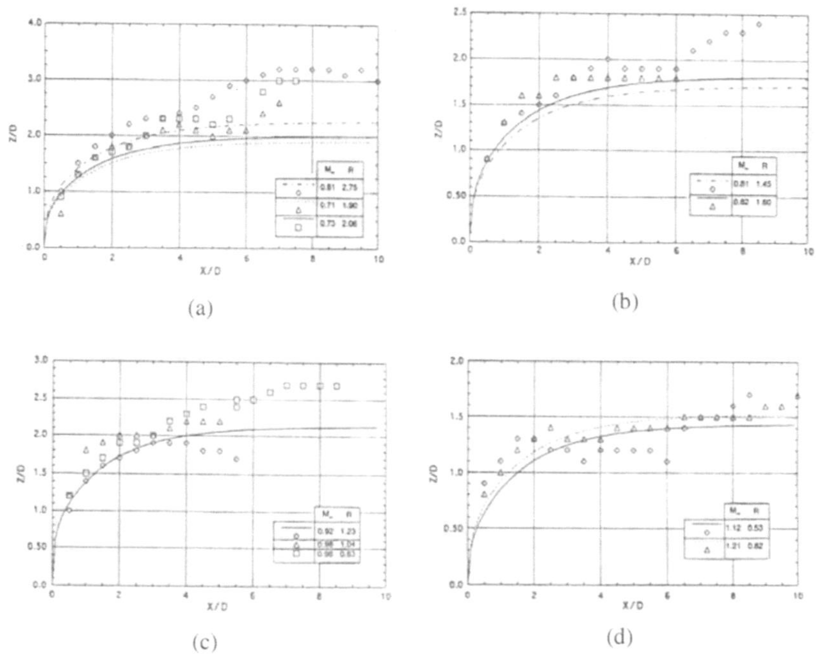

Figure 5. Comparisons for the jet in uniform crossflow of experimentally determined loci of iodine concentration maxima (symbols) with vortex pair trajectories predicted by the model (lines), as described in Section 2.2. Figures are taken from Wang, et al. (1995). In all cases, the jet Mach number M_j is unity, except for the following cases within these figures: Figure 5c (M_∞ = 0.98, $R = J$ = 0.83, and M_j = 0.85), and Figure 5d (M_∞ = 1.123, $R = J$ = 0.53, and M_j = 0.77).

tunnel, perhaps resulting from the reflection of oblique shock waves and their influence on the jet. The present studies show that this rather abrupt increase in penetration is not due to opposite wall effects, but, rather, is likely to arise from an alteration in fundamental jet structure, perhaps coincident with an alteration in the strength of the vortices which dominate the jet cross-section.

3 Gas Jets Injected into Compressible Crossflow Behind a Rearward-Facing Step

3.1 Background and Description of Experiments and Modeling

This study describes flight test experiments and modeling conducted for non-reacting, transverse gas jet injection downstream of a rearward-facing step in supersonic flow, where the experiments use PLIF imaging of iodine seeded in the jet. Details of this study may be found in Karagozian, et al. (1996). Flow conditions and injection locations which have not been extensively explored in the past were examined here: the jets were somewhat underexpanded, with freestream Mach numbers lying between 1.4 and 1.8 and with jet injection locations at one and four step heights downstream of the step. Prior wind tunnel studies have examined single non-reacting jet injection located at three (McMillin,, et al. (1994)) and four (McDaniel & Graves (1988)) step heights downstream, in addition to staged injection at three step heights (Hollo, et al. (1994)). Reasonable comparisons between experimental results and the results of the present modeling indicate that the model could be useful as a design tool in optimizing transverse jet penetration.

A gas jet injected transversely behind a rearward-facing step in supersonic flow potentially encounters several distinct, complex regions of flow. Figure 6 describes the current understanding of these flow regions (Fletcher & McDaniel (1989)), each of which can be characterized by the Mach number, static pressure, and flow direction in that region. As indicated in the figure, the

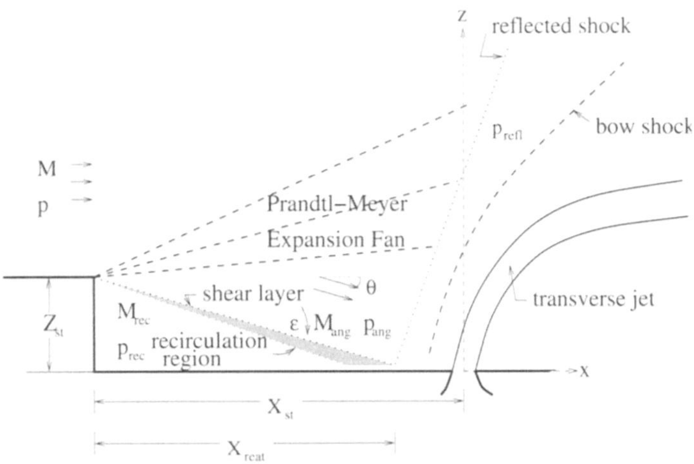

Figure 6. Schematic of flow regions associated with transverse jet injection behind a rearward-facing step in supersonic crossflow. Injection location shown is beyond the reattachment point, for clarity.

supersonic flow first passes through an expansion fan (2D in the region of present investigation) which forms at the step. Beyond the relatively uniform supersonic flow in the region past the fan (labeled with flow conditions M_{ang} and p_{ang}), a shear layer forms, across which the flow transitions from supersonic to subsonic in the low speed, recirculating flow regime. The average pressure in the recirculation cell, p_{rec}, has been found experimentally (Fletcher & McDaniel (1989), Papamoschou & Roshko (1988)) to be roughly equal to the pressure in the angled uniform flow regime, p_{ang}. The shear layer reflects from the wall (at an effective reattachment location X_{reat} from the step) as an oblique shock wave. While the reattachment point and turning angle θ are dependent on upstream Mach number, they are largely independent of wall geometry or downstream disturbances.

It is into this complex flowfield that the gas jet was transversely injected in these experiments. The jet injection location shown in Figure 6 is for the case where injection occurs beyond the reattachment point and reflected shock location. This is drawn for clarity; in the present set of experiments, injection actually occurred at two different locations that are determined to be within the recirculation region. Details for this method of determination, based on pressure and temperature measurements, are provided in Karagozian, et al. (1996). In all experimentally run cases, the jets were underexpanded, with jet-to-surrounding flow pressure ratios between 2 and 3.11. Other than the presence of the rearward-facing step and the ability to move it relative to the jet nozzle, the experiments were conducted as for the uniform crossflow experiments on the F-104G.

The present effort made use of the model described in Section 2.2. A few minor modifications to the model were made by Le (1991) in order to improve its numerical accuracy and theoretical consistency, as well as to prepare the model to accommodate the non-uniformity of upstream conditions associated with flow behind a step. These modifications include: 1) for supersonic flow about the jet cross-section, use of updated "equivalent" pressures, densities, and Mach numbers of the gas downstream of the bow shock based on output from a compressible vortex pair computation (Heister, et al. (1990)), and 2) an assumption that, before entrainment of crossflow occurs and the jet is expanded to the local pressure of the external flow, the mass flux across each cross-sectional jet slice is constant. Comparisons between the trajectories predicted by the modified transverse jet model (Le (1991)) for uniform crossflow and those predicted by the original model shown in Section 2.3 indicated a slight improvement with these refinements. While the model did not take into account the formation of a Mach disk within the jet, and thus is most appropriate for application to perfectly expanded or slightly underexpanded jets, it has been shown to be accurate to within 15% for pressure ratios p_j/p_b up to five (Heister & Karagozian (1990a)) and hence should be appropriate in the present study.

As noted above, a gas jet injected transversely behind a rearward-facing step in supersonic flow potentially encounters several distinct, complex regions of flow. In order to preserve the relative simplicity and computational efficiency of the present model, these data were extracted from present as well as prior experiments (Fletcher & McDaniel (1989), Papamoschou & Roshko (1988)), in addition to numerical data and correlations using the method of characteristics. The methodology is described in detail in Karagozian, et al. (1996).

3.2 Results

As noted in Section 2.3, two of the four flight tests conducted at NASA Dryden pertained to the nitrogen gas jet injected downstream of a rearward-facing step. Iodine PLIF images were obtained for the duration of three different experimental conditions: (a) where the jet was injected one step height downstream of the step, with $M_\infty = 1.42$, (b) where the jet was injected four step heights downstream of the step, with $M_\infty = 1.55$, and (c) where the jet was injected one step height downstream of the step, with $M_\infty = 1.77$. In all cases the jet Mach number was maintained at unity.

Sample fluorescence/concentration contour plots for iodine are shown in Figure 7 for cases (a)-(c), noting that, while the quantitative interpretation of iodine concentrations were approximate, the quantitative interpretation of overall jet structure and penetration beyond the step shear layer was accurate except just downstream of the step, where optical scattering of the laser sheet from the step for cases (a) and (c) affected the iodine fluorescence signal. The enhanced total jet penetration with injection further within the recirculation zone (i.e., closer to the step) and with increased momentum flux ratio is evident in these images.

The contours in Figure 7 may also be used to compare the degree of mixing of jet and surrounding fluid for the three cases tested. Cases (a) and (c) each had injection at one step height

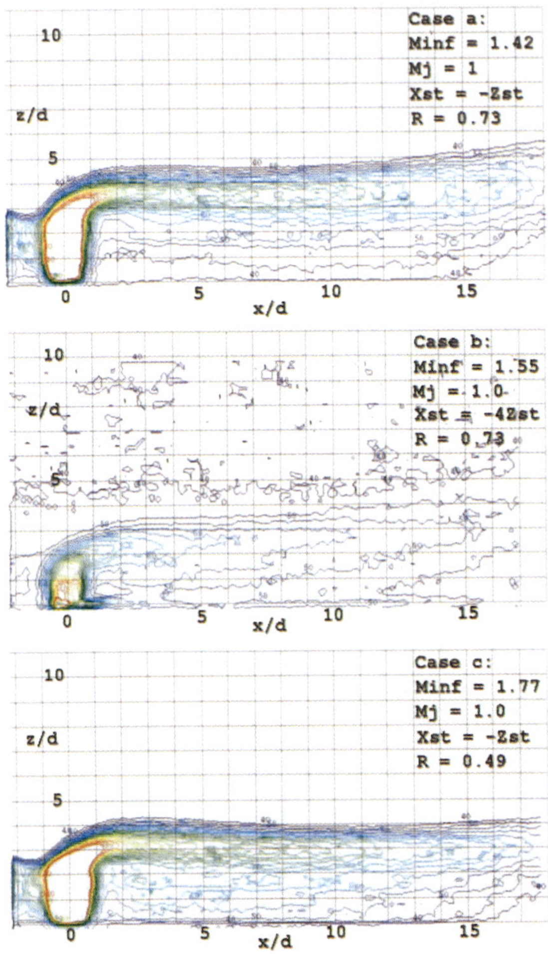

Figure 7. Iodine concentration contours for the gas jet injected behind a rearward-facing step in supersonic crossflow, for cases: (a) where the jet was injected one step height downstream of the step, with $M_\infty = 1.42$, (b) where the jet was injected four step heights downstream of the step, with $M_\infty = 1.55$, and (c) where the jet was injected one step height downstream of the step, with $M_\infty = 1.77$. In all cases the jet Mach number was maintained at unity. High concentrations of iodine appear in red and yellow; lower concentrations appear in green and blue. Each grid marking shown represents one jet diameter D.

downstream of the step, but case (a) had the higher momentum flux ratio and lower crossflow Mach number and, not surprisingly, exhibited greater penetration. Case (a) also exhibited somewhat greater mixing (i.e., a more rapid reduction in iodine fluorescence along the trajectory) than did case (c). This result confirmed prior observations concerning the enhancement of mixing with greater momentum flux ratio and with greater jet penetration (Segal, et al. (1995), McMillin,, et al. (1994)).

Cases (a) and (b), on the other hand, each had the same overall momentum flux ratio R, yet case (b) had injection at four step heights downstream with a higher crossflow Mach number and, again not surprisingly, exhibited lower jet penetration. The rate of mixing along the trajectory for case (b) was substantially greater than for case (a), however, exhibited by the sharp drop in iodine signal. Even the rate of mixing beyond the point where the jet penetrates the step's shear layer appeared to be greater in case (b) (which had a higher crossflow Mach number) than in case (a). Clearly, penetration alone cannot be used as a predictor of the degree of jet mixing, especially in the vicinity of a rearward step; in fact, extremely high rates of fuel-air mixing may actually not be desirable for scramjet applications. The very high rate of mixing in case (b) could render this condition inappropriate for autoignition due to very high strain rates which are known to delay or prevent ignition. Reduced ignition delay (resulting from reduced local rates of strain), alternatively, is known to enhance combustor performance during non-premixed supersonic combustion (Northam & Anderson (1986)).

The loci of iodine fluorescence/concentration maxima here roughly corresponded to the trajectory of the vortex pair associated with the jet cross-section, since most of the jet fluid became concentrated within a few diameters into the vortical structures. It is these estimated loci of concentration maxima which were compared with the model's prediction of the trajectory of the vortex pair. The comparisons of concentration maxima loci with model predictions are shown in Figure 8.

Results for case (a), with jet injection well within the recirculation region, indicated that the model, like the experiments, predicted enhanced penetration compared with injection further downstream within the zone but at the same momentum flux ratio (case (b)). The model underpredicted farfield penetration of the jet for case (a) by about 8% overall and overpredicted farfield penetration in case (c) by about 20% overall. These observations were likely due to inaccuracies in representing the complex flow in the recirculation zone behind the step, which was significant for this injection location. In contrast, case (b) with jet injection at four step heights downstream of the step showed model prediction to be closer to the experimental trajectory (overall, within 7%). The change in jet structure observed in the iodine contours as well as the trajectories may in fact be confirmation of the observations of Hollo, et al. (1994) that there is a transition from vortex-dominated mixing to small scale turbulent mixing along the jet trajectory.

4 Conclusions

In summary, the present flight experiments explored the behavior of a gas jet injected transversely into uniform subsonic and supersonic flow, and also behind a rearward-facing step in supersonic crossflow, in both cases in a flow regime and for injection locations previously unexplored. Iodine PLIF data gave quantitative results for jet penetration and qualitative estimates for jet mixing. A simple model for jet behavior was found to agree reasonably well with experimental data taken

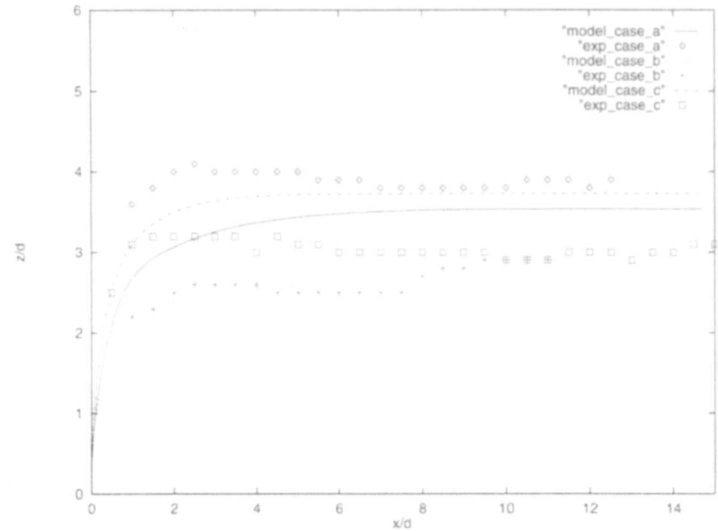

Figure 8. Comparisons of loci of iodine concentration maxima from present flight experiments (symbols) with vortex pair trajectories predicted by the model for jet injection downstream of the rearward-facing step (lines). Test conditions correspond to those given for cases a, b, and c.

in a wind tunnel (McDaniel & Graves (1988)) as well as with the present flight test experiments. While increased jet penetration may generally imply greater jet mixing and improved auto-ignition (Northam & Anderson (1986), Hollo, et al. (1994)), especially for injection at a fixed location relative to a step, as jet-to-crossflow momentum flux ratio is increased, the present experiments demonstrated that penetration alone cannot be used to predict the jet's rate of mixing with crossflow. In fact, very rapid fuel-air mixing may be undesirable for scramjet applications in that extremely high strain rates associated with rapid mixing may act to delay or prevent ignition altogether. It is clear that further controlled jet injection experiments are required, for a greater variety of compressible flow conditions, to be able to make conclusive statements concerning mixing enhancement and ignition associated with the transverse fuel jet in high speed flows.

References

Abbitt, J. D. III, Segal, C., McDaniel, J. C., Krauss, R. H., and Whitehurst, R. B. (1993). Experimental Supersonic Hydrogen Combustion Employing Staged Injection Behind a Rearward-Facing Step. In *Journal of Propulsion and Power*, 9: 472-478.

Curran, E. T. (2002). Scramjet engines: The first forty years. In *J. Propul. Power*, 17(6): 1138–1148.

Fletcher, D. G. and McDaniel, J. C. (1989). Laser-Induced Iodine Fluorescence Technique for Quantitative Measurement in a Nonreacting Supersonic Combustor. In *AIAA Journal*, 27:575-580.

Hollo, S. D., McDaniel, J. C., and Hartfield, R. J. Jr. (1994). Quantitative Investigation of Compressible Mixing: Staged Transverse Injection into Mach 2 Flow. In *AIAA Journal*, 32:528-534.

Heister, S. D. and Karagozian, A. R. (1990a). Vortex Modeling of Gaseous Jets in a Compressible Cross Flow. In J. Prop. Power, 6:85-92.

Heister, S. D. and Karagozian, A. R. (1990b). Gaseous Jet in Supersonic Crossflow. In *AIAA Journal*, 28: 819-827.

Heister, S. D., McDonough, J. M., Karagozian, A. R., and Jenkins, D. W. (1990) The Compressible Vortex Pair. In *J.Fluid Mech.*, 220:339-354.

Johnson, L. W. and Riess, R. D. (1982) *Numerical Analysis*, Addison-Wesley, Reading, MA, p. 85.

Kamotani, Y., and Greber, I. (1972) Experiments on a turbulent jet in a cross flow In *AIAA J.*, 10:1425–1429.

Karagozian, A. R. (1986b). An analytical model for the vorticity associated with a transverse jet. In *AIAA J.*, 24:429–436.

Karagozian, A. R., Wang, K. C., Le, A.-T., and Smith, O. I. (1996). Transverse Gas Jet Injection Behind a Rearward-Facing Step. In *Journal of Propulsion and Power*, **12**, 1129–1136.

Le, A.-T. (1991). Transverse Gaseous Jet Injection Behind a Rearward-Facing Step into Supersonic Crossflow. M.S. Thesis, UCLA.

McDaniel, J. C. and Graves, J., Jr. (1988). Laser-Induced-Fluorescence Visualization of Transverse Gaseous Injection in a Nonreacting Supersonic Combustor. In *Journal of Propulsion and Power*, 4:591-597.

McMillin, B. K., Seitzman, J. M., and Hanson, R. K. (1994). Comparison of NO and OH Planar Fluorescence Temperature Measurements in Scramjet Model Flowfields. In *AIAA Journal*, 32(10):1945-1952.

Northam, G. B. and Anderson, G. Y. (1986). Supersonic Combustion Ramjet Research at Langley. In AIAA Paper 86-0159.

Orth, R.C., Schetz, J.A. and Billig, F.S. (1969). The Interaction and Penetration of Gaseous Jets in Supersonic Flow In *NASA CR-1386.*

Papamoschou, D. and Roshko, A. (1988). Compressible Turbulent Shear Layer: An Experimental Study. In *Journal of Fluid Mechanics*, 197:453-477.

Rogers, R. C. and Weidner, E. H. (1983). Numerical Predictions of Residence Times Behind a Rearward Facing Step with Transverse Injection. In *Proc. of the* 20th *JANNAF Combustion Meeting*, U. S. Naval Post-graduate School, Monterey, CA.

Schetz, J. A. and Billig, F. S. (1987). Studies of Scramjet Flowfields. In AIAA Paper 87-2161.

Segal, C., McDaniel, J. C., Whitehurst, R. B., and Krauss, R. H., (1995). Mixing and Chemical Kinetics Interactions in a Mach 2 Reacting Flow. In *Journal of Propulsion and Power*, 11, 308-314.

Spaid, F. W. and Zukoski, E. E. (1966). Further Experiments Concerning Secondary Injection of Gases into a Supersonic Flow. In *AIAA Journal*, 4(12), 2216-2218.

Torrence, M.G. (1971). Effect of Injectant Molecular Weight on Mixing of a Normal Jet in a Mach 4 Airstream. In NASA TN D-6061.

Uenishi, K., Rogers, R. C., and Northam, G. B. (1989). Numerical Predictions of a Rearward-Facing-Step Flow in a Supersonic Combustor. In *Journal of Propulsion and Power*, 5:158-164.

Wang, K. S. C., Smith, O. I., and Karagozian, A. R. (1995). In-Flight Imaging of Gas Jets Injected into Subsonic and Supersonic Crossflows. In *AIAA Journal*, **33**, 2259–2263.

Jets in Crossflow - Simultaneous PIV/PLIF measurements

M. G. Mungal and L. K. Su

Mechanical Engineering Department, Stanford University, Stanford, CA 94305-3032, USA

Abstract. Simultaneous planar laser-induced fluorescence (PLIF) and particle image velocimetry (PIV) yield measurements of two-dimensional jet fluid concentration and velocity fields in turbulent crossflowing jets. The jet-to-crossflow velocity ratio is $r = 5.7$ and the jet exit Reynolds number is roughly 5000. Two flow configurations are studied, one in which the jet nozzle is flush with the tunnel wall and the other where the nozzle protrudes into the uniform region of the tunnel flow. The averaged scalar and velocity fields show a strong similarity in growth rates and centerline decay rates between the two configurations when using the centerline coordinate s. A possible approach to wake scaling is seen in the velocity fields, but is not evident in the scalar fields. Results are presented for the scalar flux components and the turbulent shear stress, as examples of the fluctuation quantities accessible by the present data.

1 Introduction

The crossflowing turbulent jet, in which a jet is injected into a perpendicular fluid stream, is of considerable practical significance in engineering systems. Combustion applications of this flow, for example, range from aeropropulsion, to gas-burning power generation, to the on-site purging of excess gases at oil drilling sites. Despite the abundance of engineering applications, however, the body of research devoted to mixing in the crossflowing jet is small. Among other issues, the complicated vortical structure of the crossflowing jet (Fric and Roshko, 1994) makes theoretical treatment difficult. Modeling efforts are described by Durando (1971), Broadwell and Breidenthal (1984), Karagozian (1986) and Hasselbrink and Mungal (1996). A problem common to modeling work is the difficulty in modeling the near field of the flow. Additionally, experimental evidence suggests that the flow is very sensitive to the ratio of jet velocity to crossflow velocity, with constrains efforts to make general conclusions about the flow configuration.

The present measurements are intended to provide a comprehensive view of the velocity and conserved scalar fields in the developing region of the flow. All measurements are made at a single jet-to-crossflow velocity ratio ($r = 5.7$). The effect of the crossflow velocity profile is considered by placing the jet exit nozzle both flush with the wind tunnel wall, and also outside the crossflow boundary layer. Velocity ratios of around 5 are of particular interest in aerospace propulsion applications. By performing the planar measurements in the center plane (i.e. the jet symmetry plane) and at various positions off the center plane, we are able to evaluate the three-dimensionality of the flow. These results are also of direct interest to ongoing efforts to compute mixing and combustion in the crossflowing jet (Schlüter and Schönfeld, 2000).

2 Experimental Arrangement

These experiments are performed in an updraft wind tunnel with air as both jet fluid and crossflow fluid. The maximum crossflow velocity in the tunnel is $v_\infty = 2.95$ m/s. The jet nozzle is a simple pipe with 6.35 mm outer diameter, inner diameter $d = 4.53$ mm, and a length of 320 mm. In the absence of a crossflow, fully developed pipe flow conditions would prevail at the jet exit; the presence of the crossflow, however, is known to affect the jet exit profile (Yuan, 1997). The average (bulk) jet velocity based on volumetric flow rate is $u_j = 16.9$ m/s, giving a velocity ratio of 5.7 and a jet exit Reynolds number of 5000. To permit assessment of the effect of the crossflow boundary layer on the flow development, two jet nozzle positions are considered, the first in which the nozzle exit is flush with the wind tunnel wall, and the second in which the nozzle protrudes 100 mm into the crossflow. The 80% point of the boundary layer profile lies 6 mm from the wind tunnel wall, so in this protruding nozzle case the jet is well outside of the crossflow boundary layer.

2.1 Optical arrangement

Particle image velocimetry (PIV) is used to measure the velocity fields while planar laser-induced fluorescence (PLIF) is used for the scalar field measurements. To provide the Mie scattering signal for particle image velocimetry, the jet flow is seeded with submicron aluminum oxide particles, while the crossflow is seeded with a glycerol-water fog produced by a commercial fog machine (Rosco 1500). A single, dual cavity Nd:YAG laser (Spectra-Physics PIV-400), with frequency-doubled 532 nm output, is used to produce two laser sheet pulses in quick succession. The time delay between pulses is as short as 8 μs. The resulting scattering signal is collected by an interline transfer CCD camera (Kodak Megaplus ES 1.0, 1k × 1k pixel resolution), fitted with a 35 mm format Nikon 105 mm f/2.8 macro lens. The frame transfer capability of this camera allows each of the two closely spaced laser sheet pulses to be captured in a separate image. This permits the use of a cross-correlation PIV algorithm (Hasselbrink, 1999), which eliminates directional ambiguity and yields improved resolution over single-image autocorrelation techniques. The algorithm used here also incorporates iterative interrogation window offset to increase vector yield.

For the laser-induced fluorescence, acetone is seeded into the jet fluid stream, to approximately 10% by volume. To excite the fluorescence signal, a XeCl excimer laser with 308 nm output (Lambda-Physik EMG 203MSC) is used. The PLIF signal is captured by a thermoelectrically-cooled, slow-scan CCD camera (Photometrics AT200), with 512 × 512 pixel resolution. The imaging lens is a 35 mm format Nikon 50 mm f/1.2 lens, fitted with a *+1* diopter close-up lens for close focusing. Two 2 mm thick BG-25 bandpass filters isolate the PLIF signal (which peaks in the range 400-500 nm) from the much brighter Mie scattering signal at 532 nm. Post-processing of the PLIF data corrects for background light levels, non-uniformity in laser sheet intensity, and local laser intensity attenuation caused by passage of the sheet through the fluorescent medium. Experiment timing and synchronization are controlled by three Stanford Research Systems DG 535 pulse generators.

3 Results

A sample PLIF image, taken in the center plane with the protruding nozzle, is shown in Fig. 1*a*. Here x is the initial jet direction, y is the crossflow direction and z is out-of-plane. In these measurements it is necessary to include the jet potential core in the imaging region, to permit the correlation of measured signal levels with jet fluid concentration. The size of the imaging region is then restricted by the desire to resolve fine scale fluctuations in the scalar field in the 512×512 pixel images. The resulting raw PLIF images span roughly 3.5 *rd* per side. In contrast, it is not required that the PIV images include the potential core for reference purposes. The PIV processing, however, inherently compromises resolution, yielding here a final vector resolution of 100×100 pixels from the original $1\text{k} \times 1\text{k}$ Mie scattering images. In order that the processed PIV results resolve fluctuations in the velocity field, the full PLIF imaging region is tiled by eight smaller PIV imaging windows, as indicated by the dashed boxes in Fig. 1*a*. For the particular scalar field shown in the figure, the simultaneous PIV field is given in Fig. 1*b*, and corresponds to the solid box in Fig. 1*b*. The essential differences that exist in the scalar and velocity fields are discussed by Hama (1962).

3.1 Mean scalar field structure and scaling

The averaged scalar field from the center plane, for the flush nozzle case, is compiled over 855 images and over 634 images for the plane at $z = 0.22\ rd$. By examining the trajectory of the points of maximum scalar concentration in the center plane, for both the flush and protruding nozzle cases, it is observed that the jet issues further into the crossflow from the flush nozzle. Using a fit of the type

$$(x/rd) = \mathrm{A}\ (y/rd)^{\mathrm{m}} \tag{1}$$

yields A = 1.95 and m = 0.302 for the flush nozzle data and A = 1.81 and m = 0.323 for the protruding nozzle data. The tendency of the flow in the flush nozzle case to penetrate further into the crossflow can be intuitively explained by noting that in that case, the jet initially flows through low-momentum fluid in the crossflow boundary layer. The analysis of velocity field trajectories suggests that the picture may be more complex, however (see following section).

The decay of the maximum scalar concentration, for both the flush and protruding nozzle cases, is shown in Fig. 2. The data are plotted in terms of the centerline downstream coordinate, s, which is determined through numerical integration of the best-fit trajectories defined by Eq. 1. Beyond the potential core, the maximum concentration $C(s)$ decays roughly as $1/s$, which is the decay dependence seen in simple jets. This contrasts with the results of Smith and Mungal (1998), for values of r ranging from 5 to 25, who found an initial decay faster than $1/s$. The results of Smith & Mungal also showed distinct branch points where the concentration decay abruptly slowed, possibly to an $s^{-2/3}$ dependence which would indicate wake-like behavior. However, the present data show a decay rate which actually increases beyond $s/rd \approx 2.5$. Also of interest is the agreement between the flush and protruding nozzle cases when the data are plotted in this centerline coordinate. The functional dependences of the decay curves are similar, as is the point at which the decay rate increases beyond $1/s$, though the potential core in the flush nozzle case penetrates further into s space.

Figure 3 shows the dependence of the flow widths on the centerline coordinate s. The flow nominal half-width for a given s is computed by first determining the maximum concentration, $C(s)$, then moving in the trajectory normal direction until the concentration drops to a specified fraction (here, 20%) of $C(s)$. Because the flow is not symmetric about the centerline trajectory in the imaging plane, it is necessary to compute these partial widths separately for both the $-x$ direction (i.e. towards the wake region) and the $+x$ direction (towards the jet outer edge). Figure 3 shows the results for δ_{-x}, δ_{+x}, and $\delta_{full} = \delta_{-x} + \delta_{+x}$. The lack of symmetry, with $\delta_{-x} > \delta_{+x}$, is clear. This asymmetry is due to jet fluid which is stripped away from the developing region of the jet by the crossflow and is deposited in the wake region (e.g. Fig. 1). For higher values of s, the $\delta(s)$ curves show evidence of an asymptotic approach to an $s \sim 1/3$ dependence, which would correspond to a wake-like scaling. Finally, the curves for both the flush and protruding nozzle cases are in good agreement for $s/rd > 1$, again showing the similarity in flow development for the two configurations when the centerline coordinate s is used. Interestingly, while the flow width, δ_{full}, appears for both the flush and protruding nozzle cases to be approaching a dependence on the centerline coordinate, s, consistent with a wake-like scaling, the concentration decay does not. These scaling laws rely upon self-similarity in the mean profiles, so it is reasonable to look to a lack of self-similarity to explain the departures from wake scaling seen in the data. The three-dimensionality of the flow will also be important.

Figure 4 shows the maximum scalar concentration as a function of s for the jet center plane, as well as for planes located 0.22, 0.45, 0.67 and 0.89 rd off of the center plane. The data are for the flush nozzle case; the protruding nozzle data are qualitatively similar. The coordinate s for all of these curves is determined from the jet centerline trajectory in the center plane, and the maximum scalar values are those in the in-plane normal direction to that trajectory. Of particular note in the figure is that the highest scalar concentration is not found in the center plane for $s/rd < 2$. Instead, the jet cross-section appears to take on a bimodal profile in which the scalar maxima lie off of the center plane. This is sufficient to explain the deviation from a wake-like concentration decay in the center plane, because the wake similarity profile has its peak on the centerline.

Figure 5 presents a different view of the three-dimensionality of the scalar field. Shown are the boundary contours for the scalar field in the center plane and in the planes $z = 0.22$, 0.45, 0.67 and 0.89 rd, together with the centerline trajectory for the center plane. The contours represent the loci of points in a given trajectory-normal plane where the local scalar value is 20% of the maximum scalar value in the center plane. On the outer boundary of the flow (i.e. the positive x side of the centerline), the deepest penetration occurs in the center plane, with progressively shallower penetration for increasing off-center position z. Meanwhile, on the wake side of the flow, with increasing y the boundaries for the off-center planes all lie further from the centerline trajectory than does the boundary in the center plane. This corresponds to a 'kidney'-like shape of the flow cross-section as one proceeds downstream. This cross-sectional shape arises because of the dominant role played by the counter-rotating vortex pair (CVP) in the large scale organization of the flow.

It is instructive that the current measurements show good agreement in decay rates and flow widths (when expressed in terms of s) between the flush and protruding nozzle cases, yet the decay rates differ from those reported by Smith & Mungal (1998), whose results were consistent over a range of velocity ratios r. The most notable difference is that Smith & Mungal used a top-hat jet exit velocity profile while the present data use a pipe flow profile. The conclusion drawn here is that the jet exit profile plays a larger role in establishing the scaling properties of the mixing field than does the jet nozzle position or the value of *r*.

3.2 Mean velocity field structure and scaling

The averaged velocity magnitude field in general reveals two distinct maxima. This bifurcated structure is shown clearly in Fig. 6, which shows the trajectory of the points of maximum velocity magnitude for the protruding nozzle case (with the crossflow speed subtracted from the vertical velocity component). The 'center streamline' plotted in the figure is found by performing a ray tracing from the center of the nozzle exit through the averaged velocity field. A fourth-order Runge-Kutta integration is used. The points of maximum velocity magnitude initially follow this center streamline quite closely. In the notation of Eq. 1, the fit to the center streamline has A = 1.92 and m = 0.342. Notably, the center streamline lies to the outside (the windward side) of the scalar centerline trajectory. At roughly $s = 1.3$ (measured along the center streamline), the points of maximum velocity magnitude abruptly shift to a trajectory lying in the wake region of the flow. The fit to this wake trajectory has A = 1.36 and m = 0.402.

It is commonly thought that the crossflowing jet can be viewed as approximating a pure jet in its near field and a wake in its far field. In attempting to identify these scalings, the choice of appropriate variables is important. For the near field, the jet is well described by the raw velocity magnitude $|\mathbf{u}| \equiv (u^2 + v^2)^{1/2}$. However, because wake scaling is defined in terms of the deficit velocity, the appropriate variable for the crossflowing jet far field is the magnitude of the velocity with the crossflow speed subtracted. In Fig. 7, the decay of this $|\mathbf{u} - v_\infty \mathbf{j}|$ term is plotted against the downstream coordinate *s*, for the flush nozzle case (the protruding nozzle case is qualitatively similar). The two decay curves represent the initial trajectory and later wake trajectory; the curves do not join smoothly because *s* is computed separately for each curve based upon the individual fits to Eq. 1. Also plotted in the figure are curves with slope -1 and -2/3. In the initial portion of the trajectory, and outside of the potential core, the magnitude decays faster than $1/s$, the pure jet rate, and does not appear to follow a power law scaling. In the wake portion of the trajectory, the magnitude appears to adhere to a power law decay with dependence very close to $s^{-2/3}$, suggesting a wake-like scaling.

The flow width of the $|\mathbf{u} - v_\infty \mathbf{j}|$ field, for both the protruding and flush nozzles, is shown in Fig. 8. Plotted are the full flow widths δ_{full}, defined as before as the distance between points, in the trajectory-normal direction, where the magnitude is 20% of the maximum value. (The flow width of the $|\mathbf{u}|$ field cannot be determined in this way because the freestream value of $|\mathbf{u}|$ is not zero.) When compared to the scalar growth rates shown in Fig. 3, the initial growth rate approximates the linear dependence of a pure jet. In Fig. 8, there is then an abrupt transition to a slower growth rate which is very close to the wake growth rate $s^{1/3}$. The curves for the flush and protruding nozzle cases are very similar, including the positions of the transition points, again pointing out the similarity in the flow scalings when expressed in terms of *s*.

The boundary contours for the $|\mathbf{u} - v_\infty \mathbf{j}|$ field in the center plane, and in the planes $z = 0.22$, 0.45 and 0.67 *rd*, for the flush nozzle case are shown in Fig. 9. The boundary contours are defined as in Fig. 5. Also plotted in Fig. 9 is the center streamline. The region enclosed by each contour is heavily weighted toward the $-x$ side of the center streamline, indicating the dominance of the wake region in the flow development. Interestingly, the kidney-shaped flow cross-section described by the scalar contours in Fig. 5 is not duplicated here. Instead, the velocity magnitude contours for the planes with higher z are contained within the contours for lower z, indicating that the cross-sectional contours of velocity magnitude are everywhere convex in shape. The absence of the out-of-plane velocity component perhaps explains the departure from the kidney-shaped cross-section. (The averaged out-of-plane velocity component in the center plane of the jet is identically zero by symmetry, and thus its absence does not affect the results shown in Figs. 6-8.)

3.3 Fluctuation quantities

Turbulence quantities involving the fluctuating velocity and scalar field terms are available due to the high resolution of the present data and are complementary to earlier data (e.g. Papaspyros et al., 1997). In particular, the simultaneous nature of the measurements permits the scalar flux components, $u'C'$ and $v'C'$, to be determined. From the evidence of previous computations, these scalar flux terms are very difficult to compute accurately (Alvarez et al. 1993). The appropriate resolution of the measurements is based upon the Batchelor scale $\lambda = \Lambda \cdot \delta\, Re_\delta^{-3/4}$ where Λ is a constant taken to be 15 following Su and Clemens (1999). Based upon this estimate the flow is resolved for $s/rd > 1$.

Scalar flux profiles

Figure 10 shows profiles of the scalar flux components $u'C'$ and $v'C'$, in the center plane for the flush nozzle case. Profiles are shown for y = 0.5, 1.0, 1.5 and 2.5 rd. The center streamline for each y position is indicated. The present data agree with the heat flux profiles reported by Andreopoulos (1983) in showing a negative correlation between $u'C'$ and $v'C'$, and in showing that the largest $u'C'$ values lie on the outer edge of the jet (larger x values). The present data differ, however, in showing distinct sign changes in the profiles, and in showing magnitudes of $u'C'$ with are noticeably larger than the magnitudes of $v'C'$. It should be noted that the data of Andreopoulos are for a flow with very low velocity ratio r = 0.5, which may be sufficient to explain the discrepancies. This notion is supported by the large-eddy simulations of Yuan (1997) at a velocity ratio, r = 3.3, closer to that of the present measurements. Yuan reported $u'C'$ profiles which show sign changes similar to those in Fig. 10.

Turbulent shear stress

Figure 11 shows the $u'v'$ turbulent shear stress field for the center plane for the flush nozzle case $u'v'$ for y = 1.0, 1.5 and 2.5 rd. In the initial region of the jet, high negative values of $u'v'$ are found on the windward $-y$ side of the jet, and high positive values are found on the wake side. These stresses are identified as resulting from the action of spanwise vorticity in the near field (Yuan, 1997), which arises from the Kelvin-Helmholtz instability. As the flow evolves, negative values of $u'v'$ begin to predominate. For increasing y, the regions of positive $u'v'$ become progressively weaker relative to the regions of negative values. In contrast to the near field, the shear stresses in the downstream portion of the flow arise from the shear between the jet and crossflow fluid.

4 Concluding remarks

Analysis of the structure and scaling of the velocity and scalar fields based upon simultaneous measurements is described here. The use of a pipe velocity profile is seen to be different than a top-hat velocity profile in the jet development. Further work includes analysis of the mean vorticity, aiming to determine the role played by the vorticity in the transition from near to far field properties in the flow. The present two-dimensional measurements yield only a limited view of the vorticity dynamics, however, because the vorticity field is dominated by three-dimensional structures. In addition, comparison with other experiments, for example those of Andreopoulos and Rodi (1984), at different velocity ratios and with different boundary conditions should prove insightful. The resolution of the present measurements also allows the determination of small scale gradients, permitting analyses of the interactions between the scalar gradient, vorticity and strain fields, for example. Finally, it is hoped that the results will serve as benchmark data for future simulations of crossflowing jets at similar velocity ratios. Additional details on this work can be found in Su et al. (2000).

References

Broadwell, J.E. and Breidenthal, R.E. (1984) Structure and mixing of a transverse jet in incompressible flow. *J. Fluid. Mech.* 148: 405-412.

Durando, N.A. (1971) Vortices induced in a jet by a subsonic cross flow. *AIAA J.* 9: 325-327.

Fric T.F. and Roshko A. (1994) Vortical structure in the wake of a transverse jet. *J. Fluid. Mech.* 279: 1-47.

Hasselbrink, E. F., Jr. (1999) Transverse Jets and Jet Flames: Structure, Scaling, and Effects of Heat Release. Ph.D. Thesis, Stanford University.

Hama, F. R. (1962) Streaklines in a perturbed shear flow. *Phys. Fluids*, 5 (6), 644-650.

Hasselbrink, E. F., Jr. and Mungal, M. G. (1996) An analysis of the time-averaged far-field properties of the transverse jet. AIAA Paper 96-0201, *34th Aerospace Sciences Meeting*, Reno, NV.

Karagozian, A. R. (1986) An analytical model for the vorticity associated with a transverse jet. *AIAA J.* 24: 429-436.

Papaspyros, J. N. E., Kastrinakis, E. G. and Nychas, S. G. (1997). Coherent contribution to turbulent mixing of a jet in cross flow. *Applied Scientific Research*, 57:291-307.

Pratte, B.D. and Baines, W. D. (1967) Profiles of the round turbulent jet in a cross flow. *J. Hydraul. Div. ASCE* 92: 53-64.

Schlüter, J.U. and Schönfeld, T (2000) LES of jets in cross flow and its application to a gas turbine burner, *Flow Turb. and Comb.*, 65 (2), 177-203.

Smith, S.H. and Mungal, M. G. (1998) Mixing, structure and scaling of the jet in crossflow. *J. Fluid Mech* 357: 83-122.

Su, L. K. (1998) Measurements of the three-dimensional scalar dissipation rate in gas-phase planar turbulent jets. In *Annual Research Briefs*, Center for Turbulence Research, Stanford, CA.

Su, L. K. and Clemens N. T., (1999) Planar measurements of the full three-dimensional scalar dissipation rate in gas-phase turbulent flows. *Exp. Fluids*, 27(6), 507-521.

Su, L. K., D. Han, R. M. Miraflor and M. G. Mungal (2000) "Measurements of Scalar and Velocity Fields in Turbulent Crossflowing Jets with Low Velocity Ratio", AIAA-2000-0815, *AIAA 38th Aerospace Sciences Meeting*, Reno, NV.

Yuan, L.L. (1997) Large eddy simulations of a jet in crossflow. PhD Thesis, Stanford University, (1997).

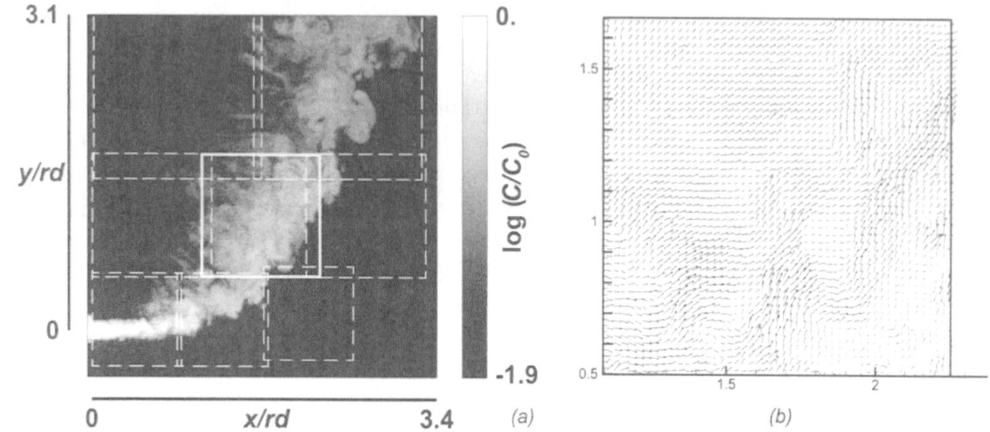

Fig. 1 A sample image pair, from the center plane of the protruding nozzle case. (a) The scalar field, showing the 8 imaging subwindows used for PIV. (b) The PIV vector field corresponding to the solid window in (a).

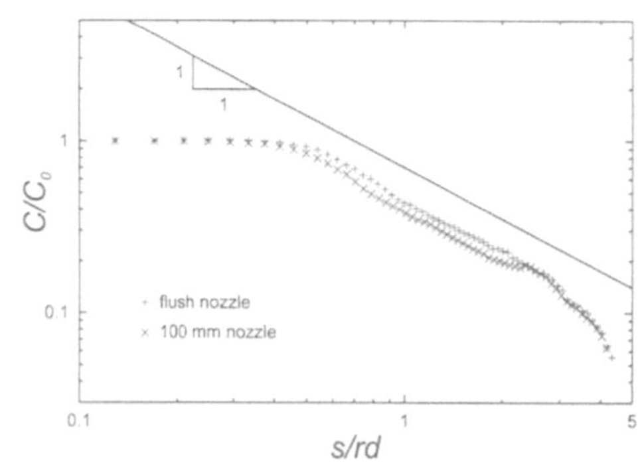

Fig. 2 . The decay of maximum scalar concentration with downstream distance in the center plane.

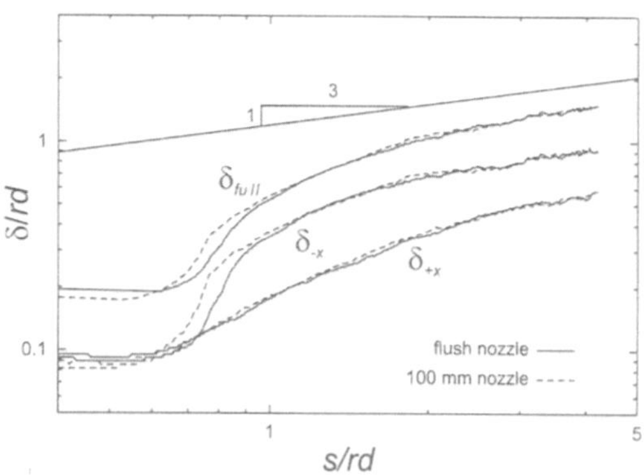

Fig. 3 Scalar field flow widths in the center plane

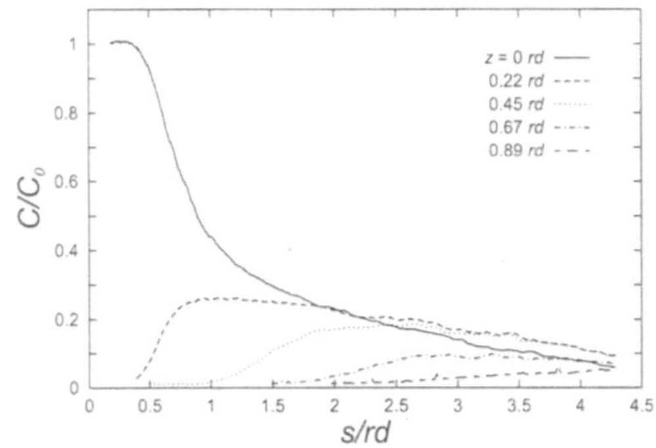

Fig. 4 The decay of maximum scalar concentration with downstream distance, for the flush nozzle case

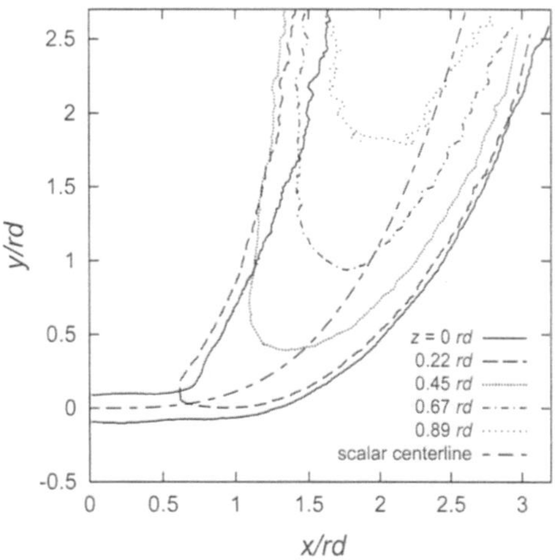

Fig. 5 Boundary contours for the scalar field

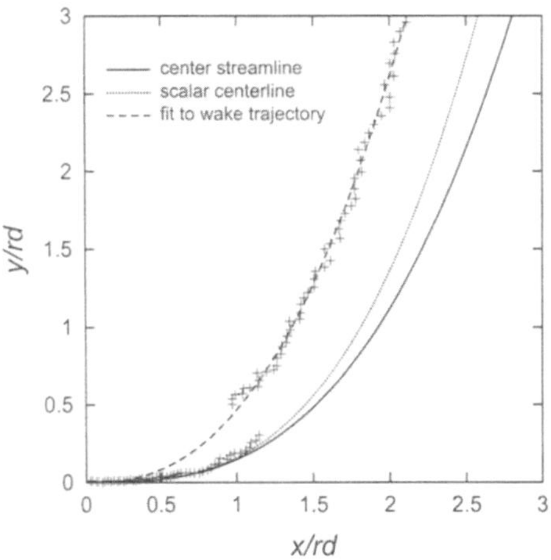

Fig. 6 The bifurcated structure of the local maxima in the crossflow-subtracted velocity magnitude field $|\mathbf{u} - v_\infty \mathbf{j}|$. Data from the protruding nozzle case.

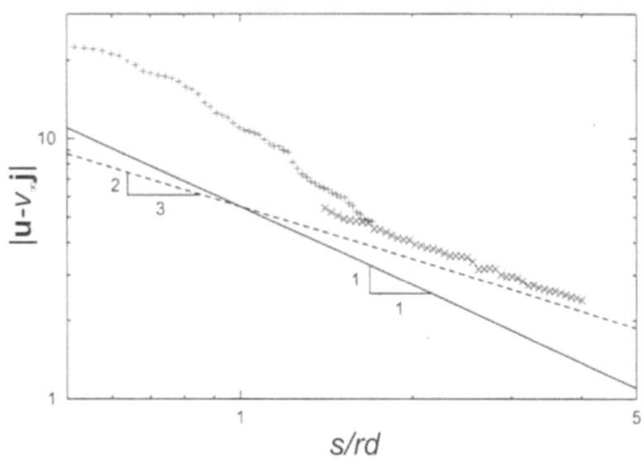

Fig. 7 Downstream decay of the crossflow-subtracted velocity magnitude $|\mathbf{u} - v_\infty \mathbf{j}|$, in the center plane for the flush nozzle case.

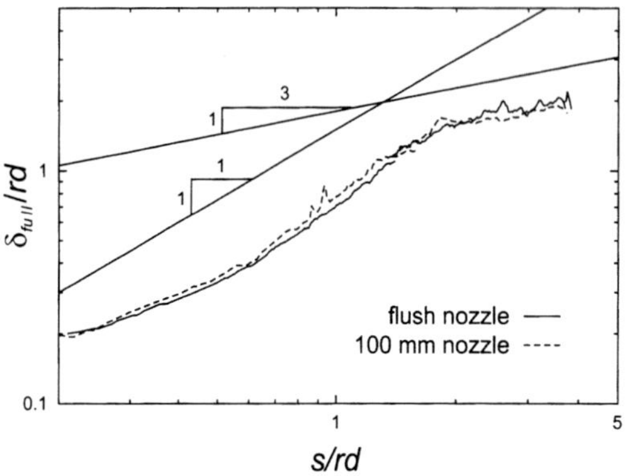

Fig. 8 Downstream evolution of the flow width of the |**u** - v$_\infty$**j**| field, in the center plane.

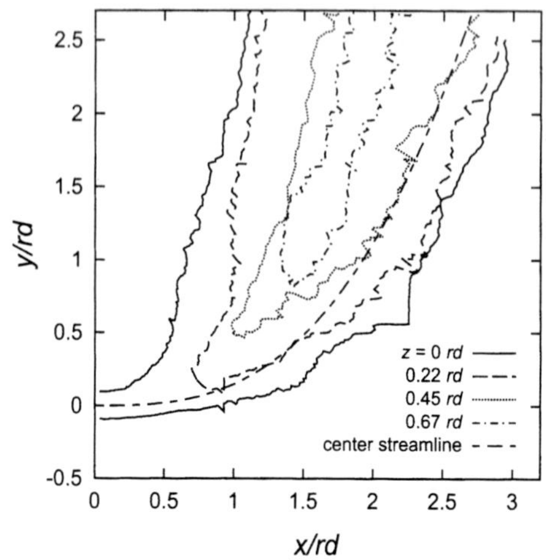

Fig. 9 Boundary contours for the |**u** - v$_\infty$**j**| field.

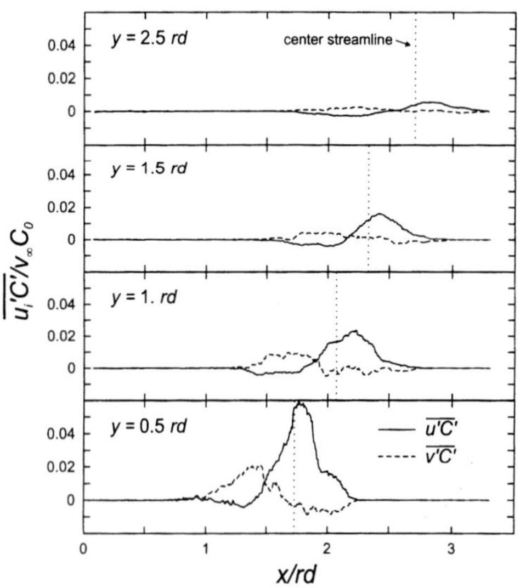

Fig. 10 Scalar flux profiles in the center plane, for the flush nozzle case.

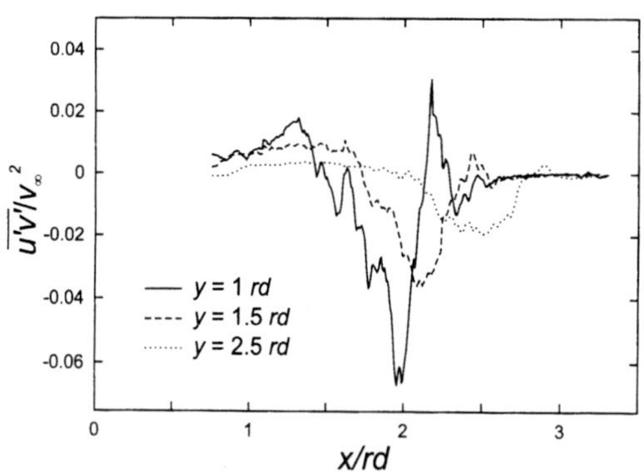

Fig. 11 Profiles of the $u'v'$ stress term, in the center plane for the flush nozzle case

Numerical appraisal of jet-to-crossflow coupling in a transverse jet

Marina Campolo, Gian Maria Degano, Alfredo Soldati * and Luca Cortelezzi[2]

Centro di Fluidodinamica e Idraulica and Dipartimento di Scienze e Tecnologie Chimiche, University of Udine, 33100, Udine, Italy

[2]Department of Mechanical Engineering, McGill University, Montreal, H3A 2K6, Canada

Abstract. The nearfield evolution of a jet in crossflow is strictly related to the coupling between the flow in the pipe and the transverse flow. In this work, we address the problem of identifying the extent of this coupling for an incompressible fluid and we examine the influence of the inlet boundary conditions on the simulation of the transverse jet. First, we validate the numerical computations using experimental data by Shandorov (Abramovich, 1963). We examine in detail the flow field calculated numerically for a jet-to-crossflow velocity ratio equal to 2.2. We evaluate in particular the evolution of the streamwise velocity profile and vorticity in the crossflow direction to characterize the structure of the jet. Second, we evaluate the effect of increasing the jet-to-crossflow velocity ratios on jet penetration. Third, we compare the numerical results obtained considering the pipe and imposing a uniform velocity profile at the jet exit. Results show that the jet-to-crossflow velocity ratio and the inlet boundary condition for the jet influence jet penetration into the crossflow and jet velocity distribution.

1 Introduction

The study of jets in crossflow is of significant importance to both engineering applications (exhaust gas cooling, combustion emission control and spray coating) and environmental applications (chimney plumes, sewerage outfalls and agricultural spraying systems). In all these applications, the object is to quantify and, if possible, control the mixing between the injectant entrained and the crossflow. Regardless of the physical or chemical nature of the injectant, as this can be heavy particles, droplets, thermal tracers or chemical species, mixing is controlled by the coherent vortical structures which characterize the jet in crossflow (Smith and Mungal, 1998). Description of the coherent vortical structures arising from the complex interaction between the jet and the transverse flow – jet shear layer vortices, horseshoe vortices, counter-rotating vortex pair (CVP) and wake vortices – can be found in Fric and Roshko (1994) and Kelso et al. (1996) among others. These articles also discuss the mechanisms proposed for formation of the coherent structures.

Early studies on jets in crossflow (Abramovich, 1963, Andreopoulos and Rodi, 1984, Broadwell and Breidenthal, 1984, Andreopoulos, 1985, Karagozian, 1986, Sykes et al., 1986, Needham

* Corresponding author. Email: soldati@uniud.it; Ph.: +390432558864; Facs: +390432558803; www: 158.110.50.22

et al., 1988) focused on the effect of the jet-to-crossflow velocity ratio on penetration and deflection of the jet. From these works, the development of the CVP system (Moussa et al., 1977) was soon identified as one of the main features determining the fluid dynamic behavior of the jet in most practical applications. Recently, investigations extended to different shapes of the jet exit (Sivadas et al., 1997, Findlay et al., 1999), to multiple jets (Leboeuf et al., 1991, Ajersch et al., 1997), and to normal impinging jets (Knowles and Bray, 1991), corresponding to practical interest and more complex configurations of transverse jets. A parallel effort has been dedicated to more fundamental investigations which are based on experiments (Kelso et al., 1996) and numerical simulations (Coelho and Hunt, 1989, Sykes et al., 1986, Hsu and Guo, 2000) aimed first at identifying and visualizing the coherent structures (Papaspyros et al., 1997, Smith and Mungal, 1998, Yuan et al., 1999, Rivero et al., 2001), then at verifying the mechanisms proposed for the generation of coherent structures (Rudman, 1996, Kelso et al., 1998, Cortelezzi and Karagozian, 2001), and finally at identifying strategies for optimal control of mixing efficiency by modulation of the crossflow (Xia and Lan, 1997) or by excitation of the jet (Chang and Vakili, 1995, Davis and Glezer, 1999, M'Closkey et al., 2002).

The extensive work on transverse jets has pointed out that the conditions at the jet-exit influence the near-field of the jet and, in turn, mixing in the near-field region (Moussa et al., 1977, Andreopoulos and Rodi, 1984, Degano et al., 2002). It is not easy, even in experiments, to know exactly the conditions at the jet-exit and this is a critical point, since most numerical simulations are performed by fixing the velocity profile at the jet-exit (Yuan *et al.*, 1999, Hsu and Guo, 2000).

The aim of the present paper is to increase the physical understanding of the mean effects of coupling between the pipe flow and the oncoming crossflow when different boundary conditions are used at the jet-exit. We use a numerical approach to simulate the flow field generated by an incompressible jet issuing from a 25 diameter long pipe perpendicular to a channel flow. Even though a number of experimental works report asymmetric velocity and scalar concentration profiles (Smith and Mungal, 1998), these results are found only at a jet-to-crossflow velocity ratio larger than 10 and in the far field, i.e. more than 25 jet diameters downstream the jet exit, where the influence of the inlet conditions is long forgotten. In this work, we focus on the jet near field, influenced by inlet conditions and on a jet-to-crossflow velocity ratio smaller than 10. In these simplified conditions, limiting our investigation to the time-averaged flow field, we will consider the jet in crossflow, time-averaged and symmetric about the jet-center plane. Furthermore, to avoid all possible complicating effects, we will consider the fluid to be incompressible. We will use the experiment by Shandorov (Abramovich, 1963) as a reference to validate our numerical simulations.

Future efforts will be focused on characterizing this interaction with fully three-dimensional, time-dependent, computer-demanding simulations and on characterizing this interaction for a compressible fluid.

2 Computational Methodology

Laboratory experiments on jets in crossflow are often conducted in wind or water tunnels. The computational domain used in this study is shown in Figure 1. The reference coordinate system is centered at the jet-exit with x, y and z representing streamwise, spanwise and vertical directions, respectively. The test section of the wind/water tunnel is represented by a box and a pipe represents the nozzle and plenum.

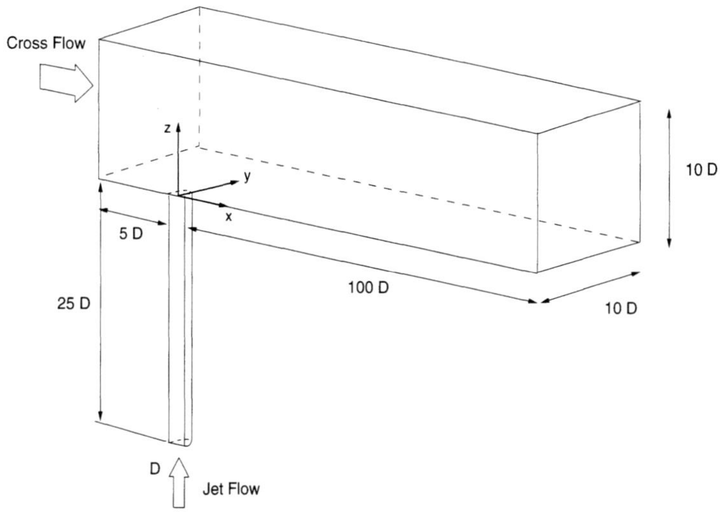

Figure 1. Geometry of considered experimental apparatus and reference coordinate system, centered at jet exit. Box represents a test section of wind tunnel and pipe models the jet outlet.

The jet in crossflow is an unsteady phenomenon (Fric and Roshko, 1994) and time-dependent numerical solutions are required to reproduce the unsteady behavior of the flow, such as oscillating wake structures (Rudman, 1996, Yuan et al., 1999, Hsu and Guo, 2000). In this work, which is a preliminary investigation into the coupling of the flow in the pipe and the crossflow, we examine the time-averaged structure of the flow – steady flow – for an incompressible fluid, with the object of testing a large number of flow conditions. This approach allows us to reduce significantly the computing time required. The steady state solution enables us to introduce a further simplification: since during the time-averaging procedure all transverse disturbances are averaged out, the flow field is symmetric with respect to the x axis. We therefore impose symmetry with respect to the jet $x - z$ plane and we solve for the flow in a geometry that is half of the entire channel.

To make the problem dimensionless, we use the orifice diameter D as a characteristic length and D/U_∞ as a characteristic time, where U_∞ is the crossflow mean velocity. Using dimensionless coordinates, the wind tunnel computational domain is 106 D long, 10 D wide and 10 D high, and the pipe is 25 D long. With reference to Figure 1, the crossflow enters the box 5 D upstream the jet orifice and the computational box extends 100 D downstream the jet orifice.

The structure of a jet issuing in a crossflow is characterized by different length scales (Smith and Mungal, 1998). We use a finite volume discretization generating nearly hexahedric cells and we refine the grid in the pipe and in the near-field of the jet adopting a hierarchical cell refinement to capture the relevant flow phenomena. We determine the final geometry and size of the computational domain through a grid sensitivity analysis. The final computational mesh is made up of 630,152 cells, of which 57,600 for the pipe and 327,680 for the region near the jet-exit, respectively. These two regions correspond to about 2‰ and 8‰ of the overall volume, respectively, but such precise grid refinement is necessary to ensure a precise calculation.

To simulate the flow, we impose (1) a uniform velocity profile of the crossflow at the box inlet and (2) a uniform velocity profile at the pipe inlet. Furthermore, we impose (3) a symmetry condition at the symmetry $x - z$ plane, (4) a pressure condition at the box outlet and (5) no-slip

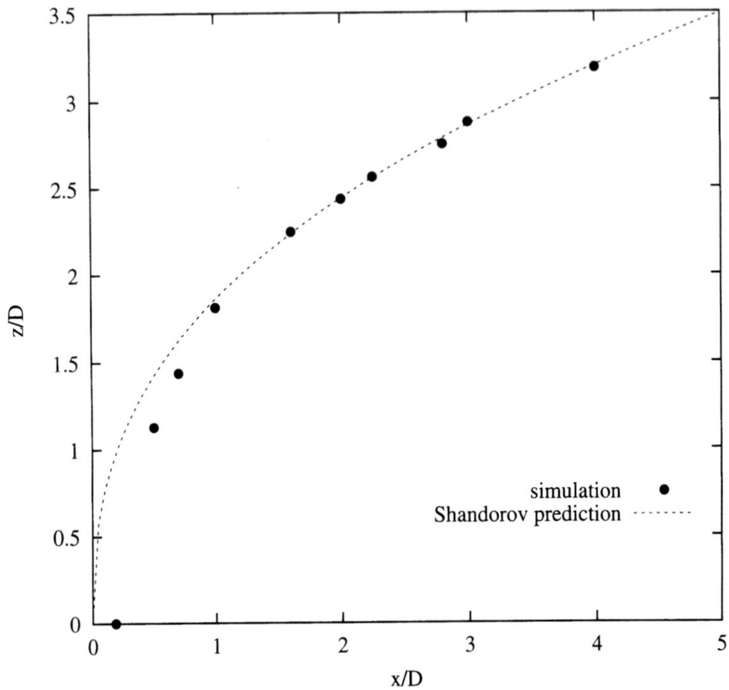

Figure 2. Comparison of axis of the jet determined experimentally by Shandorov (continuous line) and result of our computation (points). Experiments and computations match closely downstream the jet exit and differ over the first diameter, nearest to jet outlet, owing to the different type of coupling between crossflow and jet-flow near to the jet-exit.

conditions at all other boundaries. We use a pressure condition to avoid a dominant coupling between the outlet condition and the jet field evolution (Sani and Gresho, 1994).

To assess our numerical methodology, we reproduce numerically the conditions investigated experimentally by Shandorov (Abramovich, 1963). These conditions correspond to a jet Reynolds number, defined as $Re = U_{jet}D/\nu$, of more than 10^5, well into the fully turbulent regime, and by a dimensionless parameter R, defined as the jet velocity to crossflow velocity ratio, $R = U_{jet}/U_{\infty}$, equal to 2.2. We solve for the fluid balance equations in the Reynolds-averaged form using a commercial code (Star-CD) and we employ a standard $k - \epsilon$ model to reproduce the effects of turbulent fluctuations. The $k - \epsilon$ model is complemented by the algebraic *'law of the wall'* to reproduce the turbulent flow up to the near-wall regions. The $k - \epsilon$ model is simple and its accuracy can be considered comparable to that of more complex turbulence models (Renormalization Group $k - \epsilon$ models, Large Eddy Simulation models, etc.), provided that the grid is sufficiently refined near the boundaries (Knowles and Bray, 1991).

Our investigation is based on a number of simulations for different values of R, considering a fluid with fixed physical properties and changing the velocity of the jet.

(a) (b)

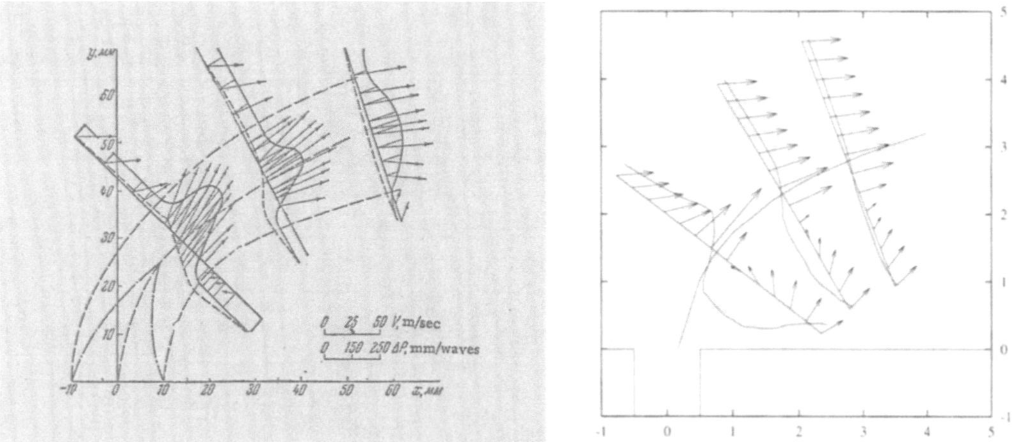

Figure 3. Velocity and pressure profiles in the plane of symmetry of a jet in crossflow: (a) measurements reported by Abramovich (1963) and (b) calculated by our simulation.

3 Results and Discussion

3.1 Analysis of pressure and velocity fields

In the first place, we aim at reproducing the experiment performed in 1947 by Shandorov, reported in the book by Abramovich (1963). These conditions correspond to a Reynolds number for the channel flow of $Re = 106,398$ and a jet-to-crossflow velocity ratio of $R = 2.2$. In his experiment, Shandorov used a jet of air injected at $U_{jet} = 77.6\,m/s$ into a crossflow of air at $U_\infty = 35.6\,m/s$ using a nozzle with an orifice diameter $D = 0.02\,m$. Since the Mach number of the jet is 0.22 and 0.10 for the crossflow, the flow can be considered incompressible and we can validate the results of our computations against the experimental data obtained by Shandorov.

The features of a jet in crossflow are conveniently discussed with reference to the jet axis. In analogy with the free jet and as reported by Abramovich (1963), the axis of the jet is defined as the locus of the points of maximum velocity. We calculate the position of the jet axis in order to establish a direct comparison with the experimental data by Shandorov and obtain a quantitative evaluation of the accuracy of our numerical simulations.

In Figure 2, the axis of the jet determined experimentally by Shandorov (dashed line) is compared against the result of our computation (solid circles). In the coordinate system used, the centerline of the pipe is at $x/D = 0$ and the pipe wall is at $x/D = \pm 0.5$. The equation of the jet axis is

$$\frac{z}{D} = R^{0.79}\left(\frac{x}{D}\right)^{0.39} \tag{1}$$

as reported by Abramovich (1963).

Experiments and computations match closely downstream the jet-exit, whereas our results differ from experiments over the first diameter, i.e. in region nearest to the jet outlet. This difference is due to the different experimental and numerical flow conditions near to the jet-exit. In the experiment, a nozzle was used to generate a top-hat similar exit velocity profile, whereas we use a pipe. This different velocity profile at the jet-exit influences significantly the initial point of the jet axis and jet development in the near-field (Antonia and Zhao, 2001).

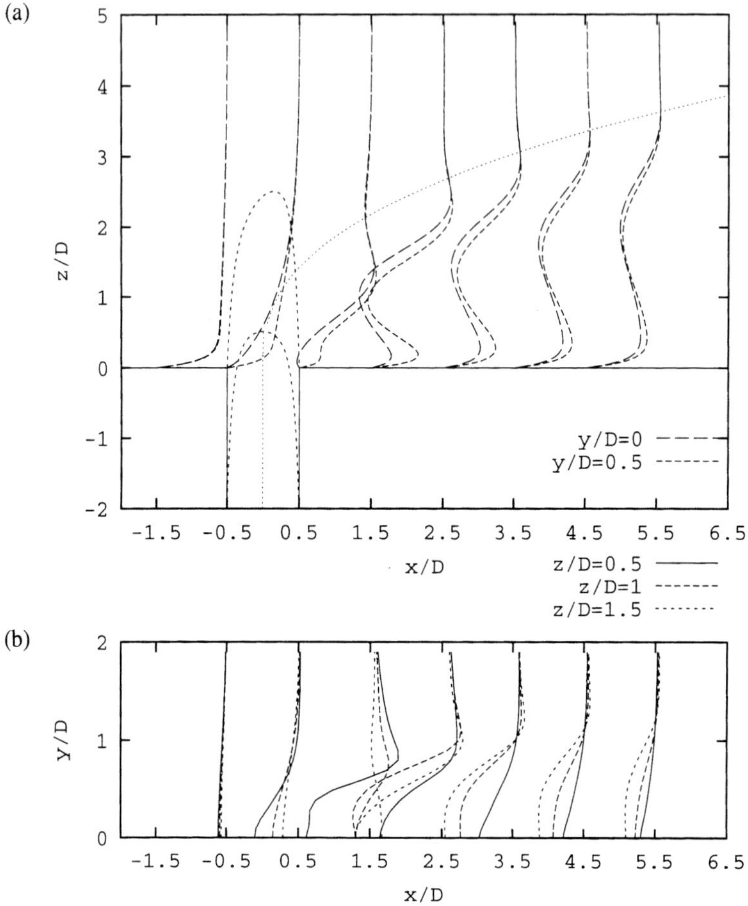

Figure 4. Interaction between crossflow and jet: (a) streamwise velocity profiles for the crossflow for $x/D = -1.5 \div 4.5$, step 1 at $y/D = 0$ (long dash lines) and $y/D = 0.5$ (short dash lines), pipe-flow profiles (very short dash lines) at $z/D = -2$ and $z/D = 0$, and axis of the jet (dotted line). (b) streamwise velocity profiles in the spanwise direction for $z/D = 0.5, 1, 1.5$.

We should remark here that the velocity profile in the pipe is fully developed already 15 D after the pipe inlet, i.e. 10 D before the exit to the channel flow. We compared velocity profile, kinetic energy profile and dissipation profile against our DNS database (Soldati and Banerjee, 1998) and we found a satisfactory agreement, with quantitative discrepancies of 15% at most.

To discuss the downstream evolution of the flow structure, it is useful to examine velocity and pressure distribution along cross-sections orthogonal to the jet axis. In Figure 3 (a), we report measurements of velocity, represented by vectors, and total and static pressure distributions, represented by solid and dashed lines, respectively as taken by Shandorov (Abramovich, 1963). We can observe that the total pressure changes noticeably going from the upper edge to the core of the jet, down to the lower side. At the upper side of the jet the pressure increases, as expected for a flow – the crossflow – encountering an obstacle – the jet issuing from the pipe. Similarly, a low pressure region shows up at the lower side. However, the jet behaves differently from a rigid obstacle, and the total pressure variation induced within the core of the jet depends on the jet velocity distribution. If we consider the velocity distribution and we remark that the axis of the jet is defined as the locus of the points of maximum velocity, we can observe that the locus of points of maximum total pressure is shifted from the jet axis toward the upper edge of the jet. Considering now the behavior of the static pressure, we observe an increase toward the upper edge of the jet. This increase is less than that expected at the wall of a solid body of the same form of the curved jet. As reported by Abramovich (1963), the cause is fluid entrainment from the crossflow. After few diameters, the distribution of static pressure is almost uniform over the cross-section orthogonal to the jet axis.

In Figure 3 (b), we show velocity (vectors) and static pressure distribution (black line) of the jet as calculated by the present numerical simulation. The behavior of the distribution of velocity vectors is similar to that obtained experimentally by Shandorov. The qualitative trend of the static pressure profile is also in agreement with the measured data. This suggests that the complex interaction between the jet and the crossflow can be correctly calculated using the current model with RANS approach.

To examine in more detail the interactions between the crossflow and the jet, we analyze the variation of velocity profiles in the region near to the jet exit. In Figure 4 (a), we show the vertical profiles of streamwise velocity probed at $-1.5\ D, -0.5\ D, 0.5\ D, 1.5\ D, 2.5\ D, 3.5\ D$ and $4.5\ D$ with respect to the pipe axis. Long dash lines represent profiles taken at $y/D = 0$, i.e. along the symmetry plane. Short dash lines represent profiles taken at $y/D = 0.5$. We show also the jet velocity profile $-2\ D$ inside the pipe and at the pipe exit (very short dash lines), and the jet axis (dotted line).

Consider first the profiles in the symmetry plane (long dash lines). We can observe that the velocity profile for the crossflow approaching the jet-exit is that of a boundary layer. The velocity profile of the jet inside the pipe up to $-2\ D$ from the jet-exit is that of a fully developed pipe-flow. When the two streams merge at the jet-exit, first, the profile of the pipe flow becomes asymmetric, and second, the profile of the wall boundary layer is modified. Apparently, the crossflow and the jet merge, with the crossflow trying to slide past the jet. Consider the behavior of the streamwise velocity profile along the symmetry plane. In the region close to the wall, we observe an initial deceleration of the crossflow. Because of mass conservation, corresponding to the deceleration at the pipe exit, there is an acceleration at the side of the pipe (short dash line), for all regions under the jet axis. This phenomenon is less pronounced than for a crossflow around a solid object. Right downstream the jet exit, we observe a region with negative streamwise velocity corresponding to the lower part of the jet and confirming Figure 3 (a) and (b). Similar profiles are also reported by Andreopoulos and Rodi (1984) who gathered data for a jet-to-crossflow velocity ratio equal to 2., very similar to the one considered in this work. They observed that for $x/D = 1$ and 2 the velocity profiles show clearly the wake behavior of the flow in the lower side of the jet.

We evaluate the wake behavior of the flow in the lower side of the jet considering the spanwise variation of streamwise velocity profiles, as shown in Figure 4 (b). We consider the same vertical planes of Figure 4 (a) and we evaluate spanwise profiles at $z/D = 0.5$ (solid line), $z/D = 1$ (short dash line) and $z/D = 1.5$ (very short dash line), respectively. We observe that upstream of the jet-exit, the streamwise velocity is uniform along the vertical, indicating an unperturbed crossflow. At $x/D = -0.5$, i.e. just upstream the jet exit, the velocity decreases near the symmetry plane, whereas the velocity profiles are almost unaffected at larger distance from from the symmetry plane. Downstream the jet exit, the streamwise velocity drops to zero in the lower plane ($z/D = 0.5$). From $x/D = 0.5$ to $x/D = 4.5$, we can observe the development of a recirculation region just below the lower side of the jet. The velocity minimum moves from the lower plane, $z/D = 0.5$, to the upper planes, $z/D = 1$ and $z/D = 1.5$ indicating that the recirculation zone bends over following the jet development.

3.2 Analysis of the counter-rotating vortex pair

The interaction between the jet and the crossflow produces a number of fluid dynamic structures. The counter-rotating vortex pair (CVP) originating at the jet exit is among the most important, since it is responsible of the crossflow entrainment in the jet and appears to dominate convective mixing. Coherent structures like CVP can be identified by regions of high vorticity and low pressure (Yuan et al. 1999).

In Figure 5 (a), we show the vorticity magnitude distribution over the $y - z$ planes located from $x = -1\ D$ to $x = 5\ D$ with step 1 D. In the plane located at $x/D = -1$, vorticity is confined close to the wall. In our experiment, we use a uniform crossflow inlet velocity and the boundary layer is not fully developed. This confines the vorticity, which is primarily shear-generated, spanwise vorticity, very close to the wall. Considering now the cross-section at $x/D = 0$, we can observe that the pipe boundary layer is advected out of the pipe mouth (Moussa et al., 1977). The action of the crossflow contributes to bend the vortex lines contributing to increase the streamwise vorticity.

To characterize better the CVP structures, we examine the distribution of the streamwise component of vorticity, as shown in Figure 5 (b). We choose a color map covering only negative values since the rotation expected for the CVP vortex is clockwise in the half plane considered. The shape of the vorticity iso-contours is in good agreement with that reported by Moussa et al. (1977). In Figure 5 (b), we observe that negative vorticity associated to the counter rotating vortex moves away from the wall. In the near wall region, downstream of the jet, a positive streamwise vorticity region develops. As discussed in Kelso et al. (1996), the flat-wall boundary layer separates owing to an adverse pressure gradient on the downstream side of the jet. This was already observed by Khan and Johnston (2000), who found two distinct vortical region downstream the jet exit. They found negative vorticity associated with the CVP vortex and positive vorticity associated with the spanwise shear layer created by the secondary flow over the wall. In Figure 5 (c), we show the iso-contours of the streamwise component of vorticity over the wall of the tunnel at $z/D = 0$. We find positive vorticity downstream the jet exit, with the separation lines, defined by $\omega_x = 0$, located along the symmetry plane (centerline reattachment point) and inclined $+60^o$ from the x axis (upstream boundary layer separation line). These positions are those already observed by Kelso et al. (1996) using dye visualizations.

3.3 Influence of the crossflow in the upstream pipe

The analysis presented in the previous section confirms the adequacy of the approach adopted in this work to investigate on the structure of the jet in crossflow. In particular, we were able to obtain qualitative results in agreement with previous experiments (Abramovich, 1963, Moussa et al., 1977, Andreopoulos and Rodi, 1984 and Kelso et al., 1996) or more sophisticated simulations (Sykes et al., 1986, Yuan et al., 1999). Our object now is to examine the influence of the pipe flow on the development of the transverse jet.

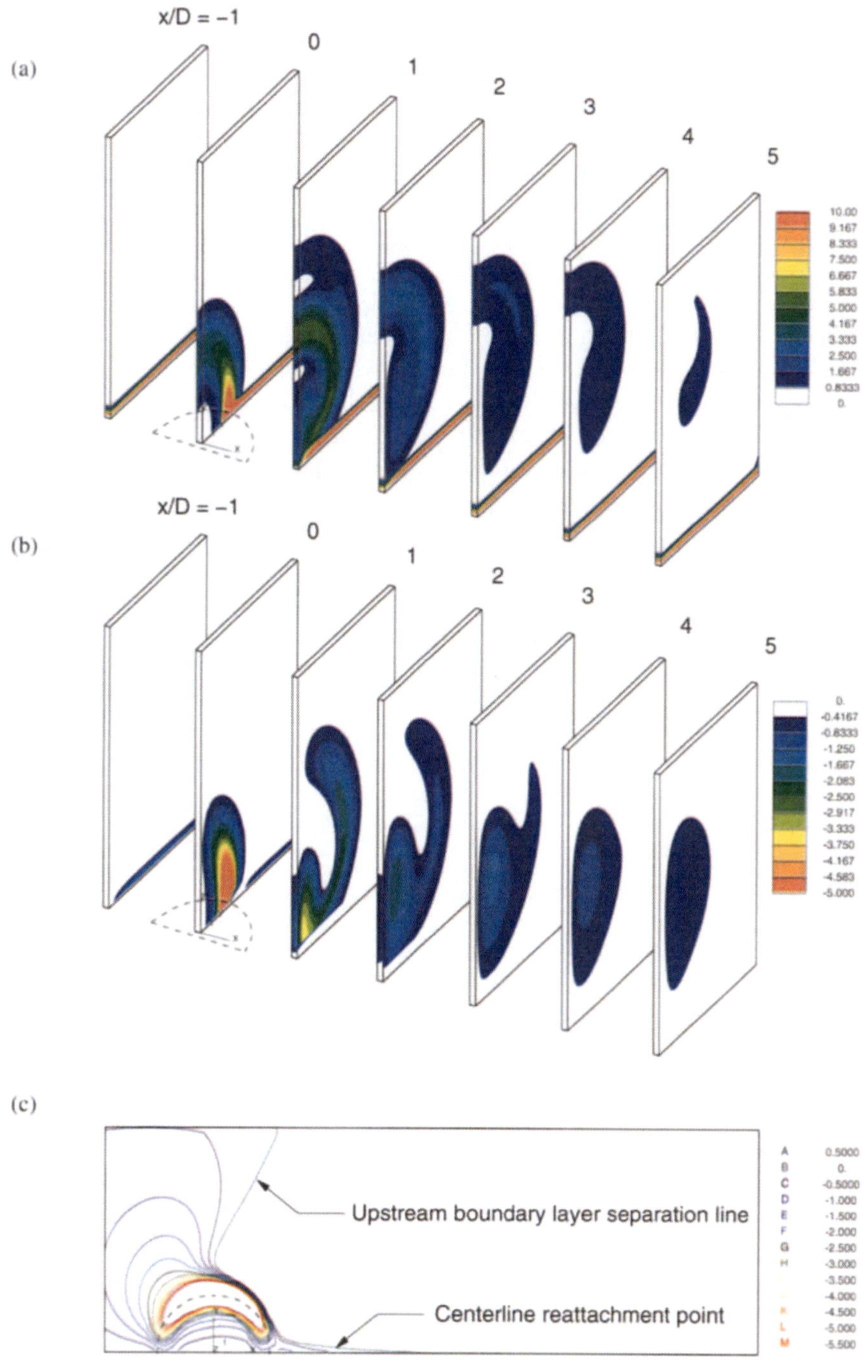

Figure 5. Iso-contours of vorticity in the near field region: (a) vorticity magnitude and (b) streamwise (x) component of vorticity for planes orthogonal to the x axis and located from $-D$ to 5 D, step 1 D, away from the pipe axis. (c) x-component of vorticity at the wall of the tunnel.

To characterize the coupling between the crossflow and the flow in the pipe, we simulate the time-averaged flow field generated by an incompressible jet entering the crossflow with different velocities. We examine different jet-to-crossflow velocity ratios, corresponding to $R = 2.2$, 3, 4, 5, 6 and 7. The averaged velocity profile inside the pipe for a jet in a quiescent fluid is axisymmetric whereas interaction with a crossflow makes this profile non-symmetric. This has been observed in experimental (Andreopoulos and Rodi, 1984, Kelso et al., 1996, Khan and Johnston, 2000) and numerical investigations (Ajersch et al., 1997).

The first objective of our analysis is to quantify the degree of asymmetry in the final stretch of the pipe and the extent of this interaction between crossflow and jet. In Figure 6, for each value of the jet-to-crossflow velocity ratio, we show the iso-contours of the axial velocity at the pipe outlet and the axial velocity distribution in the final stretch of the pipe for the different values of R examined. Owing to the crossflow – directed from left to right in Figure 6 – the axial velocity iso-contour lines lose axial symmetry with the maximum shifted downstream. For $R = 2.2$, the maximum of the axial velocity is found $x/D = 1/6$ downstream the pipe centerline. This is the initial position of the jet axis shown in Figure 2. The downstream shift of the axial velocity iso-contours at the jet-exit reduces progressively with increasing R.

To evaluate quantitatively the degree of asymmetry produced in the pipe by the crossflow, we examine the difference in velocity in the final stretch of the pipe upstream and downstream the pipe axis (Figure 6, d, e, f, j, k and l). We choose 10 lines on the symmetry plane in the pipe which are parallel to the pipe axis and $r/D = \pm 0.03$, ± 0.16, ± 0.28, ± 0.40 and ± 0.46 away from the pipe axis. Then, for each of these lines, we plot the value of the axial velocity at different axial distances from the pipe exit, using solid and dashed lines for the upstream and downstream portion of the pipe, respectively.

In a symmetric pipe flow, each pair of curves collapses onto the same straight line. In the present case, we observe that the two curves separate, with velocity on the downstream side increasing with respect to velocity on the upstream side. This effect reduces for increasing R (Andreopoulos and Rodi, 1984).

To quantify the influence of the crossflow inside the pipe, we measure where the pairs of curves separate. We calculate the z_p coordinates at which the difference between the velocity curves is more than 1% relative to the crossflow velocity for $R = 2.2$. We scale this value for the simulations at different R to keep constant the error relative to the jet velocity. The z_p coordinates calculated in this way are represented in Figure 6 (d, e, f, j, k and l) as horizontal dotted lines. For $R = 7$, the difference between the velocity curves is less than the reference scaled error everywhere inside the pipe ($z_p = 0$).

The points of maximum depth of the interaction obtained for each value of R are shown also in Figure 7. Apparently, the depth of interaction is higher at lower velocity ratios, and decreases linearly as R increases. Finally, the influence of the crossflow within the pipe is expected to vanish for R larger than 7.

3.4 Influence of the Inlet Boundary Condition

The third objective of this paper is to determine the influence of the boundary condition at the jet inlet. A common approach for the numerical modeling of jets in crossflow is to impose a given velocity profile at the jet orifice, ignoring interaction between the pipe and the crossflow. Uniform velocity profiles (Sykes et al., 1986) and fully-developed turbulent pipe flow profiles (Yuan

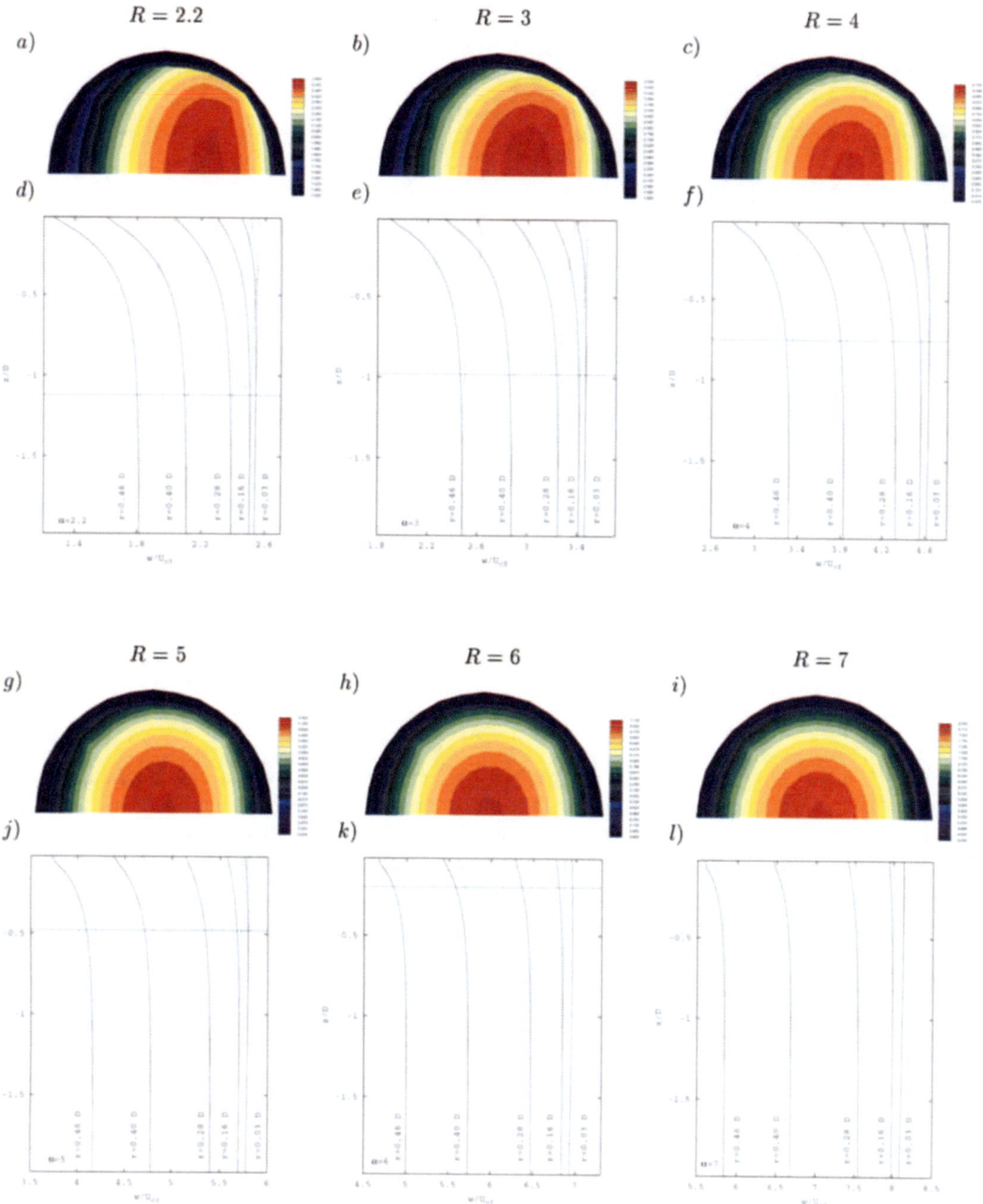

Figure 6. Interaction between jet and crossflow. Iso-contours of axial velocity at pipe outlet and axial velocity distribution in final stretch of pipe for R equal to 2.2, 3, 4, 5, 6 and 7 (from left to right). Axial velocity iso-contours become elliptical and shift downstream. This effect extends two diameters back into the pipe.

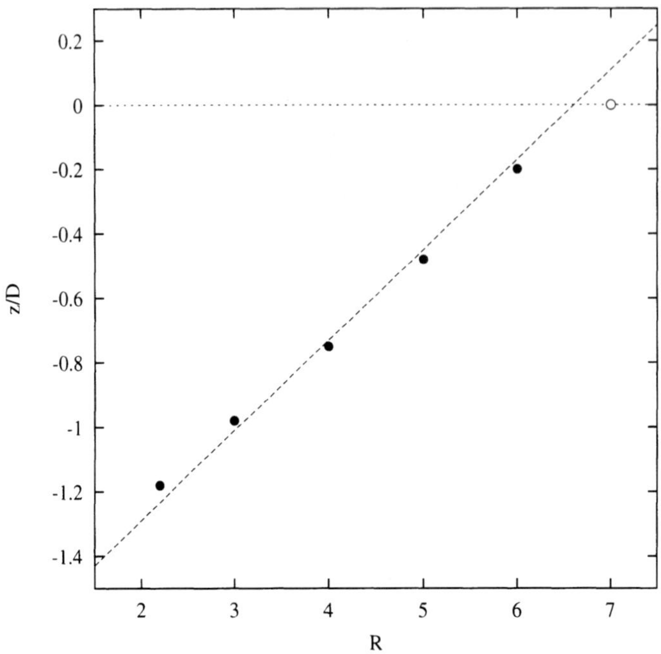

Figure 7. Depth of interaction between crossflow and pipe flow for R equal to 2.2, 3, 4, 5, 6 and 7. Depth of interaction increases linearly for decreasing R. Influence of crossflow within the pipe should vanish for values of R larger than 7.

et al., 1999, Hsu and Guo, 2000) are generally used. Nevertheless, a nonuniform distribution of the vertical component of velocity at the jet-exit is often observed experimentally, owing to interaction of the jet with the crossflow. The effect of this interaction on the numerical simulation of the jet is discussed in Coelho and Hunt (1989) and in Cortelezzi and Karagozian (2001).

To evaluate the effect of the jet profile imposed at the exit, we compare the flow field generated by a jet issuing from a pipe and by a jet obtained by imposing the inlet boundary condition directly at the jet-exit for the same values of R as examined in Section 3.3. As in Yuan *et al.* (1999) and Hsu and Guo (2000), we use a uniform velocity profile.

The axial velocity iso-contours in the $x - z$ plane corresponding to the two different conditions are compared in Figure 8. Figure 8 a, b, c, g, h and i refer to simulations with the pipe whereas Figure 8 d, e, f, j, k and l refer to simulations without the pipe. In order to use a consistent color map, we confine the color map between the minimum and maximum value found at the jet-exit for $R = 2.2$, as shown in Figure 6. This range is modified linearly for conditions corresponding to different values of R.

Simulations made with and without the pipe show that penetration of the jet in the crossflow increases with increasing values of R and that the plume of the jet is deflected downstream by the crossflow. Interesting enough, for a fixed R, the penetration is larger for the case with no pipe, although maximum axial velocity is lower. This is a consequence of the different axial velocity distribution at the jet-exit. Moussa et al. (1977) discussed the considerable variations found in the experimental results for the jet rise reported in the literature and attribute this variation to

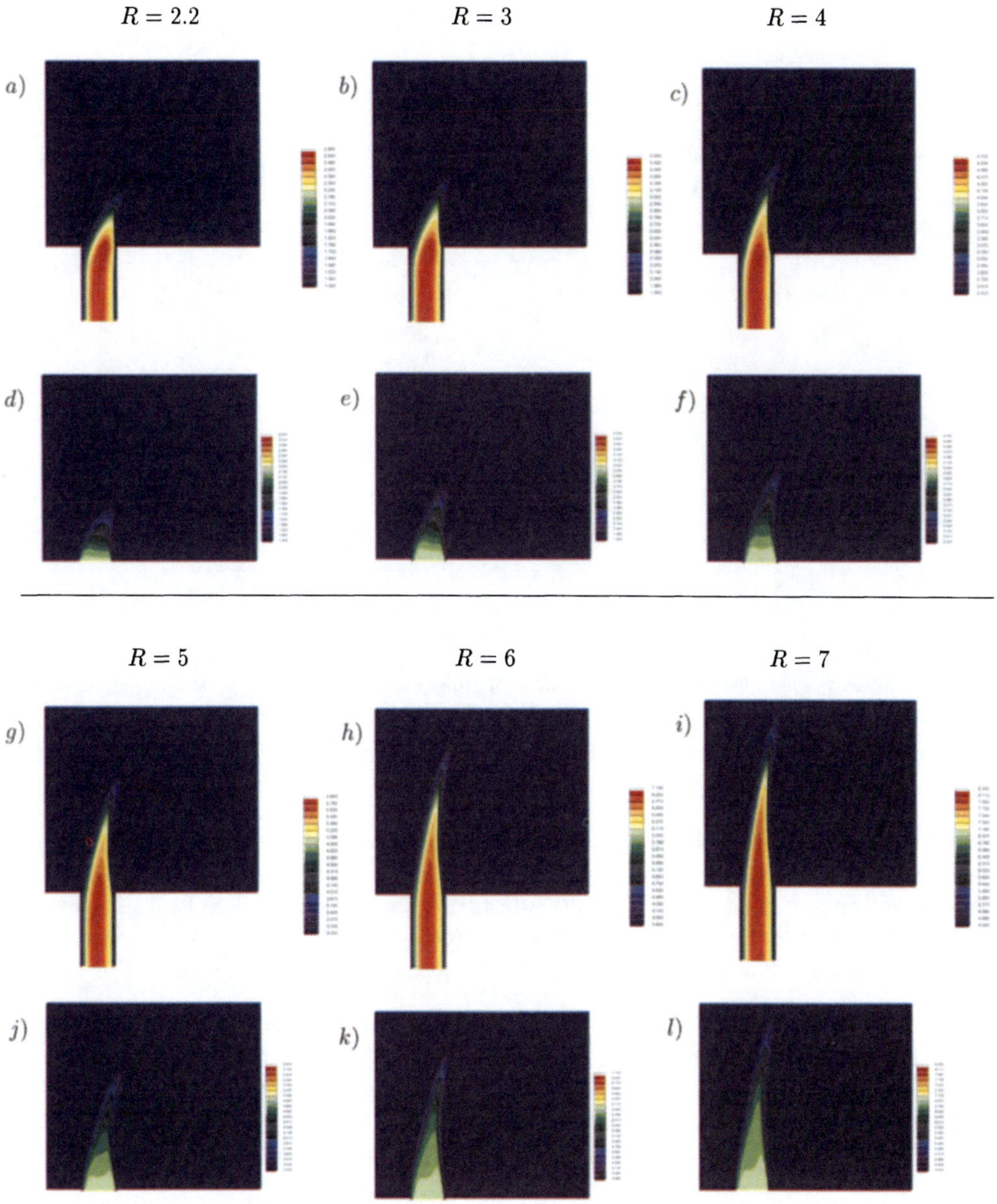

Figure 8. Shape of jets issued from pipe (a, b, c, g, h and i) or from hole in the wall (d, e, f, j, k and l) for R equal to 2.2, 3, 4, 5, 6 and 7 (from left to right). Jet penetration is qualitatively similar and increases with R. Maximum axial velocity and velocity gradients are very different for simulations with and without the pipe.

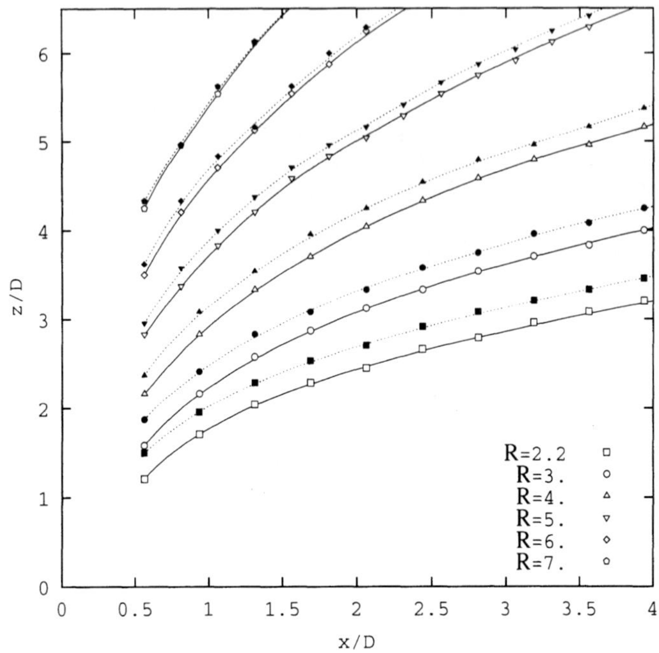

Figure 9. Penetration of the jet in to the crossflow. Starting point of jet axis and equation of jet trajectory differ for simulation made with and without the pipe. Difference is larger for smaller values of R.

differences in the flow condition at the jet-exit. From our simulation, it can be observed that the high velocity core of the jet issuing from the pipe disappears when the pipe is ignored, and that the velocity gradients decrease. These effects can play a major role on the resulting large scale convective structures responsible for mixing. Such structures are most likely related to the velocity gradients within the flow and to the flow turbulence level (Moussa et al., 1977).

We evaluate quantitatively the differences in the structure of the jet by comparing the axes of the jet calculated in the two cases. In Figure 9, the axes of jet calculated for the simulation with and without the pipe are shown as solid and dashed lines, respectively. Results differ for the position of the axis at the jet-exit (not shown in Figure) and for the penetration of the jet in the crossflow. Differences are larger for small values of R and become negligible for R equal to 7.

This suggests that the presence of a pipe should not be overlooked until R becomes sufficiently large if the interaction between the crossflow and the jet is to be captured correctly. Nevertheless, simulating numerically a long pipe is computationally demanding. Since the main effects of the interaction between the jet and the crossflow involve only the last 2 D of the pipe, as discussed in Section 3.3, we evaluate the possibility of simulating numerically this part of the pipe only. We perform the simulation with $R = 2.2$ with the last 2 D of the pipe only. At this new inlet section, we apply the velocity profile, the turbulent kinetic energy profile and the kinetic energy dissipation profile calculated from previous simulation at the same distance. We obtain nearly identical results for the velocity fields calculated in these two cases.

4 Summary and Conclusions

In this paper, we examine the influence of the pipe-flow crossflow coupling on the evolution of a transverse jet. We simulate the experimental apparatus (wind/water tunnel, nozzle and plenum) used by Shandorov (Abramovich, 1963) with a rectangular box and a circular pipe and we calculate the flow field using a commercial code and a RANS approach.

First, we validate our numerical simulation against experimental data available by Shandorov (Abramovich, 1963) to assess the ability of the RANS approach in reproducing the structure of the jet. In particular, we compare the position of the jet axis – i.e. the locus of the points of maximum velocity at various sections of the deflected jet – calculated numerically with the experimental data by Shandorov. We find good agreement except very near the jet-exit, where the experiment and the simulation differ for the velocity profile.

Then, we examine the velocity and pressure distribution for sections orthogonal to the jet axis. We find a satisfactory agreement between the experimental profiles and our calculations. We analyze also the vertical variation of the streamwise velocity profiles to understand the interaction between the crossflow and the jet. Comparison between the vertical profiles in the symmetry plane ($y/D = 0$) and in a plane at $y/D = 0.5$ shows that the crossflow slides past the jet issuing from the pipe, decelerating in the symmetry plane and accelerating in the plane $y/D = 0.5$. Furthermore, the velocity profile at the jet exit becomes asymmetric. We use also vorticity to identify the regions in which the CVP structure develops and we discuss about the interaction between the wall-generated and jet-generated vorticity. The present simulation reproduces accurately the position for which Kelso et al. (1996) observed upstream boundary layer separation in their experiments.

Then, to analyse systematically the influence of the crossflow within the pipe, we perform a number of numerical simulations varying the jet-to-crossflow velocity ratio. We are able to quantify the depth within the pipe of the interaction between the pipe-flow and the crossflow, and we find a nearly linear relation between the effect of the crossflow within the pipe and the parameter R.

We perform further simulations aimed at quantifying the difference in the near field calculations due to the imposition of a uniform velocity profile at the jet orifice instead of at the pipe inlet. A qualitative evaluation shows that the maximum velocity of the jet in the crossflow is reduced and that velocity gradients are smoothed when the uniform profile velocity is imposed at the jet orifice. Quantitative analysis shows that the initial position of the jet axis is different and that jet penetration into the crossflow is greater when the uniform velocity profile is applied at the jet-exit. The analysis suggests that differences become negligible when the value of R is larger than 7. This result can be used to decide the proper geometry and boundary condition at the jet-exit.

Finally, we evaluate the minimum length of pipe necessary to ensure precise calculations for R greater than 2.2. Limiting our conclusions to RANS complemented by $k - \epsilon$ model to account for turbulence effects, we found that a 2 D long pipe will suffice, provided that fully developed velocity, kinetic energy and dissipation profiles are applied at the pipe inlet.

Acknowledgments

The authors wish to thank Prof. Ann Karagozian for several valuable discussions pertaining to this research. One of the authors, G.M.D., thanks CATAS (UD, Italy) for support during his stay at McGill University. This study was partially supported by Regione Friuli Venezia Giulia and by NSERC under Grant RGPIN217169.

References

Abramovich, G. (1963). *The theory of turbulent jets* , Physical Sciences Engineering.

Ajersch, P., Zhou, J. M., Ketler, S., Salcudean, M., and Gartshore, I. S. (1997). Multiple jets in a crossflow: detail measurements and numerical simulations *ASME, J. Turbomachinery*, **119**, 330-342.

Andreopoulos, J., and Rodi, W. (1984). Experimental investigation of jets in a crossflow *J. Fluid Mech.*, **138**, 93-127.

Andreopoulos, J. (1985). On the structure of jets in a crossflow *J. Fluid Mech.*, **157**, 163-197.

Antonia, R. A., and Zhao, Q. (2001). Effect of initial conditions on a circular jet *Exp. Fluids*, **31**, 319-323.

Broadwell, J. E., and Breidenthal, R. E., (1984). Structure and mixing of a transverse jet in incompressible flow *J. Fluid Mech.*, **148**, 405-412.

Chang, Y. K., and Vakili, A. D. (1995). Dynamics of vortex rings in crossflow *Phys. Fluids*, **7**, 1583-1597.

Coelho, S. L. V., and Hunt, J. C. R. (1989). The dynamics of the near field of strong jets in crossflows *J. Fluid Mech.*, **200**, 95-120.

Cortelezzi, L., and Karagozian, A. R. (2001). On the formation of counter rotating vortex pair in transverse jets *J. Fluid Mech.*, **446**, 347-373.

Davis, S. A., and Glezer, A. (1999). Mixing control of fuel jets using synthetic jet technology: velocity field measurements *AIAA Paper 99-0447*, 37th Aerospace Sciences Meeting, Reno, NV, 1-15.

Degano, G. M., Campolo, M., Soldati, A., and Cortelezzi, L. (2002). Influence of inlet boundary conditions on the numerical calculation of a jet in crossflow *Int. J. Heat and Fluid Flow*, (submitted).

Findlay, M. J., Salcudean, M., and Gartshore, I. S., (1999). Jets in crossflow: effects of geometry and blowing ratio *J. Fluids Eng.*, **121**, 373-378.

Fric, T. F., and Roshko, A., (1994). Vortical structure in the wake of a transverse jet *J. Fluid Mech.*, **279**, 1-47.

Hsu, A. T., and Guo, Y. (2000). Unsteady simulation of a jet-in-crossflow *Int. J. Comp. Fluid Dyn.*, **14**, 41-53.

Karagozian, A. R. (1986). An analytical model for the vorticity associated with a transverse jet *AIAA*, **24**, 429-436.

Kelso, R. M., Lim, T. T., and Perry, A. E. (1996). An experimental study of round jets in a cross-flow *J. Fluid Mech.*, **306**, 111-144.

Kelso, R. M., Lim, T. T., and Perry, A. E. (1998). New experimental observations of vortical motions in transverse jets *Phys. Fluids*, **10**, 2427-2429.

Khan, Z. U., and Johnston, J. P. (2000). On vortex generating jets *Int. J. Heat & Fluid Flow*, **21**, 506-511.

Knowles, K., and Bray, D. (1991). Computation of normal impinging jets in crossflow and comparison with experiments *Int. J. Num. Meth. Fluids*, **13**, 1225-1233.

Leboeuf, F., Huang, G. P., Kulisa, P., and Perrin, G., (1991). Model and computation of discrete jets in crossflow", *Eur. J. Mech., B Fluids*, **10**, 629-650.

M'Closkey, R. T., King, J. M., Cortelezzi, L., and Karagozian, A. R. (2002). The actively controlled jet in crossflow *J. Fluid Mech.*, **452**, 325-335.

Moussa, Z. M., Trischka, J. W., and Eskinazi, S. (1977). The near field in the mixing of a round jet with a cross-stream *J. Fluid Mech.*, **80**, 49-80.

Needham, D. J., Riley, N., and Smith, J. H. B. (1988). A jet in crossflow *J. Fluid Mech.*, **188**, 159-184.

Papaspyros, J. N. E., Kastrinakis, E. G., and Nychas, S. G. (1997). Coherent contribution to turbulent mixing of a jet in crossflow *Appl. Sci. Res.*, **57**, 291-307.

Rivero, A., Ferrè , J. A., and Giralt, F. (2001). Organized motions in a jet in crossflow *J. Fluid Mech.*, **444**, 117-149.

Rudman , M. (1996). Simulation of the near field of a jet in a cross flow *Exp. Thermal & Fluid Sci.*, **12**, 134-141.

Sani, R. L., and Gresho, P. M. (1994). Resume and remarks on the open boundary condition minisymposium *Int. J. Num. Meth. Fluids*, **18**, 983-1008.

Schlichting, H. (1979). Boundary layer theory, Mc Graw Hill.

Sivadas, V., Pani, B. S., Butefisch, K. A., and Meier, G. E. A. (1997). Flow visualisation studies on growth of area of deflected jets *Exps. Fluids*, **23**, 105-112.

Smith, S. H., and Mungal, M. G. (1998). Mixing, structure and scaling of the jet in crossflow *J. Fluid Mech.*, **357**, 83-122.

Soldati, A., and Banerjee, S. (1998). Turbulence modification by large scale organized electrohydrodynamic flows *Phys. Fluids*, **10**, 1742-1756.

Sykes, R. I., Lewellen, W. S., and Parker, S. F. (1986). On the vorticity dynamics of a turbulent jet in a crossflow *J. Fluid Mech.*, **168**, 393-413.

Xia, L. P., and Lam, K. M. (1997). A vertical round jet issuing into an unsteady crossflow consisting of a mean current and a sinusoidal fluctuating component *J. Wind Eng. Ind. Aerodyn.*, **67&68**, 843-857.

Yuan, L. L., Street, R. L., and Ferziger, J. H. (1999). Large-eddy simulations of a round jet in crossflow *J. Fluid Mech.*, **379**, 71-104.

Nozzle Geometry Effects in the Near Field of a Round Jet

Fabrizio Tessicini[1] Roberto Verzicco[2] and Paolo Orlandi[1]

[1] DMA, Università di Roma 'La Sapienza', via Eudossiana 18, 00184 Roma, Italia
[2] DIMeG, Politecnico di Bari, via Re David 200, 70125 Bari, Italia

Abstract. Axisymmetric numerical experiments of round jets are reported with the aim of showing the nozzle geometry effects on the jet evolution. It is shown in particular that, although the simulation of the nozzle flow has a non–negligible computational cost, its effect on the jet evolution can be very important. It is also demonstrated that the flow configuration resulting at the nozzle exhaust can not be assigned at the boundary of a computational domain without considering also the flow inside the nozzle since it is a combination of radial and axial velocity components supported by an appropriate pressure distribution.

1 Motivation

Round jets are probably the most relevant free–shear flows in practical applications. The injection of fuel and oxidizer in combustors, the exhaust of chimneys and ducts in the atmosphere and sea or the mixing of coflowing streams in chemical reactors are just few examples among a wide variety.

In the literature there are many theoretical, numerical and experimental works devoted to the study of round jets expecially concerning the flow behaviour far from the nozzle. The analysis of the jet dynamics, in fact, can be roughly divided into two regions, the near field and the far field. The latter is usually fully three–dimensional and turbulent with a dynamics following 'universal' scaling laws which are only marginally affected by the initial jet details.

In the first region, in contrast, (about within 5 nozzle diameters from the jet exhaust) the flow features are dominated by the axisymmetric roll–up of the vorticity layer generated at the nozzle lips and by the pairing of the resulting toroidal structures which only eventually become azimuthally unstable and, therefore, three–dimensional.

The strength, spacing and interaction of these vortices determine, in turn, the structure of the mean field and the spreading of the jet which are fundamental quantities for characterizing the mixing between injected and ambient fluid.

Since the flow in the near field is resulting from the instability of a shear layer, it is clear that the geometry of the nozzle and the exhaust velocity profile will play a fundamental role in the determination of the jet bahaviour.

Despite the straightforward nature of this argument, the jet evolution for different nozzle geometries has been very little explored in the past. In addition, all the numerical simulations have been performed by imposing an inflow velocity profile that was evolved downstream according to the Navier–Stokes equations. This procedure clearly contradicts the physical nature of the equations that, being elliptic in space, would adjust the exhaust velocity profile not only according to the flow dynamics inside the nozzle but also to the downstream jet evolution.

In this work we will show by axisymmetric numerical experiments that indeed the simulation of the flow inside the nozzle leads to unexpected exhaust velocity profiles and to a different jet evolution. It will be shown, in addition, that the exhaust velocity is a combination of axial and radial velocity components sustained by an appropriate pressure distribution; these conditions can not be assigned at a boundary without simulating the nozzle flow.

2 Numerical set–up

The numerical solution of the problem has been performed using an adapted version of the method described in Fadlun *et al.* (1999) of which the main features are reported in the following. The viscous, incompressible, time–dependent Navier–Stokes equations have been discretized by second–order accurate finite–difference approximations in space and time. The resulting system has been solved by a fractional–step method, advanced in time using a third–order hybrid Runge–Kutta scheme with the convective terms computed explicitely and the viscous ones implicitly. The flow incompressibility is imposed at each substep through an elliptic equation which is directly solved by trigonometric expansions combined with the FISHPACK package.

The peculiarity of this method is in the introduction of a forcing $\mathbf{f}$ in order to handle complex geometries discretized on a simple grid. In particular we solve the equations

$$\frac{\partial \mathbf{u}}{\partial t} + \nabla(\mathbf{u}\mathbf{u}) = -\nabla p + \nu\nabla^2\mathbf{u} + \mathbf{f}, \qquad \nabla\cdot\mathbf{u} = 0, \tag{1}$$

with $\mathbf{f}$ time and space dependent and with a value such to prescribe a desired velocity $\mathbf{u} = \mathbf{V}$ on a boundary S immersed in the grid. The determination of $\mathbf{f}$ is easily obtained once the time–discretized version of (1) is considered

$$\frac{\mathbf{u}^{l+1} - \mathbf{u}^{l}}{\Delta t} = \mathrm{RHS}^{l+1/2} + \mathbf{f}^{l+1/2}, \tag{2}$$

with $\mathrm{RHS}^{l+1/2}$ including convective and viscous terms and the pressure gradient. From the above equation it is evident that the value $\mathbf{f}^{l+1/2}$ yielding $\mathbf{u}^{l+1} = \mathbf{V}^{l+1}$ is given by

$$\mathbf{f}^{l+1/2} = -\mathrm{RHS}^{l+1/2}\Delta t + \frac{\mathbf{V}^{l+1} - \mathbf{u}^{l}}{\Delta t}. \tag{3}$$

A particular treatment is necessary for those cells at the interface between fluid and body since the forcing has to assume an intermediate value between that given by equation (3) and the zero value. We have found that second order accurate results are obtained by linearly interpolating on the velocities at the boundary in order to determine a new value of $\mathbf{f}$ which satisfies the boundary conditions; further details are given in Fadlun *et al.* (1999) where the efficiency and accuracy of the scheme is shown by axisymmetric and three–dimensional flows.

The basic nozzle geometry is an axisymmetric cubic profile with unity exhaust radius (R_e) and a contraction $c = R_i^2/R_e^2 = 11.02$ being R_i the radius of the inflow section. The shape of the nozzle is given by $R = a_0 + a_1x + a_2x^2 + a_3x^3$ with the coefficients $a_0 ... a_3$ determined by the conditions: at $x = 0$, $R = R_i$ and $\partial R/\partial x = 0$, at $x = L$, $R = R_e$ and $\partial R/\partial x = 0$ from which it is obtained

$$R = R_i - 3(R_i - R_e)\left(\frac{x}{L}\right)^2 + 2\left(R_i - R_e\right)\left(\frac{x}{L}\right)^3,$$

with L the length of the cubic portion of the nozzle. The nozzle is completed by a constant radius section of axial length L_i before the contraction.

A second cubic law describes the external shape of the nozzle whose maximum and minimum radii are $R_{Ei} = 5R_e$ and $R_{Ee} = 1.1R_e$; the complete profile is shown in figure 1a giving a complete nozzle length equal to $7R_e$. This shape the same as used in an on going laboratory experiment (G. Buresti *Personal Communication*).

In a second nozzle geometry the exhaust section has been prolonged by a constant radius section of axial length $2R_e$. This shape is shown in figure 1b and, although the modification seems only marginal, it induces significant changes in the exhaust velocity profile and the jet evolution.

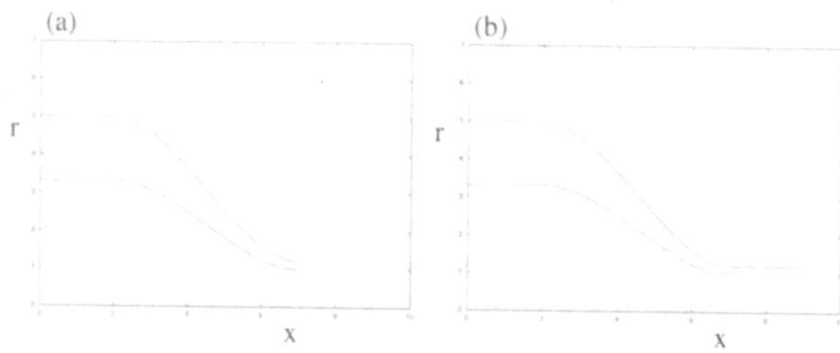

Figure 1. Nozzle geometry profiles: *(a)* basic geometry; *(b)* 'prolonged' geometry.

The Reynolds number $Re = R_e\overline{U}/\nu$ is based on the exhaust nozzle radius, the averaged velocity $\overline{U}$ at that section and the kinematic viscosity of the fluid ν; all the variables are made dimensionless accordingly.

The computational domain has been discretized by a non–uniform grid with 500×435 points in the axial (x) and radial (r) directions, respectively, with the computational nodes clustered in the dynamically active regions. The domain has a dimension $L_R \times L_x$, respectively, $30R_e \times 10R_e$ which were found enough in order to prevent artificial dynamics induced by the boundary proximity.

At the inflow of the nozzle ($x = 0$) it has been assigned an axial velocity profile $u_x(r,t) = U_i \tanh(\alpha r)/\tanh(R_i) + f(r,t)$ with α a parameter related to the thickness of the boundary layer at the nozzle wall and $f(r,t)$ a perturbation function experimentally determined and modulated in time by random numbers (M.V. Salvetti *Personal Communication*).

At the outflow of the domain $x = L_x$ standard convective boundary conditions were used (Salvetti *et al.* (1996)) since they introduce a minimal perturbation on the jet dynamics while allowing the exit of complex vortical structure from the domain.

On the lateral wall ($r = L_r$) traction–free boundary conditions have been adopted (Boersma *et al.* (1998)); they require the stresses to vanish at the boundary with the velocity components computed accordingly. In other words if n_j is the boundary normal it results

$$\sigma_{ij} * n_j = 0, \text{ with } \sigma_{ij} = -P\delta_{ij} + \nu \left(\frac{\partial u_i}{\partial x_j} + \frac{\partial u_j}{\partial x_i} \right).$$

The main advantage of this condition is the possibility of entrain fluid from outside the domain thus allowing for the correct jet growth.

All the simulations have been initiated from rest and the first 100 time units have been discarded being the initial transient; the numerical experiment has then been continued for 200 additional time units during which statistics have been collected and instantaneous velocity fields stored for the successive analysis. Each run required about 26 hours of a 800MHz Athlon–AMD processor.

3 Results

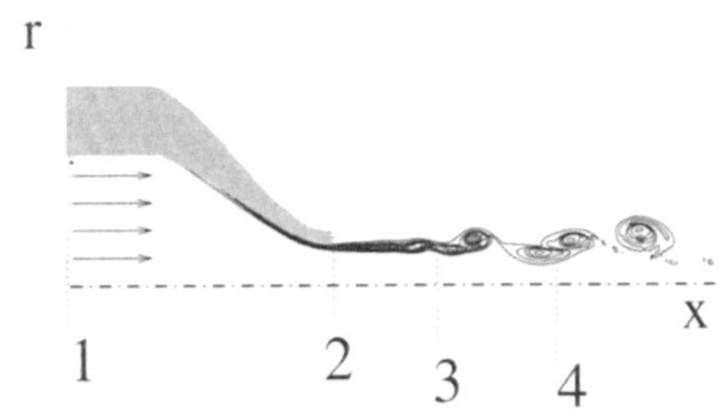

Figure 2. Typical jet flow configuration: perturbed inflow, $Re = 2000$. Vorticity contour levels $\Delta\Omega = 1$.

Before staring the discussion of the nozzle effects on the jet dynamics, we wish to briefly summarize the main features of the flow in order to aid the successive presentation of the results. In figure 2 we report a typical jet configuration with the fluid injected from section '1' at a constant flow rate; the stream converges inside the nozzle up to the exhaust (section '2') where the vorticity produced at the wall separates and is released in the ambient fluid. This shear layer initially thickens owing to viscous effect and then becomes unstable according to the Kelvin–Helmholtz instability (section '3'). This instability produces toroidal ring–like structures that pair to form larger vortices as in section '4'. It is well known that vortex rings are unstable to azimuthal perturbations, and their growth rate is related to the self–induced strain field. Since in jets each ring is subject not only to the self induced strain but also to the strain of the near structures the flow is very likely to become fully three–dimensional already after few nozzle diameters from the exhaust (about 5) and an axisymmetric simulation does not reflect anymore the flow physics.

It is very important to note that the location of the shear layer instability is extremely sensitive to the presence of perturbations in the inflow at section '1'. In preliminary simulations, in fact, we have noted that by a perturbation amplitude of the order of $10^{-3}U_i$ we were able to anticipate the onset of the instability by more than $4R_e$ (compare figures 2 and 6a). Since the main aim of this study is to investigate about the nozzle geometry effects, we decided to perform the rest of the simulation without any perturbation at the inflow.

3.1 Flow inside the nozzle

In figure 3 we show some details of the flow inside the nozzle where it is observed the convergence of the streamlines toward the nozzle exhaust and the consequent generation of a negative component of radial velocity. If this velocity would not be balanced by a radial pressure gradient the flow would continue to converge outside the nozzle owing to the fluid inertia. Contour levels of pressure are shown in figure 3e from which it is evident that the pressure is neither constant nor with zero gradient along vertical sections as usually assumed in numerical simulations.

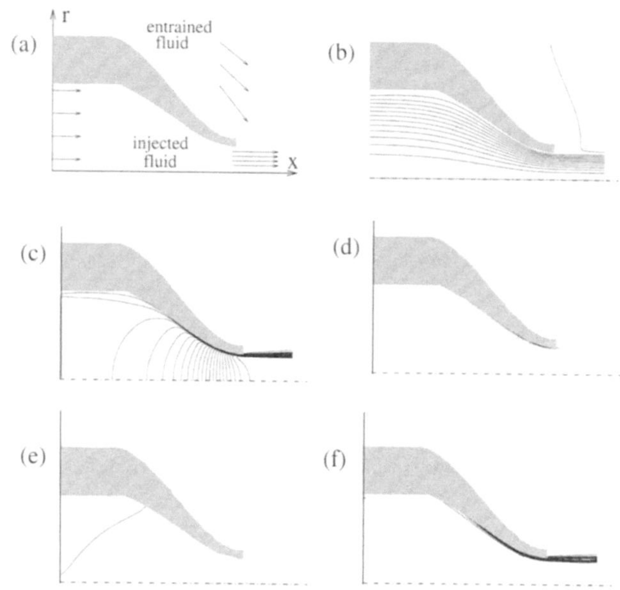

Figure 3. Flow inside the nozzle: *(a)* sketch of the flow arrangement; *(b)* streamlines, *(c)* axial velocity, *(d)* radial velocity, *(e)* pressure and *(f)* vorticity. Solid line for positive values, dotted lines for negative. $Re = 2000$.

Another important feature is the overshoot of the axial velocity profile towards the nozzle wall (figure 4a). This is easily explained when observing that in axisymmetric geometries the flow rate through a segment Δr perpendicular to the axis is linearly increasing with the distance from the axis r. Thus, on account of the mass conservation, the larger the initial distance from the axis the stronger the axial acceleration of the fluid particle; this phenomenon yields the profile of figure 4a. It is worth noting that the overshoot of the axial velocity in addition to intensifying the wall vorticity implies a change of curvature of the profile; this changes the inviscid stability properties of the jet and, therefore, its subsequent evolution.

The profiles in figure 4 also show that outside the nozzle ($r \geq 1.1$) there is a weak positive axial velocity and a negative radial velocity. This is the fluid entrained from the ambient into the main stream which causes the spreading of the jet. This is another aspect which depends on the

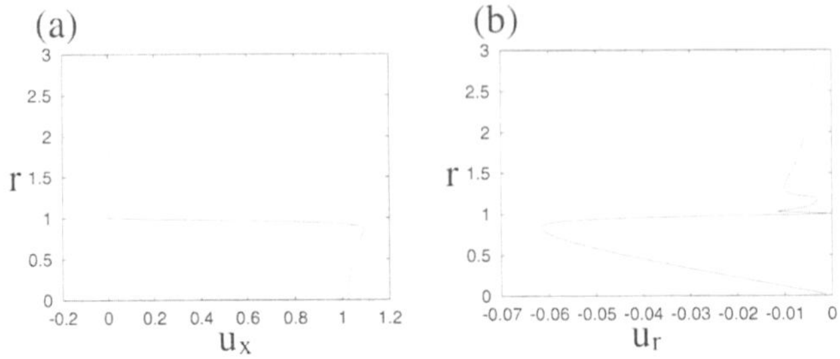

Figure 4. Velocity profiles at the nozzle exhaust: *(a)* axial component, *(b)* radial component.

external geometry of the nozzle and it can not be determined unless the complete flow set–up is computed.

With the above example we have analyzed the flow dynamics occurring inside the nozzle and its relevance to the jet evolution; we wish to mention, however, that the simulation of the nozzle flow has a non negligible computational cost (about $30-50\%$ of the total) that would become a considerable overhead in three–dimensional simulations. On the other hand, once the velocity profiles at the nozzle exit are determined, it is tempting to perform numerical simulation without computing the nozzle flow by assigning velocity distribution like those of figure 4 at a boundary of the computational domain.

This possibility has been tested in a preliminary numerical simulation where it has been observed that the jet evolution followed an unphysical behaviour. In particular, owing to the absence of the appropriate pressure distribution at the inflow, the initial negative radial velocity (figure 4b), caused the jet to converge also outside the nozzle thus producing an artificial centerline velocity peak. This is shown in figure 5a–b. respectively, by the centerline velocity evolution along the axisand by the exhaust velocity profile. When the whole nozzle geometry is computed the jet accelerates only weakly after the exhaust and then remains constant for about 6 nozzle radii, i.e. within the potential core region. For $x \geq 13$ the jet velocity steadily decreases owing to the spreading of the jet. In contrast, when the inflow condition is imposed at a boundary, there is a centerline velocity increase above 30% followed by the velocity decrease without the formation of the potential core region.

In a successive simulation the unphysical jet convergence was avoided by assigning only the axial velocity profile and setting the radial component to zero. Also in this case, the flow without the nozzle could not reproduce the results of the whole flow. It was observed in particular that, while the complete flow was able to maintain the Kelvin–Helmholtz instability without introducing any perturbation, the flow without nozzle adjusted to an almost steady state in which the vorticity layer only diffused owing to viscous effects.

The comparison of the two flows by instantaneous vorticity snapshots is shown in figure 6 and the explanation for this behaviour can be obtained looking at figure 7. In this figure it is, in fact, reported the time evolution of the centerline axial velocity at the exhaust of the nozzle. It can be observed that it oscillates in time with a frequency which is the same as that of the jet instability. This implies a feedback mechanism between the external jet evolution and the vorticity generation at the nozzle lips that, therefore, will be modulated according to the jet dynamics.

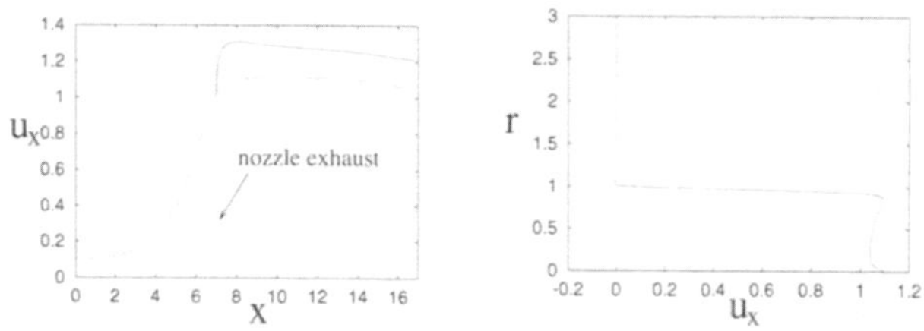

Figure 5. *left*: axial evolution of the centerline velocity, dashed line for the jet with the nozzle, solid line for the jet without nozzle. *right*: radial profile of axial velocity at a distance $0.5R_e$ from the exhaust.

Figure 6. Comparison of vorticity snapshots for the flow with and without nozzle ($Re = 2000$).

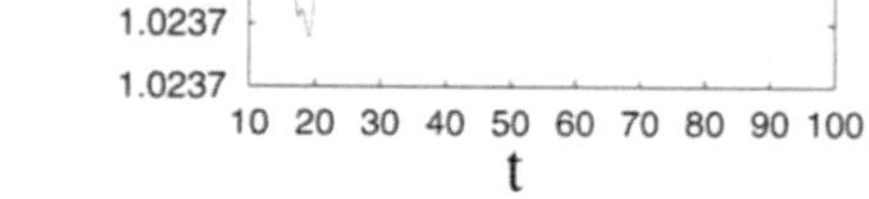

Figure 7. Time history of the centerline velocity for the jet with the nozzle; $Re = 2000$.

This effect is totally absent when the nozzle flow is not accounted for since the vorticity profile, being the spatial derivative of the imposed velocity profile, is prescribed and not computed. This prevents the vorticity modulation previously observed and therefore reduces the jet instability.

In other words, the simulation of the whole flow implies the generation of the vorticity and its evolution while the imposition of a velocity profile is in fact just a convection problem for an assigned vorticity distribution.

It is worth noting that the amplification of the Kelvin–Helmholtz instability is highly Reynolds dependent and we will show in the next section that for higher values of Re it is possible to maintain the instability also without the nozzle. This does not mean, however, that for high Re flow the effect of the nozzle becomes negligible since jets are convectively unstable (Huerre & Monkewitz (1985)) therefore a modification of the exhaust velocitiy profile changes the successive jet evolution.

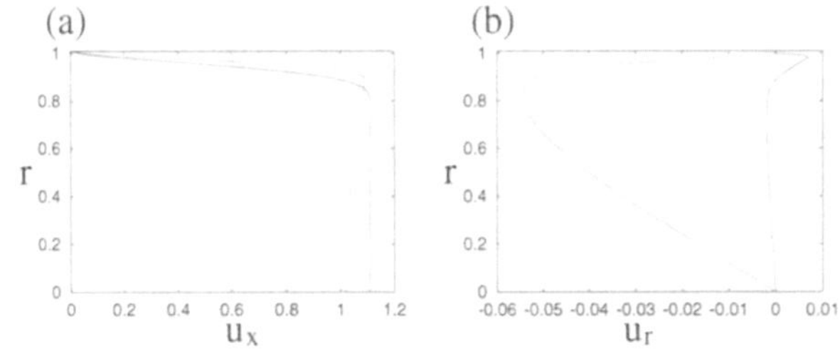

Figure 8. Comparison of the exhaust velocity profiles for the two nozzle geometry: solid line for the prolonged geometry, dashed line for the basic geometry. *(a)* axial, *(b)* radial velocity component.

3.2 Nozzle geometry effects

After having analyzed the flow dynamics inside the nozzle, in this section we will investigate the jet evolution for different geometries. In particular we will consider also the nozzle of figure 1b whose exhaust velocity profiles are given in figure 8 with the profiles produce by the geometry in figure 1a reported for comparison. We can note that the addition of a straight section at the end of the nozzle strongly reduces the overshoot phenomenon and yields a less negative radial velocity component. The third case is a jet without any geometry with the axial velocity profile of the nozzle in figure 1b assigned at the boundary of the computational domain; for all the simulation the Reynolds number is $Re = 2500$.

In figure 9 we report vorticity contour levels for the modified nozzle and for the jet without geometry. In this case, owing to the higher Reynolds number, the shear layer becomes unstable also in the case without nozzle, however the onset of the instability is delayed by about $2R_e$.

More quantitative differences in the jet dynamics can be observed from figure 10 where some integral quantities are shown; figure 10a reports the half–width radius ($R_{0.5}$), i.e. the radius where the axial velocity attains half the value of the centerline velocity. We can see that the highest values are obtained for the nozzle of figure 1a which yields the most unstable flow. Nevertheless,

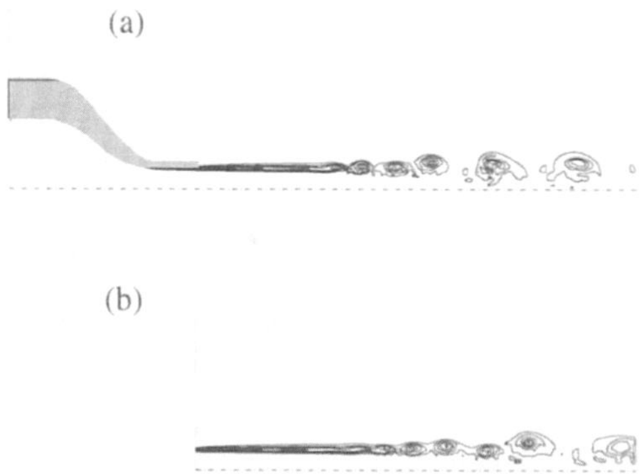

Figure 9. Comparison of vorticity snapshots for the flow with and without nozzle ($Re = 2500$).

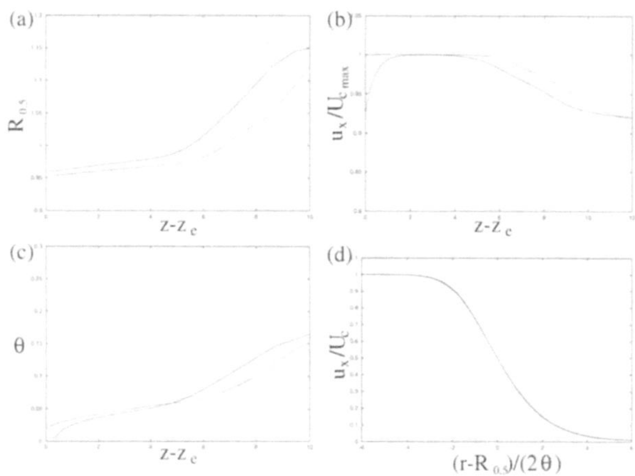

Figure 10. Comparison of integral quantities of the jet evolution; solid line for basic geometry, dashed line for prolonged geometry, dotted line no geometry: $Re = 2500$. *(a)* half–width radius, *(b)* centerline velocity, *(c)* momentum thickness and *(d)* recaled radial velocity profiles.

differences are observed also in the other two cases which have an identical exhaust axial velocity profile; in particular the jet with the nozzle eventually has a larger spreading, thus confirming the augmented jet instability. A similar indication comes from figure 10b showing the axial evolution of the centerline velocity. The first nozzle has the shortest potential core and the most pronounced velocity decrease. Once again the other two case have an almost identical initial evolution but the jet with the nozzle exhibit a different dynamics already at a distance $6R_e$.

We wish to stress that the different dynamics of the jet without nozzle does not imply that the flow does not follow any scaling law; in fact in figure 10d we report a rescaled axial velocity (Michalke (1984)) profile at a distance $8R_e$ from the exhaust showing a perfect collapse with the other two cases. This rescaling, however, is based on the local measured values of $R_{0.5}$, the centerline velocity U_c and the momentum thickness θ which evolve in a completely different way depending on the initial jet conditions.

4 Conclusions

In this paper we have explored by simple axisymmetric numerical experiments the effect of the nozzle geometry on the jet evolution. Two main results have been found, the first related to the exhaust velocity profile and the second to the jet instability. In fact, the specific shape of the nozzle (internal and external) determines the profiles of the exhaust velocity that influence the inviscid stability features of the jet.

On the other hand, these profiles are modulated in time owing to the external jet evolution thus evidencing a feed–back mechanism. This is clearly absent when the exhaust profile is imposed (instead of coming out as part of the solution) and this yields jet which are generally more stable than those obtained by the whole flow.

Another important point is that the exhaust conditions are not completely determined by a simple axial velocity profile since they generally include a radial velocity and a pressure distribution assuring the flow incompressibility. We have found that the complete flow configuration cannot be prescribed without computing the nozzle flow thus preventing the possibility of replicate an experiment without knowing the nozzle shape.

A possible extension for this study is to consider three–dimensional jets; we expect in this case even larger differences since in the concave part of the nozzle Görtler instabilities might develop and these could anticipate the azimuthal jet instability.

References

Fadlun, E.A., Verzicco, R., Orlandi, P. and Mohd–Yusof, J. (1999). Combined immersed boundary/finite difference methods for three–dimensional complex flow simulations *Journal of Comp. Phys.*, 161:35–60.

Salvetti, M.V., Orlandi, P. and Verzicco, R. (1969). Direct simulation of transitional axisymmetric coaxial jets *AIAA J.*, 34:736–743.

Boersma, B.J., Brethouwer, G. and Nieuwstadt, F.T.M. (1998). A numerical investigation on the effect of the inflow condition on the self–similar region of a round jet. *Phys. of Fluids*, 10(4):899–910.

Huerre, P. and Monkewitz, P.A. (1985). Absolute and convective instability in free shear layers *J. of Fluid Mech.* 159:151–168.

Michalke, A. (1984). Survey on jet instability theory *Prog. Aerospace Sci.* 21:159–199.

The Jet in Cross-Flow:
A Few Remarks on the Cross-Flow Structure Influence

Václav Uruba, Oton Mazur, Pavel Jonáš*

Department of Fluid Dynamics, Institute of Thermomechanics ASCR, Prague, Czech Republic

Abstract. The results of an experimental investigation of the jet in cross-flow mixing zone are presented. The jet emanates perpendicularly from a circular nozzle in a wall with parallel cross-flow, and a jet to cross-flow velocity ratio of 4 is chosen. The cross-flow was of either low turbulence or high turbulence in nature. The influence of the turbulence level on structures arising in the mixing zone has been studied in detail.

1 Introduction

The jet – cross-flow interaction represents a substantial problem for many practical applications including environmental aerodynamics – e.g. in plume dispersal from smokestacks and spreading of escaped gases. Hence the flow has attracted the attention of investigators in fluid dynamics for many years - see e.g. Margason (1993). Nevertheless, there are still several aspects of the phenomenon which are unclear, namely the influence of various operating conditions that take place in real situations. The effect of some of them is investigated in the framework of our current project, involving the influence of cross flow structure on transverse jet mixing.

In the jet – cross-flow mixing zone there are several vortex systems due to interactions that are both stationary and non-stationary in nature and highly three-dimensional in structure. The steady structures include the dominant rolled-up counter-rotating vortex pair (CVP) and the horseshoe vortex around the jet exit, while the unsteady structures include the unsteady wake vortex street shed immediately downstream of the jet exit and the free jet shear layer vortices. All those vortex structures strongly affect passive scalar exchange between the flows.

In this study an overview of our results related to the influence of cross-flow structure on the mixing zone structure is presented. Our complete experimental results may be found in prior publications listed in the references.

2 Experimental setup

The experiments have been carried out in the blow-down tunnel of the Institute of Thermomechanics. The test section was 250 x 250 x 800 mm^3, and the orifice of the nozzle was located on an inserted flat plate 80 mm downstream of its leading edge. The diameter of the jet nozzle orifice was D = 13.33 mm. The effective jet – to – cross-flow velocity ratio is defined in the case of two identical fluids (air of the same temperature) as $R \equiv V_j / V_\infty$, with the value R =

* Václav Uruba, Oton Mazur, Pavel Jonáš, Institute of Thermomechanics, ASCR, Dolejškova 5, 182 00 Praha 8, Czech Republic; Phone: +420 2 8658 8547, Fax: +420 2 8658 4695, E-mail: uruba@it.cas.cz

4.00±0.01 (V_j = 40.8 m/s, V_∞ = 10.2 m/s) used here. The situation is shown schematically in Figure 1, where the coordinate system is introduced which includes the components (U, V, W) of the instantaneous velocity vector $\mathcal{V}$ at a given point. A detailed description of the measurement setup is presented in Uruba et al. (1999).

Several experimental techniques were used to explore the mixing zone 4 to 30D downstream of the jet orifice. To quantify the 3D steady flow pattern, a 5-hole pressure probe was chosen

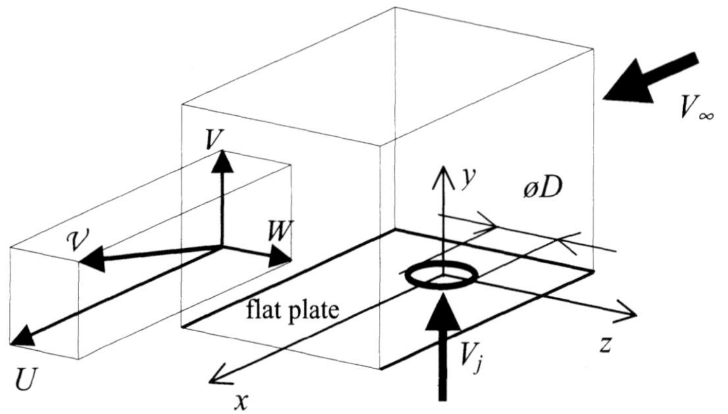

Fig.1 – Experimental setup and coordinate system

which incorporated an intelligent pressure scanner, PSI model 9010. The external diameter of the probe was 2.6mm. Miniature total and static pressure probes have also been used.

For unsteady measurements the CTA anemometry technique was used, using both a single hot wire perpendicular to the cross-flow direction (probe type DANTEC 55P01) and an x-probe (type DANTEC 55P51), indicating two velocity components simultaneously (U and V or U and W). The CTA anemometer was a DISA type M, used with a National Instruments data acquisition system. The measurements were carried out in the region $10D < x < 30D$, $0.5D < y < 10D$, $0 < z < 7D$ (the symmetry for negative values of z was supposed and checked). The frequencies of fluctuations in the signals ranged up to 10kHz, so an analog low-pass filter at 20kHz and sampling frequency of 50kHz was used. The statistical characteristics were evaluated from 10 second records acquired at each particular position. Uncertainties were found to lie within 1% of the mean velocities and within 3% of the fluctuating velocity values for hot wire measurements, and within 2% for pressure measurements and 0.5° for flow angle measurements, as obtained using the 5-hole probe.

Two conditions with different cross-flow structures were examined. The "low intensity turbulent cross-flow" case (hereafter called LT) corresponded to a turbulence intensity associated with longitudinal velocity fluctuations for the cross-flow of 0.05%, while the "high intensity turbulent cross-flow" case (hereafter called HT) corresponded to a turbulence intensity of 6% in the x-direction relative to the jet orifice center. In both cases the wall boundary layer thickness was about 0.1D at this position. The turbulence intensity associated with longitudinal velocity is defined as the ratio of the longitudinal velocity signal's standard deviation to its mean value, written as a percentage.

3 Results

3.1 Pressure measurements

The pressure measurements were carried out in the *y-z* planes for $x = 10, 15, 20, 25$ and $30D$ with determination of static and total pressure distributions as well as mean velocity vector directions. Consequently the mean vorticity vectors were evaluated as well.

The velocity vector fields were studied in detail in the *y-z* planes. The dominant counter-rotating vortex pair (CVP) is clearly visible from the results, as seen in Figure 2. Moreover, almost perfect symmetry was observed from the mean velocity vector distribution. Hence all subsequent results will be presented only for one half of the jet trajectory or mixing zone.

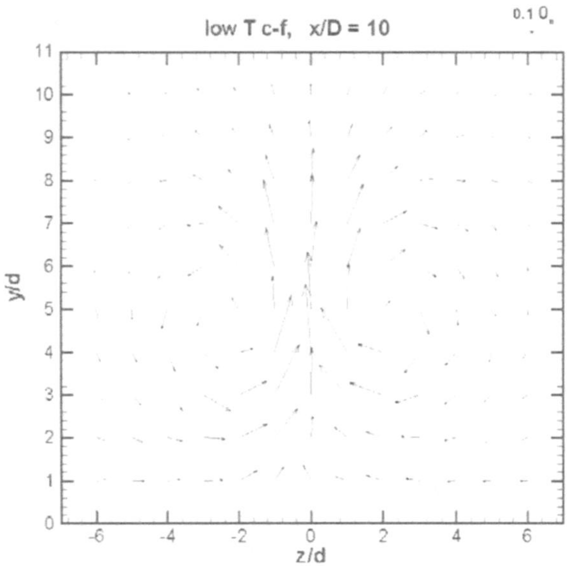

Fig.2 – Mean velocity vectors distribution in the *y-z* plane at x/D=10, LT case

The CVP was studied in detail for the both cases of cross-stream structure (LT and HT). In Figure 3 distributions of the mean longitudinal vorticity component are given. The vorticity was put into non-dimensional form using the jet diameter D and the cross-flow velocity V_∞. The vortex structures were found to be stronger in the HT cross-flow case than for the LT case in Figure 3. The mean vorticity distribution, which has its maximum in the center of the vortex, shows a maximum which is about 30% higher in the HT case than in the LT case.

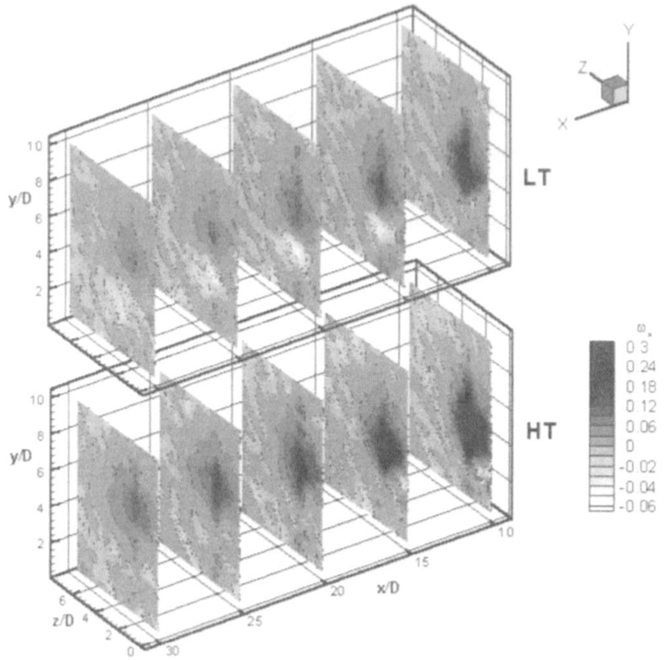

Fig.3 – Vorticity distribution comparison

Furthermore, another vortex pair was found in the LT case. This "associated" vortex pair was significantly weaker and opposite in orientation compared with the CVP. It was localized below the main vortex pair arising in the region x=15 to 20D. No associated vortex pair was observed in the HT case. In the upper half of Figure 3, corresponding to the LT case, the distinct minima were visible in the associated vortex location. It seems there was no connection between this associated vortex pair and the horseshoe vortex around the jet exit, although the orientations were the same. Yet the horseshoe vortex is typically weak in the case of a very thin upstream wall boundary layer, and moreover its location is usually different from that of the "associated" vortex observed here.

In Figure 4, a comparison of the mean flow patterns in the x/D = 30 plane for the LT and HT cases is given. For clarity, stream traces are added. The centers of the CVP and the associated vortices were, respectively, (y/D, z/D) equal to (6, 3) and (3, 3). An explanation of the associated vortex pair formation mechanism may involve entrainment of ambient air towards the low-pressure region in the wake behind the jet near its orifice, which was found to be much stronger for the LT case. The associated vortex pair is discussed further in Uruba et al. (2000).

The properties of the mean velocity vector field were also studied in detail. It was found that the vector field was of a divergent nature (i.e. fluid particles were pushed out from the vortex center) for small x values, while the flow became gradually convergent further downstream. This changeover process began at x = 15D and ended at 25D, however in the HT case it was somewhat faster and shifted upstream. In Figure 4 the residue of a divergent region may be seen near the CVP center (thick elliptic line) in the LT case only. This divergent-convergent feature of the flow field plays an important role in the jet diffusion process.

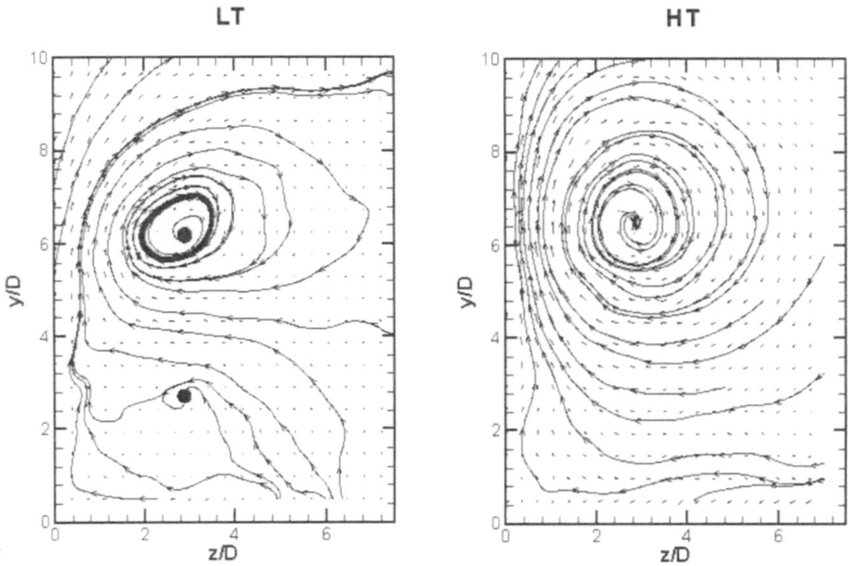

Fig.4 – Velocity field in $x/D = 30$ for LT (left) and HT (right) cross-flow

The centerline trajectory of the jet was studied for the LT and HT cases. The centerline is defined as the set of points of maximal mean velocity modulus in the plane of symmetry. The shape of the centerline was found to lie within published data, see Margason (1993). However, the centerline for the HT cross-flow case was found to be slightly higher than in the LT case, as seen in figure 5, but the difference is not very significant. Thus the high turbulence cross-flow ap-

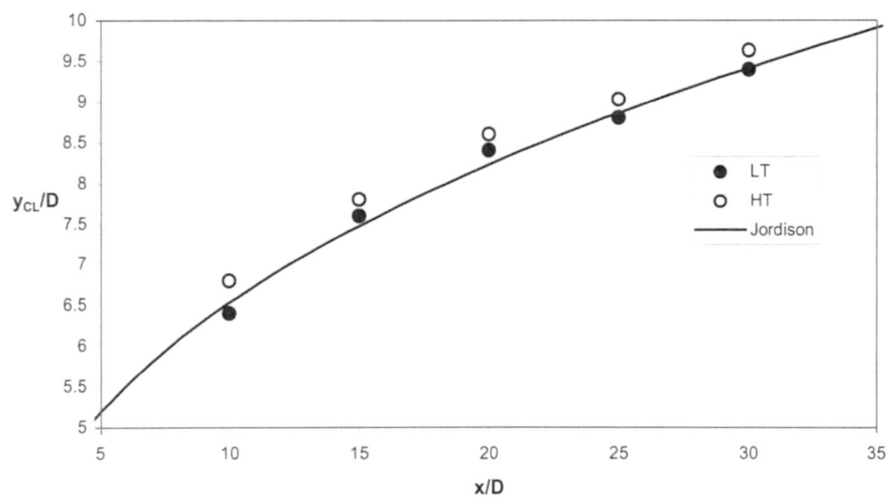

Fig.5 – The jet centerline shape

peared to be easier for the jet to penetrate than did the low turbulence crossflow. The solid line in Figure 5 represents Jordison's approximation of the jet centerline for $R = 4$ (referenced by Margason, 1993).

3.2 Hot wire measurements

Several statistical characteristics were evaluated from the hot wire data. The voltage data records were converted into velocity time series and then processed. The central moments of order up to 4 in both velocity and its time derivative were calculated.

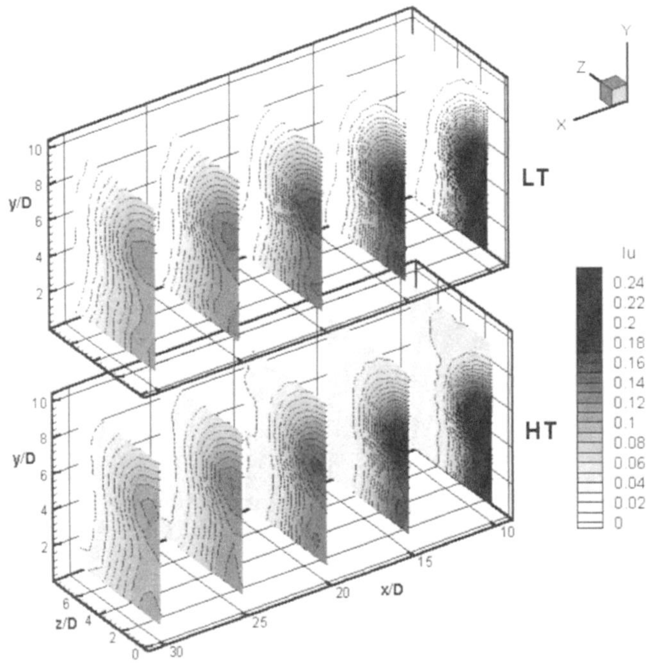

Fig.6 – Distribution of intensity of longitudinal velocity fluctuations

In Figure 6 a comparison of the intensity of longitudinal velocity fluctuation distributions is given for the two cases. No significant difference between the LT and HT cases could be found within 30 jet diameters of injection. The maximal value of the intensity of fluctuations (about 25%) could be found near the plane of symmetry on the level of the centerline (i.e., the maximum of the mean velocity modulus). The only difference in these values was in the fluctuation intensities outside of the mixing zone, as expected.

In Figure 7, a comparison of flatness factor distributions is given. The flatness factor is defined as the fourth central statistical moment divided by the signal's standard deviation with respect to relevant power. A factor or 3 is subtracted so that the flatness factor is equal to 0 for a Gaussian process. We observed a big difference in flatness factor between the LT and HT cases. While in the HT case the value of the flatness factor was nearly constant in the entire mixing zone, in the LT case, very high maxima were distinct along the jet mixing zone. The same behavior was found in the skewness factor, which is defined similarly to the flatness factor but in terms of the third central statistical moment. This difference between LT and HT cases was related to

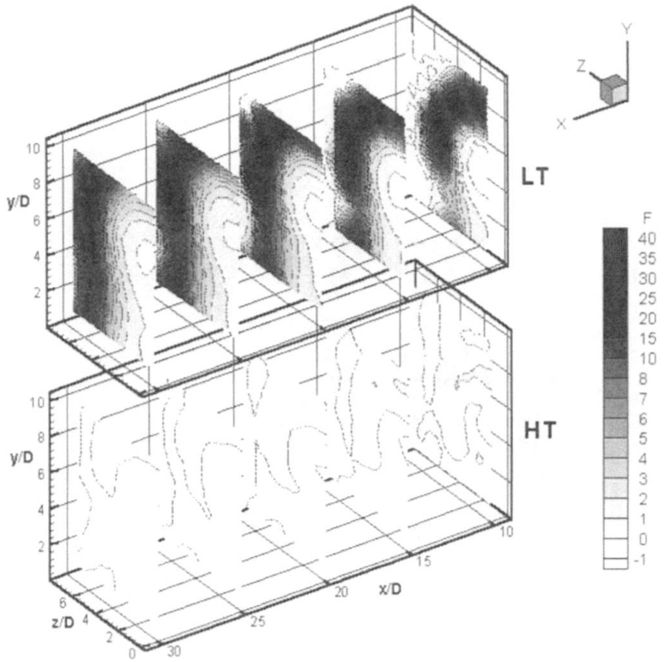

Fig.7 – Distribution of the flatness factor of longitudinal velocity fluctuations

the intermittent character of the flow. Near the edges of the jet mixing zone, the intermittency should be the same in each case. While in the LT case the classical intermittency signal, with laminar and turbulent sections, is evident, in the HT case there is only more or less homogenous turbulent signal. In the LT case one fluid region appeared laminar and the second appeared turbulent in the intermittency region, while in the HT case both fluids were in the turbulent state.

A detailed intermittency analysis in the LT case was carried out. The velocity time series were high-pass filtered and the energetic detection function was calculated: $Det(t) = |\partial u/\partial t * u|$, where u is the fluctuating part of the longitudinal velocity component. The threshold value was set slightly higher than that corresponding to the points outside of the mixing zone. The result is given in Figure 8. The intermittency function γ is defined as the probability that the flow at a given point is turbulent. For the fully turbulent case, γ is equal to 1, while for the fully laminar case γ is equal to 0. The turbulent regions (black) were distinguished inside the bent jet, while the intermittent regions (gray) were found near the jet edges and below the jet, and the laminar regions (white) were observed outside the mixing zone. The intermittent regions coincide well with the high flatness factor regions in Figure 7. In the HT case no such analysis could be carried out because the entire domain was always in the turbulent state. A detailed analysis of the intermittency is given in Uruba et al. (2001).

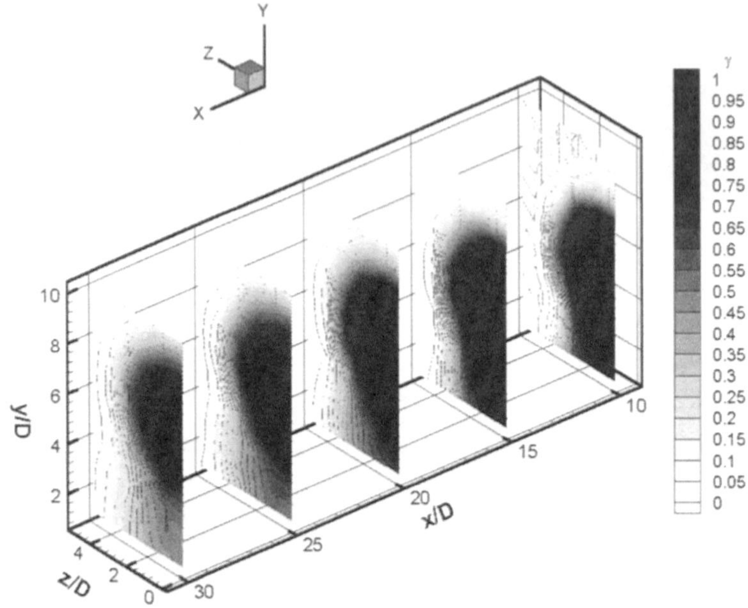

Fig.8 – Distribution of intermittency function, LT case

To explore the non-stationary coherent structures behind the near field region of the jet, which is similar to the well-known von Kármán vortex street arising behind a bluff rigid body, we examined the plane of symmetry in detail. Time series of all three velocity components were evaluated there. In Figure 9 one sees an example of spectra for all three velocity components at the point $x/D = 4$, $y/D = 1$, $z/D = 0$. In the w velocity component spectrum the most pronounced peak is clearly visible. Its value is higher in the LT case than in the HT case. In Figure 10 the downstream dependence of the Strouhal number, Sh, based on the jet diameter D and the cross-flow velocity V_∞ and corresponding to the peak frequency in the x direction, is shown. If we compare this result with vortex shedding behind a circular cylinder ($Sh = 0.2$), the data match if the diameter of the equivalent circular cylinder D_e is set to lie in the range $1.57D$ to $1.77D$, with a slightly higher value for the LT case. Hence the transverse jet wake structures behave similarly to vortices shed from a rigid cylinder with the diameter scaled as indicated for D_e.

4 Conclusions

We conclude that, in addition to the dominant counter-rotating vortex pair, an associated vortex pair may be observed in the low turbulence cross-flow case. The counter-rotating vortex pair was stronger in the high turbulence cross-flow case than in the low turbulence cross-flow case.

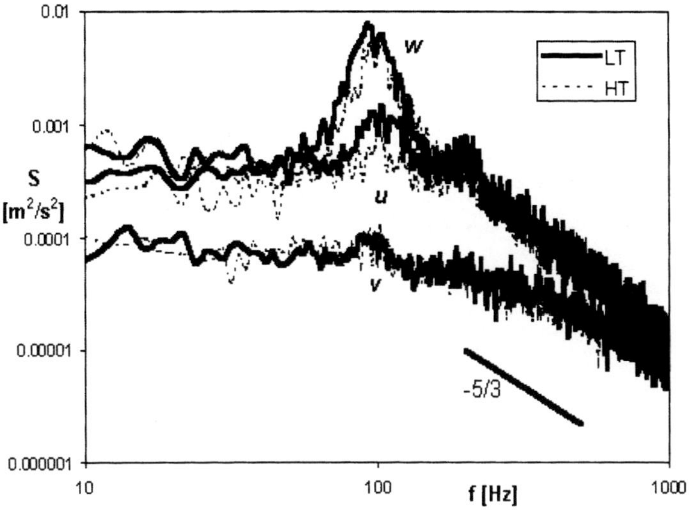

Fig.9 – Spectra example ($x/D = 4$, $y/D = 1$, $z/D = 0$)

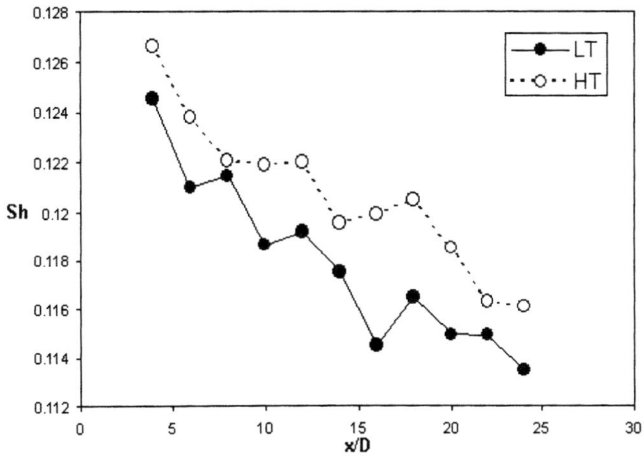

Fig.10 – The peak Strouhal number vs. x-position comparison

The non-stationary structure of the mixing zone was very similar in both LT and HT cases as concerns the intensity of fluctuations, but there were significant differences between the cases in their higher moment distributions. High values were found near the boundary of the mixing zone only in the low turbulence cross-flow case. The differences are related to the intermittent character of the signal near the mixing zone boundary in the LT case. The unsteady structure of the zone behind the bent jet is very similar to the wake behind a cylindrical bluff body. The diameter of the equivalent cylinder is equal to about 1.5 to 1.8 jet diameters and it is greater for the low turbulence cross-flow case.

5 Acknowledgement

Presented work was supported the grant from budget of the Czech Republic by means of the Ministry of Education, Youth and Physical training of the Czech Republic, the project COST 715, No. OC 715.90 and the grant of the Grant Agency of the Czech republic No. 101/98/K001.

References

Fric, T.F., Roshko, A. (1994). Vortical structure in the wake of a transverse jet. *J.Fluid Mech.*, vol. 279, pp. 1-47.

Hemrle J., Uruba V. (1998). On the Steady 3D Pattern of a Jet – Cross-Flow Mixing Zone. In: *Proceedings of colloquium Dynamics of Fluids '98*, Praha 20-22 October 1998, pp.19-20.

Luxa, M., Uruba V. (2000). Visualization of Jet in Cross-Flow. In: *Proceedings of colloquium Dynamics of Fluids 2000*, Praha 24-25 / 10 / 2000, pp.75-76.

Margason, R.J. (1993). Fifty Years of Jet in Cross Flow Research. *72nd AGARD Panel Meeting and Symposium*, Winchester, 19-22/4/1993.

Uruba V., Jonáš P., Mazur O. (1999). Facility for Experimental Investigation of Jet-Cross flow Interaction. In: *Proceedings of Engineering Mechanics '99*, Svratka, 17 - 20 May 1999, vol.3, pp 737-740.

Uruba V., Jonáš P., Mazur O. (1999). On the Unsteady Behaviour of a Jet – Cross Flow Mixing Zone. In: *Book of Abstracts, GAMM Annual Meeting,* Metz, 12-*16* April 1999, p 157.

Uruba, V., Jonáš, P., Mazur, O. (2000). On the Coherent Structure Behaviors in the Jet – Cross Flow Interaction. In: *Book of Abstracts, 4th EUROMECH Fluid Mechanics Conference*, 19-23 November 2000, Eindhoven, p.83.

Uruba, V., Mazur, O., Jonáš, P., Hemrle, J. (1999). On the Unsteady Characteristics of the Jet – Cross Flow Interaction. In: *Proceedings of colloquium Dynamics of Fluids '99*, Praha 19 - 20 October 1999, pp.219-224.

Uruba, V., Mazur, O., Jonáš, P., Hemrle, J. (1999). On the Jet - Cross Flow Interaction. In: *Book of abstracts, EUROMECH Colloquium 391 "Wind tunnel modeling of dispersion in environmental flows"*, Praha, 13-15 September 1999, pp.56-57.

Uruba, V., Mazur, O., Jonáš, P. (2000). The Instabilities in Jet-Cross Flow Interaction. In: *Proceedings of Engineering Mechanics 2000*, Svratka, May 15-18 2000, vol.IV, pp 129-134.

Uruba, V., Mazur, O., Jonáš, P. (2000). On the Influence of the Cross Flow Structure Interacting with a Perpendicular Jet. In: *Book of abstracts of Annual Scientific Conference, GAMM*, Göttingen, 2-7 April 2000, pp.155-156.

Uruba, V., Mazur, O., Jonáš, P. (2001). Intermittency Analysis in a Jet - Cross-Flow Interaction. In: *Proceedings of Engineering Mechanics 2000*, Svratka, 14-17 May 2001, CDROM, 8 pages.

Uruba, V., Mazur, O., Jonáš, P. (2001). Coherent structures in Jet – Cross-Flow Interaction. In: *Book of abstracts, GAMM 2001 Annual Scientific Conference*, February 15 – 15, 2001, Zurich, pp. 151-152.

Uruba, V., Mazur, O., Jonáš, P., Luxa, M. (2001). On the Jet – Cross-Flow Interaction. In: *Proceedings of the conference Topical Problems of Fluid Mechanics '2001*, Praha 21/2/2001, pp. 125-128.

Part II

Background on Modeling, Dynamical Systems, and Control

Three-dimensional Vortex Modeling of Unforced Transverse Jets

Luca Cortelezzi[1] and Ann R. Karagozian[2] *

[1]Department of Mechanical Engineering, McGill University, Montreal, H3A 2K6, Canada

[2]Department of Mechanical Aerospace Engineering,
University of California, Los Angeles, California 90095-1597, U.S.A.

Abstract. The present computational study focuses on the mechanisms for the dynamical generation and evolution of vorticity associated with the jet in crossflow, with special focus on nearfield flow evolution and the generation of the counter-rotating vortex pair (CVP) observed to dominate the jet cross-section. Transient numerical simulations of the flowfield are performed using three-dimensional vortex elements. Vortex ring rollup, interactions, tilting, and folding are observed in the nearfield, consistent with the ideas described in the experimental work of Kelso *et al.* (1996), for example. The time-averaged effect of these jet shear layer vortices, even over a single period of their evolution, is seen to result in initiation of the CVP. Insight into the topology of the flowfield, the formation of wake vortices and the entrainment of crossflow is also provided in this study. Further details on the complete study may be found in Cortelezzi & Karagozian (2001).

1 Introduction

Among the important physical phenomena associated with the jet in crossflow is the formation of vortical structures, in particular the counter-rotating vortex pair (CVP), which are observed to strongly influence jet behavior. These time-averaged vortical structures are suggested to be associated with enhanced overall mixing efficiency for the transverse jet as compared with the free jet or mixing layer (Broadwell & Breidenthal, 1984). While the *global* features of the transverse jet can be understood and predicted in terms of the dynamics of the CVP, the *generation* and *evolution* of these vortical structures is not completely understood. A schematic diagram of the global or mean flow features of the single jet in crossflow is provided in Figure 1. The reader is referred to Karagozian (2002) for the background on and applications of jets in crossflow.

The simulation of a transverse jet constitutes a challenge for any numerical scheme. This flowfield is in general three-dimensional, unsteady, and, for many applications, compressible. The *formation* of the jet, i.e., the initial transient, is of crucial importance in studying the evolution of the jet. Past numerical simulations of transverse jets (e.g., Uenishi *et al.*, 1989) have incorporated turbulence models such as $k - \epsilon$, while the recent three-dimensional Navier-Stokes calculations by Rudman (1994) successfully capture many details of the laminar evolution of

* This work has been supported by NASA Dryden Flight Research Center under Grant NCC 2-374, by the California Energy Commission/EISG under Grant 51240A/99-01-56, by the University of California Energy Institute, and by NSERC Grant RGPIN217169.

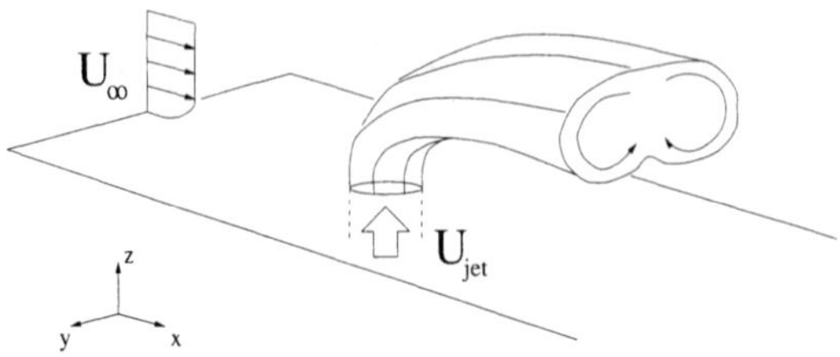

Figure 1. Flow geometry for a jet of velocity U_{jet} in a crossflow of freestream velocity U_{∞}. Figure taken from Cortelezzi & Karagozian (2001).

the flow. Another recent computational study of the jet in crossflow is that of Yuan & Street (1998) and Yuan *et al.* (1999), in which a large eddy simulation (LES) of the incompressible turbulent flowfield is performed. Computed features of the transient jet evolution as well as turbulent stresses are seen to compare well with experimental data, although direct quantitative comparisons of detailed flow features are possible when the computational input conditions (jet-to-crossflow velocity ratio, jet Reynolds number) are made to be different from those stated in the experiments, possibly due to different upstream pipe flow conditions. These differences could also arise due to the fact that, in some experimental studies, the jet flow characteristics are measured with the crossflow turned off, with the assumption that the jet flow at the exit plane and within the nozzle is not strongly influenced by the crossflow. This assumption is not entirely accurate, as discussed by Campolo *et al.* (2002).

Vortex methods are particularly useful in simulating unbounded, incompressible, unsteady flows where the fluid containing vorticity occupies only a small fraction of the total fluid volume. Vortex methods permit accurate, gridless simulation of the evolution of the vorticity field. The velocity field can then be reconstructed from the vorticity field. In the past 25 years, many vortex schemes have been proposed for the simulation of two and three-dimensional flows (e.g., Leonard, 1985; Sarpskaya, 1994; Gharakhani & Ghoniem, 1997; Cottet & Koumoutsakos, 2000). Several features are common among these schemes: the vorticity field is discretized, the position of the vortex elements is identified by Lagrangian coordinates, and the evolution of the vortex elements is computed. The velocity field associated with a vortex element is often referred to as the "induced" velocity field. Vortex elements move under the velocity field induced by themselves, by other vortex elements, and by the background velocity field. The nonlinear process of mutual induction constitutes the foundation of vortex methods.

Vortex methods have been used to simulate some aspects of the evolution of jets. A number of investigators have studied the evolution of a cylindrical vortex sheet in space (Ashurst, 1983; Martin & Meiburg, 1991; Martin & Meiburg, 1992; Dahm *et al.* 1992). The characteristics of a bifurcating jet at high Reynolds numbers have been studied by Parekh *et al.* (1988), while more recently, Nitsche & Krasny (1994) simulated the formation of a vortex ring at the edge of a circular tube, with excellent corroboration with experimental observations. An inviscid 3D vortex sheet model was used to study the nearfield flow in the vicinity of the orifice of a transverse jet by Coelho & Hunt (1989). The vortex sheet model included a distribution of sinks as well as vortices, used to simulate the complex distortion of the vorticity field as well as entrainment just above the injection wall. This complex distortion will be explored in the present study as well.

The present study focuses on the characterization of the nearfield penetration and evolution of the transverse jet. Vortex methods are used to simulate and analyze the generation and evolution of the vorticity field. We confirm the mechanism proposed by Kelso *et al.* (1996) and others describing the formation of the CVP. Furthermore, we suggest an approach which reconciles results obtained from instantaneous measurements or flow visualizations and ensemble averaged measurements. Finally, we characterize the nearfield entrainment of the crossflow into the jet and suggest mechanisms for the initiation of the CVP as well as the initiation of downstream columnar vortices as observed experimentally.

2 Mathematical formulation

The flowfield induced by a jet in crossflow is in general three-dimensional and unsteady. We model this flow using 3D vortex filaments to characterize the generation and evolution of the vorticity field. A series of assumptions is made in order to isolate the mechanisms for the production of vorticity in this problem. First we consider the injectant and crossflow to be incompressible, with the same density ρ. We neglect the coupling between the jet and the wall boundary layer upstream of the jet. The influence of the crossflow on the jet flow at the orifice is computed here (via postprocessing of data), but the effect of the crossflow on the flow deep within the jet nozzle is not represented; this effect is explored in a separate study (Campolo *et al.*, 2002). The present problem is made dimensionless using the radius of the jet, $D/2$, as a characteristic length, and the characteristic time, $D/2U_\infty$, where D is the diameter of the jet.

The assumption that the crossflow does not affect the flow deep within the pipe allows us to build an underlying potential flow in the problem by superimposing a mean jet flow perpendicular to a mean crossflow. The mean crossflow velocity U_∞ is in the x-direction over the jet plane x-y in the upper half space $z \geq 0$ (see the geometry indicated in Figure 1). The crossflow is prescribed to have a laminar boundary layer of constant thickness δ, with a cubic profile (i.e., $U = U_\infty \left[\frac{3}{2}\frac{z}{\delta} - \frac{1}{2}\left(\frac{z}{\delta}\right)^3\right]$) for $0 < z \leq \delta$ upstream of the jet. The mean jet flow, i.e., the flow through a semi-infinite pipe that discharges into the half space $z \geq 0$, is represented by a semi-infinite, cylindrical vortex sheet of diameter D and circulation γ per unit length. The axis of the cylinder coincides with the negative z-axis (see Figure 1). An analytic solution that relates the bulk or average jet velocity U_{jet} to the circulation per unit length γ of a semi-infinite, cylindrical vortex sheet was derived by Parekh *et al.* (1988). This is used as a preliminary estimate in the present model, yet, as will be discussed below, the influence of the evolving vorticity in the flowfield (above the injection wall) affects the specific value of U_{jet} present in the vicinity of the exit plane.

We model the evolution of the jet shear layer by introducing a vortex filament near the jet exit at each time step dt. The vortex filaments used in this simulation are three-dimensional vortex rings of finite core size. The ith vortex filament is represented by a set of N_i nodes interpolated by a cubic spline. The velocity field induced by all vortex filaments at a point $\mathbf{r}$ in space may be computed using the following modified Biot-Savart integral:

$$\mathbf{u}(\mathbf{r}) = -\sum_{i=1}^{N_f} \frac{\Gamma_i}{4\pi} \int_{C_i} \frac{[\mathbf{r} - \mathbf{r}_i(\eta, t)]}{[|\mathbf{r} - \mathbf{r}_i(\eta, t)|^2 + \alpha\sigma_i^2]^{3/2}} \times \frac{\partial \mathbf{r}_i}{\partial \eta} d\eta, \tag{1}$$

where N_f is the total number of vortex filaments, and $\Gamma_i \equiv \gamma^2 dt/2$, σ_i, and C_i are the circulation, the core radius, and the contour of the ith vortex filament, respectively, situated at position $\mathbf{r}_i$. The parameter α controls the vorticity distribution within the vortex core. For the computation shown here we set $\alpha = 0.413$ to impose a Gaussian distribution within the core, whose size σ_i changes in time due to vortex stretching (see below). The computation is advanced in time via a second order predictor/corrector scheme. Since the vortex core size is not considered to be affected by viscous diffusion in the present computations, the simulation may be regarded to be one examining the essentially inviscid evolution of vorticity in the flowfield, e.g., as done for three-dimensional shear layers by Ashurst and Meiburg (1988).

The circulation, the initial position, and the number of nodes of the vortex filaments are crucial to an accurate simulation of the transverse jet. Recently, experimental work conducted by Kelso *et al.* (1996) showed that the vorticity lines near the jet exit are tilted with respect to the jet plane. The circulation assigned to the filaments at a given time step may thus be deduced from the vorticity convected through the orifice. The initial number of nodes is chosen so that the cubic spline that interpolates them accurately approximates a circle of radius $D/2$. For the computations shown here, the initial number of nodes is 24; further refinement in the computation with respect to the initial number of nodes yields virtually identical results. The initial distance of the filament from the jet plane is determined by the velocity of the fluid in the pipe. The initial tilt of the filament with respect to the jet plane is chosen so as to minimize the radial velocity of the fluid at the jet exit. This tilt models the result of the interaction between the jet and the crossflow at injection, although there are clearly physical mechanisms associated with this asymmetry which are not represented.

It is important to note that a vortex filament does not stretch uniformly as it is affected by the crossflow. Filaments actually stretch differently depending on their position along the vortex sheet. As will be seen, and as originally proposed by Kelso *et al.* (1996), the downstream portion of a filament whose plane becomes aligned with the direction of the jet stretches the most because this part of the filament wraps around the upstream part of the secondary vortex ring formed previously. Stretching and deformation of the filaments is resolved in the present computation using the cubic spline interpolation algorithm, which permits accurate and efficient addition and redistribution of nodes as stretching and deformation occurs in order to maintain an accurate simulation. Hence a typical computation might yield as many as 200 nodes per vortex filament by the time the filament has traveled to about four jet radii downstream of injection. The spline scheme also permits computation of the velocity induced at a point, accounting for the contribution due to the curvature of the filaments. The tradeoff in using a spline scheme, however, is that the velocity induced by the spline filament must be calculated by quadrature, which adds to the computational expense.

Maintaining an appropriate level of overlap between adjacent vortex filaments is also important for an accurate simulation. Here the initial core size is assigned to the filament based on the thickness of the boundary layer in the pipe. The time step is chosen so as to guarantee good overlap between adjacent filaments near the jet exit. Vortex stretching tends to reduce the level of overlapping, however; stretching tends to separate the filaments along the jet axis and to increase their length, thus decreasing the size of the filament cores. The change in core radius of the ith filament, σ_i, due to stretching, may be tracked by the following equation:

$$\frac{d}{dt}(\sigma_i^2 L_i) = 0 \tag{2}$$

where L_i is the instantaneous length of the ith filament. This approach, which maintains a constant vortex filament volume, has been used in simulations of 3D shear layers (Ashurst and Meiburg, 1988), for example. Although the detrimental effects of stretching could be balanced using a computationally expensive redistribution of the vorticity field, this is not implemented in the present strategy. The level of overlap for the duration of the nearfield computations performed here is found to be satisfactory because computations performed with larger core sizes yielded virtually identical results to those shown in the present study.

Appropriate boundary conditions for the simulation of the jet in crossflow are difficult to implement because of topological complexity arising from the presence of the pipe. There were several alternative boundary conditions examined in the present simulations. First, image vortex filaments were imposed in order to represent the jet injection wall boundary. Second, panel methods were used to represent the injection wall and the jet orifice. Third, image vortex filaments and panel methods were used jointly. Panels were placed within the jet orifice in order to cancel the effect of the images and produce the specified jet velocity. The result of the refinements implemented by panel methods alone or in conjunction with the image vortex filaments were less accurate in the vicinity of the jet orifice because of the poor behavior of the panels near the jet rim. Simulations with only image vortex filaments produced the most reasonable results, although the nonlinear interaction of the vortex system with the incoming shear layer vorticity right at the injection plane is not precisely represented. Yet in all simulations, irrespective of the type of boundary conditions imposed, the same jet flowfield is generated in time for $z > D/8$, attesting to the robustness of the vortex system generated for the jet in crossflow. Results shown in the present study are for the case where image vortex filaments are used to represent the boundary conditions.

The nonlinear, time-dependent evolution of and interactions within the vortex system in fact create a different evolving centerline jet velocity at and in the vicinity of the injection plane than is predicted by the relation of Parekh *et al.* (1988). The initial transient in the flow, as the jet is started, causes an increase in the jet velocity in time until sustained vortex rollup and tilting occurs, after which the jet centerline velocity essentially remains constant in time, only oscillating slightly in magnitude at the frequency of rollup of the vortex rings. For example, many of the baseline cases shown in the following sections consider a transverse jet with a nondimensional boundary layer thickness of $\delta/D = 0.5$, and with an initial distribution of circulation for a vortex sheet which would correspond to a non-dimensionalized maximum jet velocity of 2 according to the relation of Parekh *et al.* (1988). But with this input condition, after the evolution of the transverse jet has taken place, the vorticity in the flowfield has a chance to affect the jet exit plane conditions such that the velocity in the vicinity of the jet exit plane increases. The baseline case actually produces a peak jet velocity (at the centerline) of $U_{jet} = 10.8$, or a mean nondimensional jet velocity (which also corresponds approximately to the velocity at $(x, y, z) = (D/8, 0, D/8)$) of $U_{jet} = 5.4$. The present computation thus requires a postprocessing of the vorticity field to compute the actual mean or peak jet velocity.

3 Time-resolved nearfield evolution of the jet in crossflow

In this section we present results for the nearfield evolution of a single circular jet in crossflow. Emphasis in the investigation is placed on cases where the ratio between the mean jet and free-stream velocities lies between 2 and 11. These 3D vortex simulations thus may be regarded to represent the low speed flow dynamics seen in the experimental studies of Kelso *et al.* (1996) and Lim *et al.* (1998) for the jet in crossflow. The reader is referred to Cortelezzi & Karagozian (2001) for more comprehensive results.

Figure 2 shows the side view of the evolution of a jet in a crossflow when the mean jet-to-crossflow velocity ratio (or non-dimensional jet velocity) is $U_{jet}/U_\infty = 5.4$ and the boundary layer thickness is $\delta/D = 0.5$. As noted in Section 2, the nonlinear interaction of the vortex system with the jet flow creates a higher mean jet velocity than that initiated at the beginning of the computation (with the velocity computed according to Parekh *et al.*, 1988), so that the actual mean jet velocity near the exit plane is computed by postprocessing the vorticity field. The evolution of the jet is represented by plotting the position of all of the nodes of the vortex filaments at a given time, projected onto the $x - z$ plane at $y = 0$. At nondimensional time $t = 1.5$ only a large vortical structure generated by the initial roll-up of the separating shear layer dominates the flow. The starting vortex ring is tilted with respect to the jet plane and the roll-up of the vortex sheet is tighter on the upstream side, as indicated in Figure 2(a). At time $t = 2.0$, the size of the starting vortex ring, the tilt angle, and the asymmetry between upstream and downstream roll-up have increased. The tilt angle and the asymmetry are responsible for the deformation of the cylindrical vortex sheet that connects the starting vortex ring to the jet exit (Figure 2(b)). The cylindrical vortex sheet stretches on the upstream side while it starts to fold on the downstream side. Folding is the first indication of the formation of the counter-rotating vortex pair (CVP), as observed by Kelso *et al.* (1996). At time $t = 2.5$, this folding has triggered a roll-up of the cylindrical vortex sheet starting the formation of a secondary vortex ring (Figure 2(c)). At time $t = 3.0$, the formation of the secondary vortex ring is completed. The secondary vortex ring is bent nearly at a right angle with respect to its initial orientation (Figure 2(d)). Thus, as suggested by Kelso *et al.* (1996), this secondary vortex ring has folded into two parts: " ... the upstream part whose plane tilts with the mean curvature of the jet, and the downstream part whose plane becomes aligned with the direction of the jet, and so the sides of the downstream part contribute to the circulation of the CVP." Figure 2(d) also shows that the side of the secondary vortex ring aligned with the direction of the jet begins to wind around the downstream part of the (new) starting vortex ring. As a consequence, the asymmetry of the secondary vortex ring is responsible for the deformation of the cylindrical vortex sheet that connects the secondary vortex ring to the jet exit, in turn generating a new fold. The process of folding and the consequent generation of a new secondary vortex ring repeats itself. At time $t = 3.5$, a new secondary vortex ring is formed and a new fold is recognizable. The side of the new secondary vortex ring aligned with the direction of the jet rolls up, in a sort of a spiral, around the side aligned with the direction of the jet of the previous secondary vortex ring, as indicated in Figure 2(e). Note that the winding process stretches the vortex filaments, increasing the vorticity in the direction of the jet. Finally, at time $t = 4.0$, a new secondary vortex ring is forming (see Figure 2(f)), which repeats the process of rollup, folding, and interaction with prior and new vortex ring structures. There appears to be a precise period of formation of the secondary vortex rings. For the conditions shown in Figure 2 the nondimensional period T is about 0.68, where T is made nondimensional by the jet radius and the crossflow velocity. At flow times beyond 4.0, the interaction between the starting vortex ring and the first and second secondary vortex rings generates a complex topology of the vortex filaments, indicating a breakdown of the vortical structures.

We also computed the evolution of a jet in a crossflow when the velocity ratio, or nondimensional mean jet velocity, is $U_{jet}/U_\infty = 2.5$ and 10.8, respectively, and the boundary layer

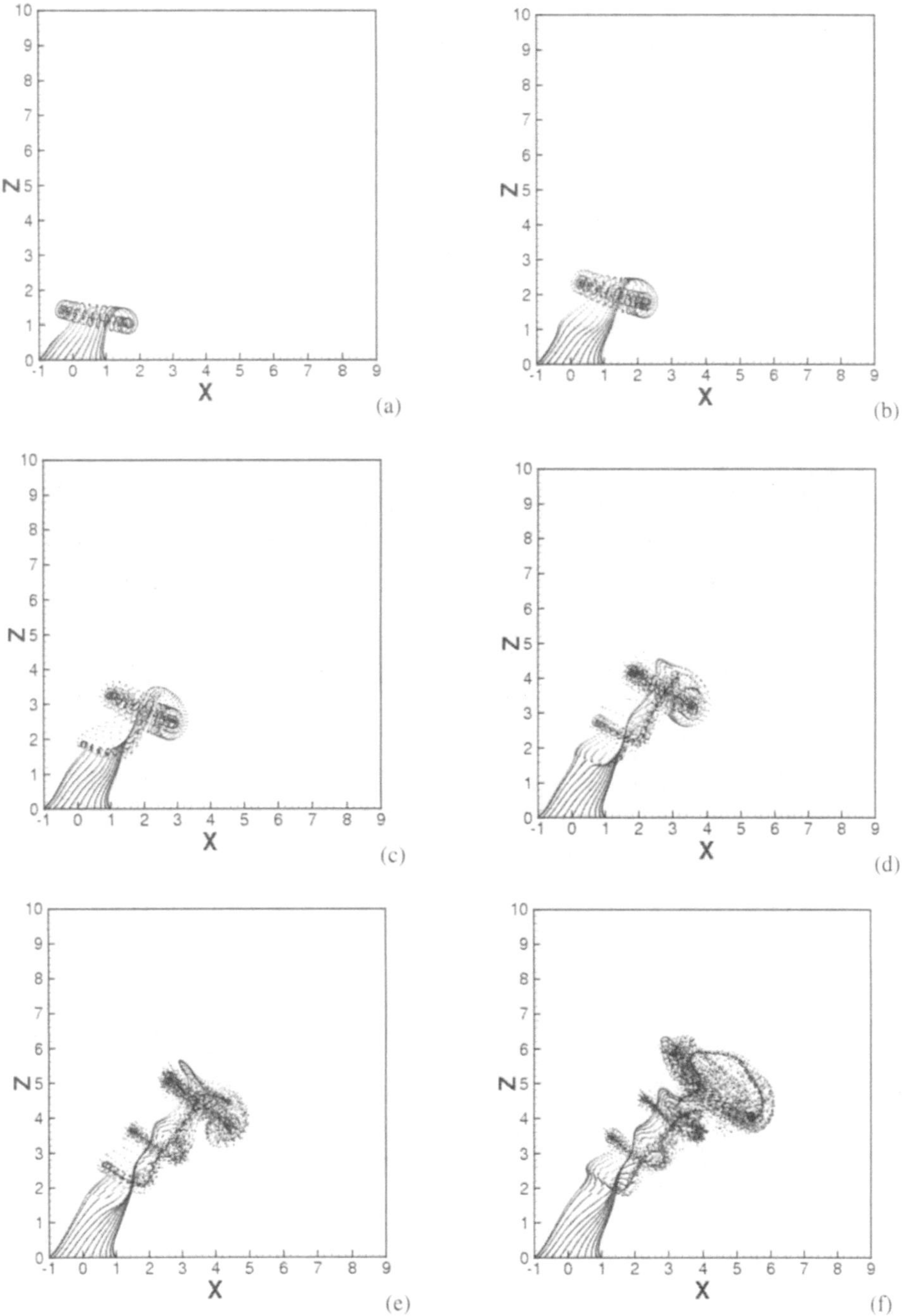

Figure 2. Side view of the evolution of the transverse jet: $U_{jet}/U_\infty = 5.4$, $\delta/D = 0.5$, t=1.5(a), 2.0(b), 2.5(c), 3.0(d), 3.5(e), and 4.0(f). Vortex filaments with initially 24 nodes and core size $\sigma^2 = 0.1$ are introduced in the flow at each time step $dt = 0.01$. Figure taken from Cortelezzi & Karagozian (2001).

thickness is $\delta/D = 0.5$. The mechanisms of vortex rollup and bending described for the case $U_{jet}/U_\infty = 5.4$ apply as well to the case $U_{jet}/U_\infty = 2.5$ and 10.8. Folding and generation of secondary vortex rings alternates as in the previous case, although the temporal and spatial scales are different. The characteristic periods of formation of the secondary vortex rings are approximately $T = 1.0$ and 0.38, respectively. These observations suggest that the frequency of rollup increases with increasing jet velocity, at least in the regime explored here.

Figure 2 shows good agreement between vortex filament simulations and the experimental observations pertaining to vortex roll-up in the transverse jet as presented by Kelso *et al.* (1996). On the other hand, Figure 2 does not provide strong evidence for the sustained existence of the global CVP nor for its formation. In Section 4 we will provide evidence that the formation and existence of the CVP is at least in part an outcome of the averaging process inherent to most experimental studies, consistent with the experimental observations of Smith & Mungal (1998).

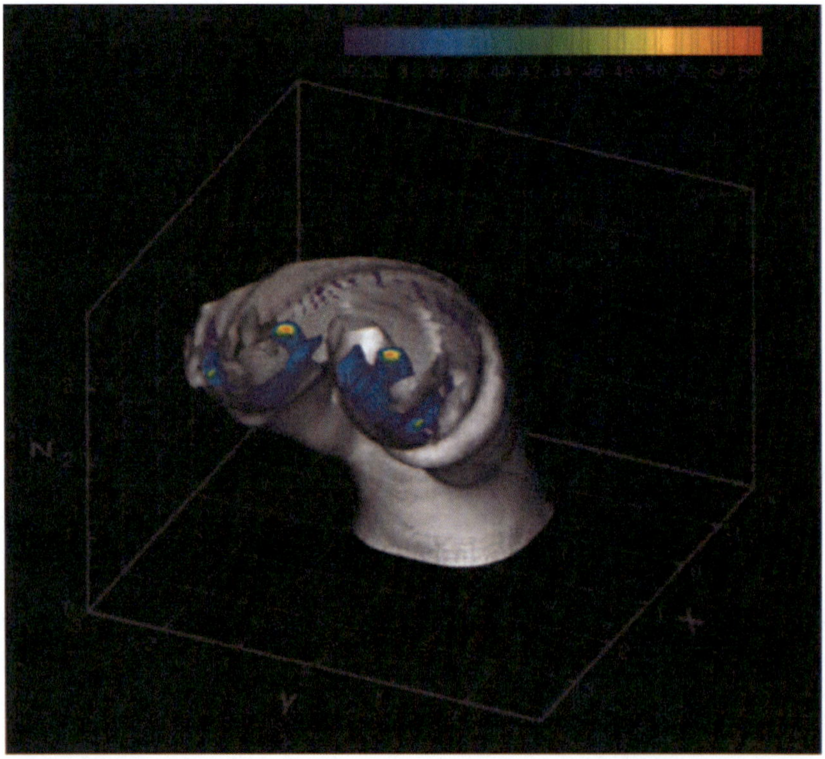

Figure 3. Ensemble-averaged magnitude of the vorticity field for the case $U_{jet}/U_\infty = 5.4$, $\delta/D = 0.5$. The light grey outer isosurface corresponds to average vorticity magnitude $\bar{\omega} = 4.0$, while colored isosurfaces correspond to values of $\bar{\omega}$ of 30 and above, as delineated in the color map shown. The cross-hatching at the lower vorticity values (30-38) is an artifact of the plotting routine used here. Figure taken from Cortelezzi & Karagozian (2001).

4 Formation of the counter-rotating vortex pair

To reconcile the temporal evolution of the vorticity field with experimental observations of the CVP, we attempt to simulate the averaging process commonly used in experiments. We compute the three-dimensional velocity field at different times during a given cycle associated with the formation of a secondary vortex ring. Subsequently, we compute the ensemble average of the velocity fields over this cycle and, from the averaged velocity field, we compute the associated averaged vorticity field. The ensemble over which we perform the average, 13 measurements, is small with respect to the ensembles usually used in experiments, which may involve hundreds of measurements, yet due to the present use of 3D Lagrangian vortex dynamics, the current averaging process is in fact able to reconstruct an accurate time-averaged structure of the jet. When the number of realizations within each ensemble average is either doubled or halved, there is no apparent difference in results.

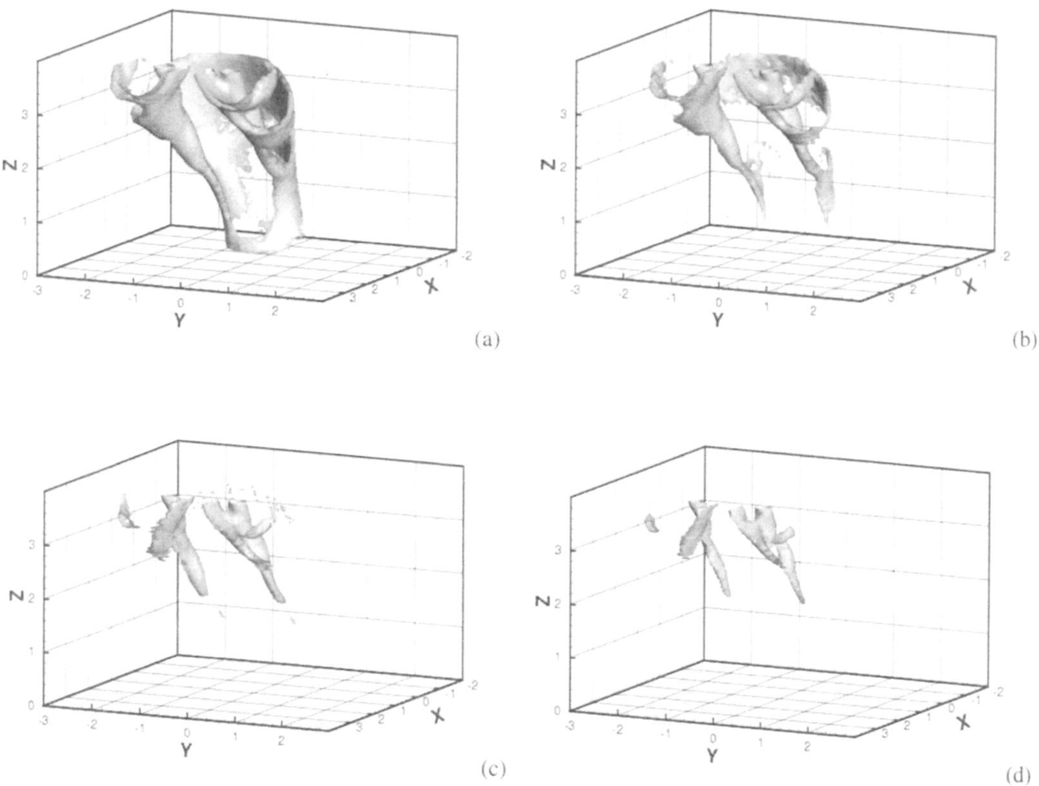

Figure 4. "Exploded view" of the ensemble-averaged magnitude of the vorticity field for the case $U_{jet}/U_\infty = 5.4$, $\delta/D = 0.5$. The average vorticity isosurfaces correspond to (a) $\bar{\omega} = 25$, (b) $\bar{\omega} = 27$, (c) $\bar{\omega} = 30$, and (d) $\bar{\omega} = 32$. Figure taken from Cortelezzi & Karagozian (2001).

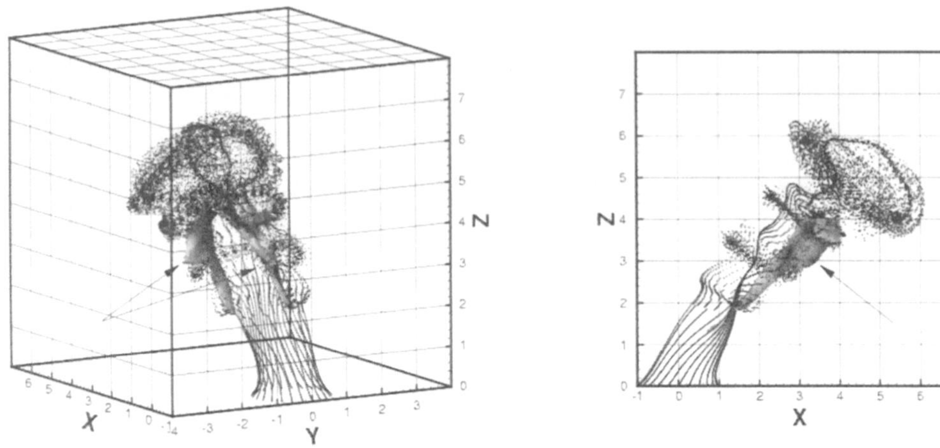

(a) (b)

Figure 5. Instantaneous positions of the transverse jet: $U_{jet}/U_\infty = 5.4$, $\delta/D = 0.5$, at time t=4.0, with three-dimensional view (a) and side view (b), as in Figure 2. Superposed on these plots is the ensemble-averaged 3D vorticity isosurface $\bar{\omega}$ = 30, shown in grey and indicated by the arrows. Figure taken from Cortelezzi & Karagozian (2001).

To visualize the three-dimensional nature of the initiation of the CVP, we compute the averaged vorticity from the time (ensemble) averaging of the velocity vector field. Figure 3 shows a three-dimensional view of the magnitude of the averaged vorticity field for the jet in crossflow where $U_{jet}/U_\infty = 5.4$. The light grey surface corresponds to the isosurface of the magnitude of the averaged (nondimensional) vorticity field, $\bar{\omega} \equiv \sqrt{\omega_x^2 + \omega_y^2 + \omega_z^2}$ equal to 4.0. This isosurface may be designated as that which separates regions of "high" and "low" vorticity for these specific operating conditions. This light grey isosurface is in fact nearly identical to the isosurfaces where the RMS value of the cross product of velocity and vorticity vectors is minimized. Hence the $\bar{\omega} = 4.0$ isosurface is likely to be a reasonable representation of the "jet" region, although at $z = 0$ this isosurface lies somewhat external to the actual jet orifice due to the boundary conditions employed. The regions in which the magnitude of the vorticity vector is greater than 30 (shown in darker grey in Figure 3) have strong similarity to the jet structure observed experimentally when jet fluid is seeded with dye (Kelso *et al.*, 1996). High-magnitude vorticity is clearly contained within two slim columns which resemble the CVP cores. Figures 4(a)-(d) show "exploded" views of some of the separate vorticity isosurfaces seen in darker grey in Figure 3, for the values $\bar{\omega}$ = 25, 27, 30, and 32, respectively. The origins of the initiation of the CVP are evident as rollup and bending of the vortex sheet occurs (Figure 4(a)), leading to tight columnar vorticity isosurfaces (Figure 4(b)) which originate even within one radius of the jet orifice. Strengthening of this rollup is evident as one moves further away from the injection plane (Figures 4(c)-(d)). These robust vortical structures, which form independently of the alternative computational boundary conditions explored, appear to be the initiation of the fully developed CVP structure seen experimentally in the farfield. This nearfield initiation of the CVP is suggested by a number of recent experiments (Kelso *et al.*, 1996; Smith & Mungal, 1998).

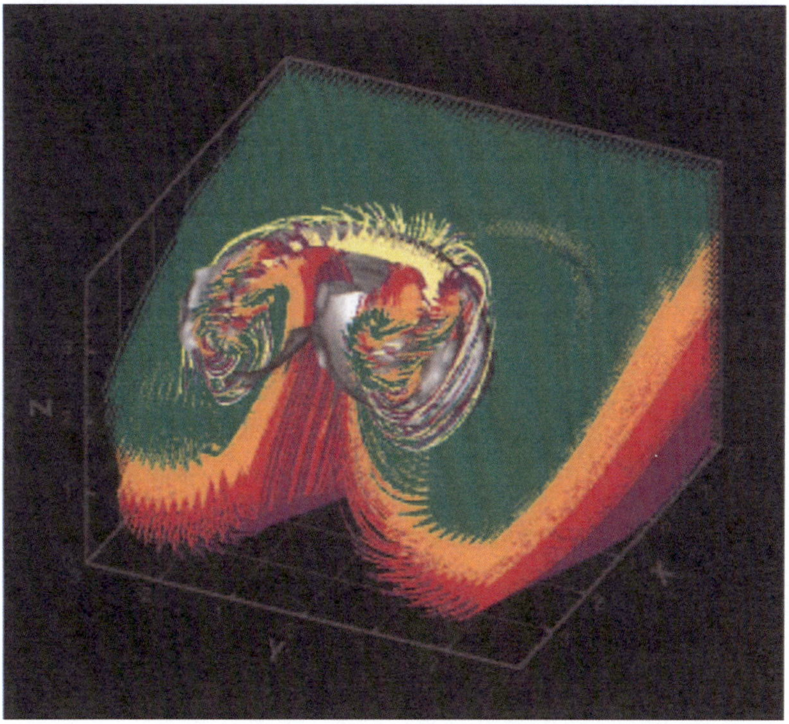

Figure 6. Averaged magnitude of the vorticity field $\bar{\omega}$ = 4.0 for the case $U_{jet}/U_\infty = 5.4$, $\delta/D = 0.5$, on which is superposed the evolution of horizontal arrays of "smoke" tracers. The tracer streaklines are computed from the ensemble-averaged velocity field. The yellow streaklines are associated with fluid originating within the jet nozzle, while the other colored streaklines represent crossflow fluid originating upstream of the jet, at $x = -2$, within one radius (purple), two radii (pink), three radii (orange), and four radii (green) of the injection wall. Figure taken from Cortelezzi & Karagozian (2001).

The ensemble averaging here also enables validation of the ideas of Kelso *et al.* (1996) with respect to initiation of the CVP via folding and tilting of the secondary vortex ring which enters the flowfield. Figures 5(a)-(b) show 3D and side views, respectively, of the instantaneous positions of vortex filaments for the $U_{jet}/U_\infty = 5.4$ case, as done in Figure 2(f) at time $t = 4.0$, on top of which are superposed the grey, ensemble-averaged 3D vorticity isosurfaces $\bar{\omega}$ = 30. As noted in Figures 3 and 4, the $\bar{\omega}$ = 30 contour appears to enclose the cores of the evolving three-dimensional CVP; in Figures 5(a)-(b) this contour is clearly coincident with the "downstream part" of the tilted secondary vortex ring, similar to concentrated dye evolution visualized in the experiments of Kelso *et al.* (1996) and more recently in Lim *et al.* (1998). Thus there is strong evidence that, as proposed by Kelso *et al.* (1996), the initial bending of the vortex sheet is the first sign of the initiation of the CVP structure.

5 Crossflow entrainment by the transverse jet

Among the benefits of using the jet in crossflow in a gas turbine engine environment is the enhancement of the entrainment of crossflow into the vicinity of the jet and the subsequent enhancement of the molecular mixing of the two fluids that occurs. Entrainment by a transverse jet or flame is typically seen to be much greater (see Broadwell & Breidenthal, 1984; Brzustowski, 1977; Karagozian, 1986b; Smith & Mungal, 1998) than that by a free jet (Hottel & Luce, 1953) or a buoyant turbulent jet (Steward, 1970), for example. Smith & Mungal (1998) suggest that, in light of the evolving development of the CVP in the nearfield of the jet, the enhanced mixing in the jet's nearfield corresponds to and likely results from the structural *formation* of the CVP rather than its fully-developed *existence*. This nearfield enhancement of entrainment and the potential for enhancement of molecular mixing may be explored in the present vortex element simulations.

The present simulations allow a means for detailed examination of nearfield entrainment because the ensemble averaging of the velocity vector field is by definition time-independent. Consequently, the entrainment of the crossflow can be characterized using stream-tracers imposed in the ensemble-averaged flowfield here, since streamlines and streaklines coincide. Figure 6 shows the light grey vorticity isosurface $\bar{\omega} = 4.0$ for the jet in crossflow with velocity ratio $U_{jet}/U_{\infty} = 5.4$, as done in Figure 3, on which there is superposed the evolution of a densely packed array of smoke tracers initially placed two radii upstream of the jet centerline. The yellow streaklines enclosed within the upstream portion of the vorticity isosurface are associated with fluid originating within the jet nozzle, which was initially distributed uniformly within the circular jet orifice at $z = 0$. The other colored streaklines represent crossflow fluid originating upstream of the jet, initially situated within one radius (purple), two radii (pink), three radii (orange), and four radii (green) of the injection wall. These streakline patterns indicate that the crossflow is deflected about the jet, then can be strongly entrained into the jet on its lee side, with streaklines winding into the vortical structures which form in the ensemble-averaged flowfield. The concentration of the yellow jet fluid within the upstream portion of the $\bar{\omega} = 4.0$ vorticity isosurface, above the bulk CVP structure, is consistent with experimental findings dating from the work of Kamotani & Greber (1972). These and subsequent experimental studies suggest that the local center of the CVP structure spatially "lags" or lies below the local maximum in axial jet velocity. Most of the crossflow streaklines shown in Figure 6, even those originating close to the injection wall (purple), are uplifted on the lee side of the jet into the centerline region of the evolving vortex pair structure, later winding into the vortices themselves.

This latter observation is further consistent with those of Fric & Roshko (1994) as well as Smith & Mungal (1998), which suggest that the upright wake vortices forming downstream of the transverse jet originate in the wall boundary layer. As the purple streaklines originating in the wall boundary layer are deflected about the jet, they fan outward and upward into the evolving

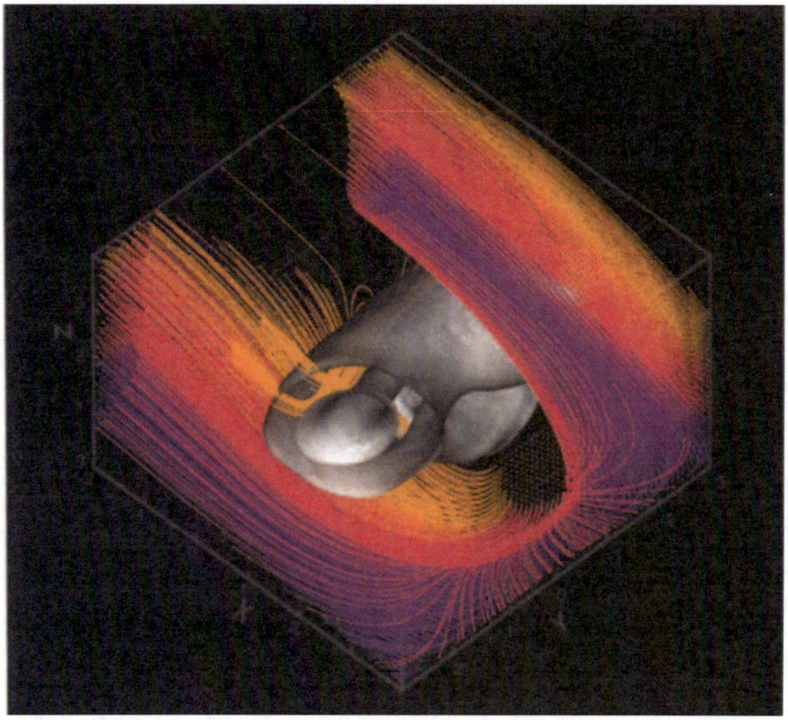

Figure 7. Averaged magnitude of the vorticity field for the case $U_{jet}/U_{\infty} = 5.4$, $\delta/D = 0.5$, on which is superposed the backward evolution of arrays of "smoke" tracers originating outside of the jet at $x = 4$. The tracer streaklines shown are those, computed from the averaged velocity field, which have bypassed the jet structure and are not entrained into the CVP. Figure taken from Cortelezzi & Karagozian (2001).

CVP associated with the jet fluid. This phenomenon is also similar to the uplifting of boundary layer dye into the jet observed in the experiments of Kelso *et al.* (1996), which led these researchers to postulate the existence of a saddle-node point on the lee-side of the jet exit. This three-dimensional critical point is a saddle in the jet plane and a node in the symmetry plane. As the streamlines approach the critical point, they stretch in the jet plane due to the saddle point and fan out in the symmetry plane due to the node creating a stream-surface. Since the region of our computed saddle point is far enough downstream of jet injection, we believe that the present simulations provide an accurate rendering of this phenomenon. A more comprehensive topological map of the flow near the jet exit may be found in the discussion by Kelso *et al.* (1996).

Figure 6 also suggests that, far enough away from the jet nearfield structure (in the y-direction), smoke tracers can completely bypass the jet, apparently avoiding any substantial degree of entrainment into the jet. This observation is further demonstrated in Figure 7, which shows smoke tracers initially situated four radii *downstream* of and external to the jet, at $x = 4$, with the same color scheme as in Figure 6 in the region $0 \leq z \leq 3$. A "backward" tracing of the origination of the smoke is made, for this ensemble-averaged flowfield, to the location $x = -2$. Virtually all of these "external" streamtracers correspond to streaklines which originated in the crossflow, indi-

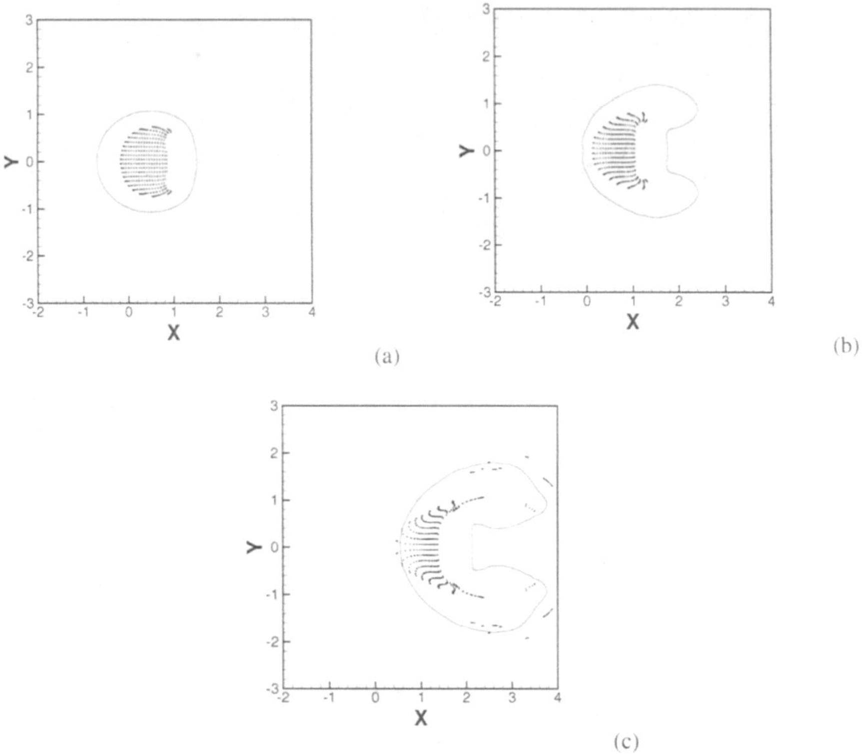

Figure 8. Slices in the $x-y$ plane of the transverse jet evolution: $U_{jet}/U_{\infty} = 5.4$, $\delta/D = 0.5$, at z-locations 1.0(a), 2.0(b), and 3.0(c). A slice of the average vorticity isosurface $\bar{\omega} = 4.0$ is shown. Also shown, in black, are the positions of streamtracers which originated within the jet, corresponding to tracer positions shown in yellow in Figure 6. Figure taken from Cortelezzi & Karagozian (2001).

cating that there has been a complete bypassing of the jet. Such streaklines show evidence of a "bathtub"-shaped stream-surface surrounding the jet nearfield which contains crossflow fluid that is entrained into the jet. All the crossflow fluid outside of this "bathtub"-shaped stream-surface completely bypasses the jet.

The evolution of streamtracers within the ensemble-averaged transverse jet flowfield also provides data from which entrainment of crossflow by the jet may be visualized. Figures 8(a)-(c) show a series of slices of the averaged jet flowfield in the $x-y$ plane, at different z-locations, which measure nondimensional distance from the injection wall (z = 1, 2, and 3, respectively). These slices, of thickness Δz =0.04, reveal the $\bar{\omega}$ = 4.0 vorticity isosurface, which at the injection wall lies somewhat external to the jet orifice. Also shown are the positions of streamtracers which originated within the jet (shown in black; these were yellow in Figure 6). The evolution of the flowfield in the z-direction provides further visual evidence for the entrainment of crossflow as the transverse jet and associated CVP develop. As z increases, the region containing black (jet) streamtracer particles proportionately decreases within the area enclosed by the vorticity isosurface.

Consistent with the visualization in Figure 6, as the crossflow is entrained into the jet in Figure 8, principally on the lee side, jet fluid tends to become concentrated within the upstream portion of the deformed "jet slice". Figure 8(c) further suggests that a small amount of jet fluid can be ejected out of the vorticity isosurface downstream of injection, as more and more of the "jet" region is comprised of crossflow fluid. At the location $z = 3$, the jet fluid itself comprises less than one third of the fluid contained within this slice of the vorticity isosurface. The association of vortex tilting and folding with entrainment into the jet is further visualized in Figures 8(a)-(c). In fact, the jet fluid's nearfield deformation in the $x - y$ plane, evolving from a circle to the cusped structures seen by the tracers in Figures 8(a)-(c), is remarkably similar to that shown by Kelso *et al.* (1996) as an interpretation of the reorientation processes for the jet shear layer.

6 Summary and conclusions

The nearfield evolution of the jet in crossflow has been simulated here using temporally evolving, 3D vortex elements. This simulation allows the examination of the instantaneous flow structure of the jet as well as an ensemble-averaged view of the flowfield. Instantaneous images exhibit clear rollup of vortical structures, in addition to tilting and folding of the evolving rings. Folding of the vortex rings is seen, consistent with the ideas of Kelso *et al.* (1996), to be responsible for the initiation of the counter-rotating vortex pair as well as the stretching and deformation of later-forming vortical structures. Jet penetration as well as periodicity associated with the rollup, folding, and interaction with prior and new vortex ring structures are seen to be dependent on specific flow conditions, including the jet velocity and crossflow velocity. The influence of these parameters on penetration corresponds well with experimental observations (Kamotani & Greber, 1972; Abbitt *et al.*, 1993; Karagozian *et al.*, 1996). If increasing vortex rollup period is seen to suggest increased jet-crossflow interaction and hence potential for entrainment and mixing, then the present results are further consistent with experimental findings (Karagozian *et al.*, 1996).

While instantaneous images provide some evidence for the potential initiation of the CVP through the downstream part of the folded vortex ring structure, ensemble-averaged images (averaged over a single vortex rollup period) provide much stronger evidence of the initiation of a robust CVP. Moreover, the ensemble-averaged velocity field may be used to predict the evolution of streaklines placed upstream of the jet, so that entrainment of crossflow by the jet may be visualized. There is clear evidence of entrainment of crossflow on the lee side of the jet, with a concentration of injectant in the upstream portion of the jet, similar to experimental observations (Kamotani & Greber, 1972). There is also visualization of a fanning of the deflected upstream boundary layer flow about the jet and upward into the lee side of the jet, which suggests origins of the experimentally-observed wake vortices within the boundary layer, again consistent with experiments (Fric & Roshko, 1994; Smith & Mungal, 1998).

Entrainment of crossflow into the nearfield of the jet is also visualized in these simulations by taking slices of the jet, in the $x-y$ plane, at different z locations. Clear evidence of the increasing proportion of crossflow fluid (and the decreasing proportion of injectant) in downstream jet slices is observed. Entrainment of crossflow appears to be clearly enhanced as the jet is deformed and as vortex rollup results in the formation of the CVP. While corresponding enhancement of fluid mechanical mixing as a result of the entrainment cannot be quantified in the present study, since molecular diffusion is not explicitly represented, the conditions for which entrainment increases can be used to identify conditions for which potentially greater mixing of injectant and crossflow can occur.

The authors wish to thank Professors Anthony Leonard, Anatol Roshko, and Frank Marble for several valuable discussions pertaining to this research.

References

ABBITT, J. D. III, SEGAL, C., MCDANIEL, J. C., KRAUSS, R. H., AND WHITEHURST, R. B. 1993 Experimental supersonic hydrogen combustion employing staged injection behind a rearward-facing step *J. Prop. Power* **9**, 472–478.

ASHURST, W. T. 1983 Large eddy simulation via vortex dynamics *AIAA* Paper No. 83-1879-CP.

ASHURST, W. T. AND MEIBURG, E. 1988 Three-dimensional shear layers via vortex dynamics *J. Fluid Mech.* **189**, 87-116.

BROADWELL, J. E. AND BREIDENTHAL, R. E. 1984 Structure and mixing of a transverse jet in incompressible flow *J. Fluid Mech.* **148**, 405–412.

BRZUSTOWSKI, T. A. 1977 Hydrocarbon turbulent diffusion flame in subsonic cross flow. *Turbulent Combustion: Prog. Astro. Aero.* **58**, 407–430.

CAMPOLO, M., DEGANO, G.M., CORTELEZZI, L. AND SOLDATI, A. 2002 Numerical appraisal of jet-to-crossflow coupling in a transverse jet. *Manipulation and Control of Transverse Jets*, Springer-Verlag.

COELHO, S. L. V. AND HUNT, J. C. R. 1989 The dynamics of the near field of strong jets in crossflows *J. Fluid Mech.* **200**, 95–120.

CORTELEZZI, L. AND KARAGOZIAN, A. R. 2001 On the formation of the counter-rotating vortex pair in transverse jets *J. Fluid Mech.* **446**, 347–373.

COTTET, G-H. AND KOUMOUTSAKOS, P. D. 2000 *Vortex methods - Theory and Practice* Cambridge University Press.

DAHM, W. J. A., FRIELER, C., AND TRYGGVASON, G. 1992 Vortex structure and dynamics in the near field of a coaxial jet *J. Fluid Mech.* **241**, 371–402.

FRIC, T. F. AND ROSHKO, A. 1994 Vortical structure in the wake of a transverse jet *J. Fluid Mech.* **279**, 1–47.

GHARAKHANI, A. AND GHONIEM, A.F. 1997 Three-dimensional vortex simulation of time dependent incompressible internal viscous flows. *J. Comp. Phys.* **134**, 75–95.

HOTTEL, V. O. AND LUCE, R. G. 1953 Burning in laminar and turbulent fuel jets. *Fourth Symposium (Intl.) on Combustion*, 97.

KAMOTANI, Y., AND GREBER, I. 1972 Experiments on a turbulent jet in a cross flow. *AIAA J.* **10**, 1425–1429.

KARAGOZIAN, A. R. 1986b The flame structure and vorticity generated by a chemically reacting transverse jet. *AIAA J.* **24**, 1502–1507.

KARAGOZIAN, A. R. 2002 Background on and applications of jets in crossflow. *Manipulation and Control of Transverse Jets*, Springer-Verlag.

KARAGOZIAN, A. R., WANG, K. C., LE, A.-T., AND SMITH, O. I. 1996 Transverse gas jet injection behind a rearward-facing step *J. Prop. Power* **12**, 1129–1136.

KELSO, R. M., LIM, T. T., AND PERRY, A. E. 1996 An experimental study of round jets in cross-flow. *J. Fluid Mech.* **306**, 111–144.

LEONARD, A. 1985 Computing three-dimensional incompressible flows with vortex elements. *Ann. Rev. Fluid Mech.* **17**, 523–559.

LIM, T. T., KELSO, R. M., AND PERRY, A. E. 1998 A visual study of vortex rings fired transversely into a crossflow. *13th Australian Fluid Mechanics Conference*, Monash University, Melbourne, Australia.

MARTIN, J. E. AND MEIBURG, E. 1991 Numerical investigation of three-dimensionally evolving jets subject to axisymmetric and azimuthal perturbations. *J. Fluid Mech.* **230**, 271–318.

MARTIN, J. E. AND MEIBURG, E. 1992 Numerical investigation of three-dimensionally evolving jets under helical perturbations. *J. Fluid Mech.* **243**, 457–487.

NITSCHE, M., AND KRASNY, R. 1994 A numerical study of a vortex ring formation at the edge of a circular tube. *J. Fluid Mech.* **276**, 139–161.

PAREKH, D. E., LEONARD, A. AND REYNOLDS, W. C. 1988 Bifurcating jets at high Reynolds numbers. *Report TF-35, Thermosciences Division, Department of Mechanical Engineering, Stanford University.*

RUDMAN, M. 1994 Numerical simulation of a jet in crossflow. *International Colloquium on Jets, Wakes, and Shear Layers* Melbourne, AU. CSIRO.

SARPSKAYA T. 1994 *Adv. Appl. Mechanics* **31**, 113–247.

SCHULLER, T., KING, J., MAJAMAKI, A., AND KARAGOZIAN, A. R. 1999 An Experimental Study of Acoustically Controlled Gas Jets in Crossflow *Bull. Amer. Phys. Soc.* **44(8)**, 111.

SMITH, S. H. AND MUNGAL, M. G. 1998 Mixing, structure and scaling of the jet in crossflow. *J. Fluid Mech.* **357**, 83–122.

STEWARD, F. R. 1970 Prediction of the height of turbulent diffusion buoyant flames. *Combust. Sci. Tech.* **2**, 202–212.

UENISHI, K., ROGERS, R. C., AND NORTHAM, G. B. 1989 Numerical predictions of a rearward-facing-step flow in a supersonic combustor *J. Propul. Power* **5**, 158–164.

YUAN, L.L. AND STREET, R. L. 1998 Trajectory and entrainment of a round jet in crossflow *Physics of Fluids* **10**, 2323–2335.

YUAN, L. L., STREET, R. L., AND FERZIGER, J. H. 1999 Large-eddy simulations of a round jet in crossflow *J. Fluid Mech.* **379**, 71–104.

Large-Eddy Simulation of Jets: State of the Art and Open Problems

Maria Vittoria Salvetti

Department of Aerospace Engineering, University of Pisa, Italy

Abstract. The mathematical formulation of large-eddy simulation is introduced and the closure problem (subgrid scale modeling) is briefly reviewed. Some fields which need further investigation, in order that large-eddy simulation becomes a completely reliable tool, in particular for the analysis of complex flows, are identified and discussed. Specific problems for application to jet flows are also illustrated. Finally, examples of large-eddy simulations of jets in cross flow are presented.

1 Introduction

The direct numerical simulation (DNS) of jet flows at realistic Reynolds numbers is at present impossible, because of the enormous amount of computational resources required. On the other hand, since the dynamics of jet flows is highly unsteady and three-dimensional, simulations based on the numerical discretization of the Reynolds-averaged Navier-Stokes equations encounter significant difficulties in predicting this type of flows. Conversely, the large-eddy simulation (LES) approach, in which only the large scale motion is simulated, while the small scale effects are modeled, provides a compromise which appears to be particularly promising for the simulation of jet flows. Indeed, flow unsteadiness and three-dimensionality are automatically accounted for in LES. Furthermore, the flow dynamics, at least in the region close to the jet exit, is dominated by the evolution of vortical structures of quite large dimensions, which are thus directly simulated in LES.

2 Mathematical Formulation and Subgrid-Scale Modeling

In the LES approach, the governing equations are filtered in space over subdomains $C_1, ..., C_N$, with $\bigcup_{i=1,\ldots,N} C_i = \Omega$, Ω being the computational domain. Filtered quantities are obtained by the following convolution:

$$\overline{f}(X,t) = \int_{C_i} f(X',t)G(X'-X)dX' \qquad (1)$$

where $G(X'-X)$ is a normalized filter function. In 3D the filter function may be written as follows:

$$G(X'-X) = \prod_{\alpha=1}^{3} G_\alpha(x'_\alpha - x_\alpha) \qquad (2)$$

$G_\alpha(x'_\alpha - x_\alpha)$ being a filter in the direction x_α.

Several types of filters can be used. Classical ones are the Gaussian filter:

$$G_\alpha(x) = \left(\frac{6}{\pi \Delta_\alpha^2}\right)^{1/2} exp\left(-\frac{6x^2}{\Delta_\alpha^2}\right) \tag{3}$$

or the *top-hat* or *box* filter:

$$G_\alpha(x) = \frac{1}{\Delta_\alpha} \text{ if } -\Delta_\alpha/2 \leq x \leq \Delta_\alpha; \ 0 \text{ otherwise} \tag{4}$$

in which Δ_α is the filter width. A cut-off filter can be also applied in the wave number space:

$$\widehat{G}_\alpha(k) = 1 \text{ if } -\pi/\Delta_\alpha \leq k \leq \pi/\Delta_\alpha; \ 0 \text{ otherwise} \tag{5}$$

Recently, discrete filters in physical space have been proposed specifically for application to LES (see, for instance, Vasilyev et al., 1998).

The LES governing equations are obtained by applying the filter to the Navier-Stokes equations. Let us consider, for instance, the incompressible case; by using the commutativity between the filter and the differential operators, the following filtered equations are obtained (in Cartesian coordinates and using the Einstein notation):

$$\frac{\partial \overline{u}_i}{\partial t} + \frac{\partial \overline{u}_i \overline{u}_j}{\partial x_j} = -\frac{\partial \overline{p}}{\partial x_i} + 2\nu \frac{\partial \overline{S}_{ij}}{\partial x_j} - \frac{\partial \tau_{ij}}{\partial x_j} \tag{6}$$

in which $\overline{u}_i$ is the component in the i direction of the filtered velocity, $\overline{p}$ is the filtered pressure, $S_{ij} = \frac{1}{2}\left(\frac{\partial \overline{u}_i}{\partial x_j} + \frac{\partial \overline{u}_j}{\partial x_i}\right)$ is the resolved strain rate tensor and $\tau_{ij} = \overline{u_i u_j} - \overline{u}_i\, \overline{u}_j$ is the *subgrid scale stress tensor*, which must be modeled to close the problem. Subgrid scale (SGS) modeling is indeed a critical point for a successful LES.

The most widely used model in early LES was the eddy-viscosity Smagorinsky model (1963), in which the subgrid scales are assumed to have a purely dissipative effect. Thus, the subgrid scale stress tensor may be expressed as a function of the resolved strain tensor as follows:

$$\tau_{ij} - \frac{\delta_{ij}}{3}\tau_{kk} = -2C\bar{\Delta}^2 |\bar{S}| \bar{S}_{ij} \tag{7}$$

in which $|\bar{S}| = (2\bar{S}_{ij}\bar{S}_{ij})^{1/2}$ and $\nu_s = C\bar{\Delta}^2|\bar{S}|$ may be considered as a SGS eddy viscosity. However, this model revealed some important drawbacks; first of all, it contains an input coefficient C to be specified *a priori*, which can not be assumed to be flow independent unless the filtered scales are contained well inside an ideal inertial range of locally isotropic and homogeneous turbulence. Furthermore, once the coefficient C is specified, the model limiting behavior near a wall and in laminar flow is incorrect and the backscatter of energy from small scales to large scales can not be accounted for. Finally, the principal axes of τ_{ij} are assumed to be parallel to those of the resolved strain rate tensor $\bar{S}_{ij}$, which is not generally correct. Among the proposed remedies to some of the previously cited problems are the use of *ad-hoc* damping functions (Van-Driest functions) near the solid walls, or procedures to obtain analytical expressions of the subgrid scale viscosity (see, for instance, Kraichnan, 1976 and Yakhot et al., 1989).

However, the first model which overcame most of the previously listed limitations is the *dynamic* subgrid model proposed by Germano et al. (1991). Equation (7) is retained, but the coefficient C becomes a parameter $C(X,T)$ that is explicitly computed by using the information from the smallest resolved scales. This model automatically has the correct behavior near solid walls and for laminar flow ($C \rightarrow 0$), and permits backscatter of energy from small scales to large scales ($C < 0$). However, the coefficient C computed following the dynamic procedure shows strong *unphysical* oscillations. In correspondence with high negative peaks, the subgrid-scale viscosity (negative) may become larger than the molecular viscosity and this leads to numerical instabilities. To avoid this problem, the C parameter may be averaged over homogeneous directions (if they exist), or local averaging and/or clipping may be used. Modifications of the dynamic procedure have also been proposed. Ghosal et al. (1995) proposed an integral formulation of the algebraic identity used in the dynamic computation of C, which led to a nicer mathematical formulation of the dynamic procedure. However, it requires the solution of an integral-differential equation at each time step and implies a significant increase in computational time. Furthermore, the application in LES showed only a marginal damping of the unphysical oscillations of C. More recently, Meneveau et al. (1996) proposed to average C over flow pathlines by backwards time integration, with a weighting function that gives decreasing weights to past events. In this way, a better behavior of the model parameter is obtained (see also Sarghini et al., 1999), with a moderate computational overhead. The dynamic procedure may also be generalized to account for variation of C with scale, as done, for instance, by Porté-Agel at al. (2000).

Scale similarity models (Bardina et al., 1980, Liu et al., 1994, Akhavan et al., 2000) provide an alternative class of SGS parameterizations. They are based on the assumption of similarity between the resolved and the unresolved field, and they can be broadly represented as:

$$\tau_{ij} = C_{sim} \, \widetilde{\overline{u}_i \overline{u}_j} - \widetilde{\overline{u}}_i \widetilde{\overline{u}}_j \tag{8}$$

in which the filter denoted by a tilde may be either the grid filter itself (Bardina et al., 1980) or a different one (Liu et al., 1994, Akhavan et al., 2000), and C_{sim} is a model coefficient. Similarity models naturally account for backscatter and show high correlation with actual SGS stresses in *a priori* tests. Their main drawback is that, when implemented in simulations, they do not provide enough energy dissipation and, thus, lead to inaccurate results. A possible remedy is to combine the scale similarity model (8) with an eddy-viscosity term (7), obtaining in this way the so called *mixed* models (Bardina et al., 1980, Zang et al., 1993, Liu et al., 1994, Salvetti and Banerjee, 1995, Akhavan et al., 2000). Mixed models that combine eddy-viscosity with an additional term different from Eq. (8) have also been devised (see, for instance, Horiuti (1997) and Winckelmans et al. (1998)). The dynamic procedure can be used to compute one (Zang et al., 1993, Akhavan et al., 2000, Winckelmans et al., 1998) or both model coefficients (Salvetti and Banerjee, 1995, Horiuti, 1997, Sarghini et al., 1999).

More recently, various deconvolution procedures (Leonard, 1974, Domaradzki and Saiki, 1997, Geurts, 1997, Stolz and Adams, 1999) have been used to recover the resolved part of the unfiltered velocity field from the filtered one. Thus, the subgrid scale term can be evaluated by using the approximated unfiltered field. This approach has given good results with a limited computational cost; however, deconvolution must be supplemented by regularization (Stolz and Adams, 1999) or by an ad-hoc model, to account for the effects of the remaining subgrid scales (Domaradzki and Saiki, 1997).

Multi-scale concepts have also been applied to SGS modeling by Hughes et al. (2000). By variational projection, the LES governing equations are split into large scale and small scale equations and a Smagorinsky type SGS model is applied only to the small scales. This idea is very simple to be implemented for spectral methods, for which good results have been obtained (Hughes et al., 2001).

For compressible flows, the LES equations are obtained by introducing the density weighted or Favre filter. Besides the subgrid-stress tensor, additional unclosed terms appear in the filtered equations; not much work has been done up to now for the physical modeling of these terms. We refer to Erlebacher et al. (1992), Vreman et al. (1995) and Lesieur and Comte (1997) for details.

3 Open Problems in LES

Before describing specific applications of LES to jet flows, we briefly discuss some points which need further investigation, in order that LES becomes a completely reliable tool, in particular for the analysis of complex flows, such as those encountered in many practical applications.

As briefly reviewed in the previous section, although remarkable advances have been made in SGS modeling, a model of general validity is not yet available and the limits of many of the proposed models still need to be established. The situation is worsened for compressible flows or when additional physical phenomena, such as combustion, scalar mixing or particle tracking, are considered.

Other points of controversy concern numerical discretization. For instance, there is no general agreement on the spatial accuracy needed for numerical schemes in LES. Some studies indicate that highly accurate schemes (more than 6th order or spectral) are needed (see, for instance, Ghosal (1996) or Kravchenko and Moin (2000)). On the other hand, several examples of successful LES with 2nd order schemes exist in the literature (see, for instance, Akselvoll and Moin, 1996, Jones and Wille, 1996, Voke and Gao, 1998, Yuan and Street, 1998, Yuan et al., 1999, Schlüter and Schönfeld, 2000, Zhou et al., 2001, McIlwain and Pollard, 2002 for jet flows). It is clear that with high order schemes applications to complex flows are difficult.

Numerical dissipation is often unavoidable, in particular for high Reynolds number and compressible flows. Although there are some studies in the literature which claim that the numerical dissipation given by monotone flux treatments may be an accurate SGS model (see Fureby and Grinstein, 1999, Grinstein et al., 1995, Grinstein and De Vore, 1996, Grinstein at al., 1996 and Grinstein, 2001 for applications to jet flows), it is common opinion in the LES community that the numerical viscosity can interact with the SGS model and significantly deteriorate the results (see, for instance, Mittal and Moin, 1997 and Garnier et al., 1999).

Commonly, no explicit filtering is applied, since it is assumed that the numerical discretization acts as a filter (grid filter). However, in general, it is not clear which type of filter corresponds to a given numerical discretization. This is not only an academic problem because the definition of the filter enters in many SGS models and the nature of the SGS terms depends on the used filter (see, for instance, Salvetti and Beux, 1998). If an explicit filter, larger than the spatial discretization, is applied, a cleaner formulation is obtained. However, this implies the use of an actual resolution that is higher than the nominal one, leading to a significant increase of computational cost.

In general, our understanding of the interaction between filtering, numerical discretization and SGS model is at present quite limited.

As far as LES of jet flows is concerned, more specific problems arise. In most cases, the computational domain is only a part of a laboratory or real configuration. Thus, appropriate boundary conditions must be supplied.

Early LES have been carried out for temporally-evolving jets, i.e. periodic boundary conditions were used in the axial direction. This has the advantage of leading to *simple* simulations, in which only a small box of fluid is followed. However, to obtain a description closer to reality, spatially-evolving jets must be simulated, so that inflow and outflow boundary conditions are needed.

The inlet of the computational domain usually coincides with the jet outlet. The specification of suitable inlet conditions is critical since it is known that they significantly influence the jet dynamics. Therefore, in principle, the real instantaneous behavior of the incoming flow should be reproduced in the simulations, especially as regards the shear-layers. For that, instantaneous values of velocity (and, probably, of pressure) should be available all over the jet outlet section. Such an information is extremely difficult, or even impossible, to be obtained in practice; thus, some approximations have to be made. The simplest choice is to assume a constant inlet velocity profile, e.g. to use mean flow data usually available from experiments. However, it has been shown (see, for instance, Salvetti et al., 1996) that this is generally not enough to correctly reproduce the flow dynamics, not even qualitatively. Thus, a fluctuating profile, which approximates the actual velocity fluctuations in the incoming flow, must be superposed to the mean one. In the case of forced jets, a harmonic time perturbation at the excitation frequency (or frequencies) can be used (Ollson and Fuchs, 1996, Pruett, 2000, Grinstein et al., 1995, Grinstein and De Vore, 1996, Grinstein at al., 1996, Grinstein, 2001, Zhou et al., 2001, McIlwain and Pollard, 2002), while in the unforced case it is not trivial to construct such a fluctuating profile. A choice adopted in the literature (Jones and Wille, 1996, Gamet and Estivalezes, 1998, Le Ribault et al., 1999) is to use a perturbation having a broadband spectrum in time and a given shape in space, possibly obtained from experimental first order moments. Alternatively, the instantaneous velocity obtained in separate LES of fully-developed pipe or channel flows may be given as inlet conditions (Akselvoll and Moin, 1996, Jones and Wille, 1996, Voke and Gao, 1998). Clearly, all these are only approximations of real situations, in which the jet flow originates from nozzles, pipes or ducts, and the jet dynamics itself may affect the inflow velocity and pressure. Recent direct numerical simulations of a complete jet configuration, including the flow in the nozzle (Verzicco, personal communication), indicated that these *feedback* effects are far from being negligible. However, large-eddy simulations of complete configurations appear at present very difficult to be carried out because of the huge computational cost and memory required and because of the inadequacy of most of the classical numerical methods in dealing with complex geometries.

As for the outflow boundary, *non-reflecting* boundary conditions, i.e. conditions allowing vortical structures to exit the computational domain without significant distortion or reflection, are needed. For instance, for incompressible flows conditions of convective type can be used (Akselvoll and Moin, 1996, Ollson and Fuchs, 1996, McIlwain and Pollard, 2002):

$$\frac{\partial u_i}{\partial t} + C\,\frac{\partial u_i}{\partial n} = 0 \tag{9}$$

where n denotes the normal to the outflow boundary and C, which represents a convection velocity, may be assumed constant (given a priori) or be computed numerically at each time-step. However, these non-reflecting boundary conditions work only if either the computational domain is long enough to avoid feedback effects on pressure at the inflow (Salvetti et al., 1996) or a buffer zone is also used after the non-reflecting boundaries (e.g. Hu, 1996). In both cases, there is a noticeable increase in computational cost. For compressible flows, more complex non-reflecting boundary conditions exist, which take flow characteristics into account (Thompson, 1987, Giles, 1990, Poinsot and Lele, 1992, Grinstein, 1994). They are used, for instance by Grinstein et al. (1995), Le Ribault et al. (1999), Gamet and Estivalezes (1998), Pruett (2000) and Grinstein (2001).

In the radial or lateral direction, conditions allowing entrainment to be correctly accounted for are required. Free-slip conditions, commonly used in LES, do not satisfy this requirement even if the domain is very large, while *traction free* conditions, specifically proposed for jet flows (Boersma et al., 1998), seem to satisfy this requirement quite well.

Finally, for jets in cross flow or impinging jets, boundary conditions at the solid wall must be specified. At high Reynolds numbers it is unpractical to resolve near wall structures, which have dimensions that scale with the viscous sub-layer height, and approximate boundary conditions or near wall models should be devised. The wall treatment in LES of high Reynolds number flow is far from being a resolved problem (see the review by Cabot and Moin (2000)).

4 Applications to Jets in Crossflow

Several examples of LES of jet flows can be found in the literature, e.g. Pruett (2000) for a 2D axisymmetric jet configurations, Grinstein at al. (1996), Ollson and Fuchs (1996), Urbin and Metais (1997), Gamet and Estivalezes (1998), Hong and Mingde (1999) and McIlwain and Pollard (2002) for round jets, Akselvoll and Moin (1996) for round coaxial jets, Rizk and Menon (1988), Gao and Voke (1995), Cziesla et al. (1997), Olsson and Fuchs (1998) and Voke and Gao (1998) for impinging jets, Grinstein et al. (1995), Grinstein and De Vore (1996), Fureby and Grinstein (1999) and Grinstein (2001) for square and rectangular jets, Dai et al. (1995), Weinberger et al. (1997), Le Ribault et al. (1999) and Akhavan et al. (2000) for plane jets, Desjardin and Frankel (1998), Wall et al. (2000) and Steiner and Bush (2001) for reacting jets and Nadaoka et al. (1999) and Zhou et al. (2001) for multi-phase jets and plumes.

To our knowledge, only three large-eddy simulations of jets in crossflow are documented in the literature. The earliest one was the simulation of a *plane* jet in a crossflow by Jones and Wille (1996). The Navier-Stokes equations for incompressible flows were discretized by a finite-volume method on a non staggered mesh (velocity and pressure computed at the cell centers). Spatial discretization was obtained through centered finite differences (no numerical dissipation) while a multi stage implicit difference algorithm was used for time advancing. The resulting numerical method is second-order accurate both in space and time.

Three SGS models were tested, viz. the Smagorinsky model, the eddy-viscosity dynamic model and a one-equation eddy-viscosity model (Schmidt and Schumann, 1989).

The analyzed configuration, sketched in Fig. 1, reproduced the one of the experiments by Chen and Hwang (1991); the Reynolds number, based on the channel height H and the bulk velocity u_b, was 1.9×10^4, while the ratio between the jet velocity and u_b was 7.34 and $H/D = 24$, D being the nozzle width. Two different $87\times30\times30$ grids were used: a rectangular mesh refined in the vicinity of the jet exit and a grid adapted on the basis of the solution on the previous one, i.e. refined in the mean shear layers developing around the jet.

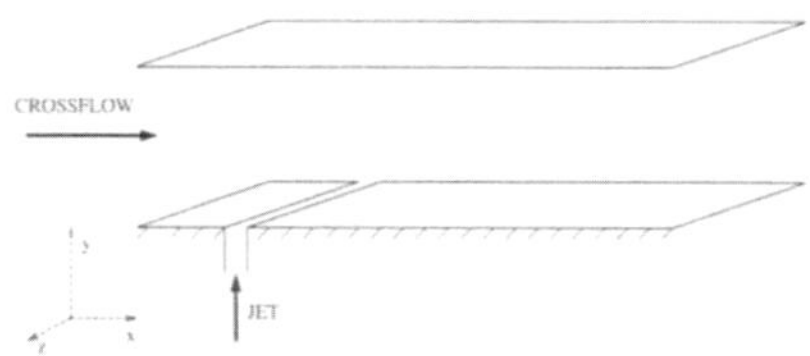

Figure 1. Sketch of the configuration simulated in Jones and Wille (1996).

Two different inflow conditions were also considered: in the first case, a perturbation, having a broadband spectrum in time and a r.m.s. profile typical of channel flow in the wall normal direction, was superposed to the mean inlet velocity profile, while in the second one the instantaneous velocities deriving from a channel flow simulation were used. Non-reflective convective conditions were used at the outflow, while periodicity was assumed in the spanwise direction. At the solid walls ($y = 0$, H), approximate conditions (Grötzbach, 1993) were prescribed because both grids were too coarse to permit the resolution of the viscous sublayer.

A reasonable overall agreement with experiments was obtained, although significant discrepancies were observed inside the recirculation region (even for mean quantities). Mesh adaption did not improve the results. On the authors' opinion, this was due both to the adaption criteria, which was based on the mean velocity field, and to the increased anisotropy and reduced resolution in the recirculation region for the adapted mesh. Results were insensitive to the inflow conditions and relatively insensitive to the SGS modeling, although different models provided very different values for the SGS eddy-viscosity.

The main criticism raised by Yuan et al. (1999) to the simulations by Jones and Wille was that "they used approximate boundary conditions at the wall rather than resolving the turbulent wall boundary layer, which introduced inaccuracies in their computed results that are difficult to evaluate".

Yuan et al. (1998, 1999) carried out large-eddy simulations of a round jet issuing normally into a crossflow (Fig. 2). The Navier-Stokes equations for incompressible flows were projected into a generalized coordinate system and discretized by a control-volume method on a non staggered co-located mesh, in which cartesian velocities and pressure are computed at the center of the cells and contravariant volume fluxes at cell faces. Spatial discretization was second-order accurate and a QUICK method was used for convective terms (numerical viscosity). Time advancing was based on a semi-implicit fractional-step algorithm (second-order accurate). Domain decomposition was used to efficiently discretize the domain geometry. The dynamic mixed model of Zang et al. (1993) was used for SGS terms.

Simulations were carried out for different values of the jet-to-crossflow velocity ratio, $R = U_j/U_c = 2.0, 3.3$, of the Reynolds number based on crossflow velocity and jet diameter, $Re_D = (U_c D)/\nu = 1050, 2100$ and of the Froude number, $Fr = U_j/[gD(\rho_c - \rho_j)/\rho_c]^{1/2} = 10, \infty$, ρ_c and ρ_j being the density of cross and jet flows. Comparisons with experiments by Sherif and Pletcher (1989) ($R = 4$, $Re_D = 4820, 9066$, $Fr = 36$) were made. The computational domain had the following dimensions (in the directions x, y and z in Fig.2 respectively): $12.7D \times 8D \times 12D$ for $R = 3.3$ and $13.7D \times 8D \times 9D$ for $R = 2$. In both cases, the jet exit plane was centered $2.7D$ downstream of the crossflow inlet plane. At the inflow, a laminar boundary layer velocity profile was assumed. On the lateral and top surfaces, free slip conditions were prescribed, while no slip was enforced on the bottom. At the outflow, a zero gradient boundary condition was imposed on all flow variables. The incoming pipe velocity profile was prescribed at a distance of $1D$ below the wall and was obtained from an additional simulation of temporally-evolving pipe flow.

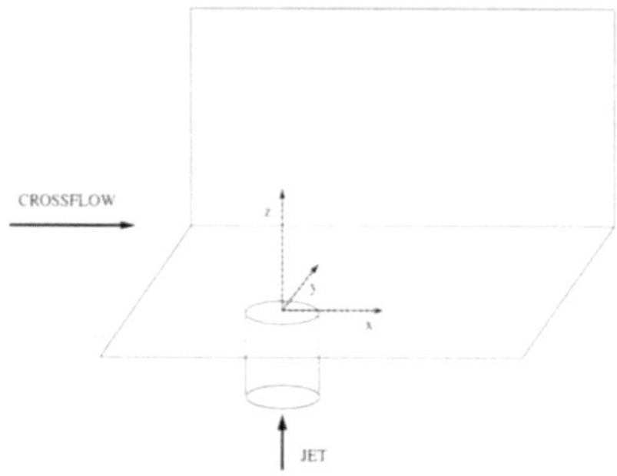

Figure 2. Sketch of the configuration simulated in Yuan et al. (1998, 1999).

As stated previously, domain decomposition was used, with thirteen grids and a total of 1.34×10^6 control volumes (approximately 45% of which placed downstream of the jet exit), resulting in a very fine resolution of the circular boundary condition at the jet exit and near the solid wall; the first control volume near the jet exit had the following dimensions, in wall units: $\Delta x^+ = 1.4$, $\Delta z^+ = 2.2$). Note that, to reach such a resolution, the Reynolds number in the simulation had to be kept lower than in the reference experiments. The sensitivity of the solution to grid refinement was not investigated, because of the high computational cost of the simulations. Indeed, each computation took approximately 2000h on a single processor of a Cray J90.

An overall satisfactory agreement with the experiments was found, but some discrepancies were observed. The authors attribute these discrepancies mainly to differences between experiments and simulations in the jet inflow conditions and in the Re_D, leading to a different status of the crossflow boundary layer. In particular, in the experiments the incoming turbulent pipe flow was not fully developed. A sensitivity analysis carried out numerically showed significant effects of the outlet conditions on the flow dynamics. As regards the crossflow boundary layer, in the experiments, transition to turbulence probably occurred upon encountering the perturbation by the jet, while the boundary layer was completely laminar in the simulations (lower Re_D). In our opinion, possible additional sources of inaccuracy might be the boundary conditions used in the simulations, in particular at the outflow. Moreover, other studies in the literature (Andreopoulos, 1982, Schlüter and Schönfeld, 2000 and Degano et al., 2001) indicate that, at least for $R = 2$, the flow in the pipe which supplies the jet should be simulated on a length larger than that considered by Yuan et al. ($1D$). These highly resolved simulations were used by the authors for a detailed analysis of the flow and new structures were identified in the near field of the jet. Jet trajectories and spread mechanisms were also examined.

Finally, Schlüter and Schönfeld (2000) simulated a jet in crossflow, including the plenum chamber, the pipe and the nozzle of the jet. The standard and filtered (Ducros et al., 1997) Smagorinsky models were used for SGS terms, while the compressible Navier-Stokes equations were discretized by the AVBP parallel solver (Schönfeld and Rudgyard, 1999). This solver is based on the cell-vertex finite volume method with a central difference Lax-Wendroff discretization, combined with a linear preserving artificial viscosity; spatial accuracy ranges from $O(1.5)$ to $O(1.6)$. Time advancing is carried out by an explicit multistage Runge-Kutta scheme.

Three different configurations were considered, reproducing the experiments by Andreopoulos and Rodi (1984) (R=2, Re=82000), by Smith and Mungal (1998) (R=5 and Re=16400) and by Toy et al. (1993) (twin jets, R=6 and Re=31800).

Multi-block structured grids were used, having a much lower number of points than in the LES by Yuan et al. (9×10^4 and 2×10^5 grid nodes, corresponding to a near wall resolution at the jet exit of Δz^+= 90 and 50 respectively). The authors identify the used grid resolution as the lower limit to obtain acceptable results. Unfortunately, not many details are available in the paper about domain dimensions and boundary conditions.

Despite thc coarse grid resolution, the agreement with the experiments was rather satisfactory, especially with the filtered Smagorinsky model and for the Smith and Mungal test case, which is characterized by the lowest Re_D among those considered. Thus, the authors conclude that the LES approach has the capability to be used as research tool in the design process of industrial equipments, such as gas turbine burners.

One may conclude that for jets in crossflow, as in general, LES could play two different roles. In an academic context, LES can be carried out at moderate Reynolds numbers and with highly resolved grids to study in detail the physics of the flow as for direct numerical simulation, but at slightly higher Reynolds numbers (roughly less than an order of magnitude higher than in DNS). The calculations by Yuan et al. (1998, 1999) may be considered an example of this kind of use of LES. On the other hand, LES could be an alternative to RANS, in engineering applications, thus characterized by high Reynolds numbers and more complex geometry. The simulations by Schlüter and Schönfeld (2000) are a valuable step in that direction. However, before LES may actually take this second role, satisfactory answers to most of the previously cited open problems should be given.

References

Akhavan, R., Ansari, A., Kang, S., and Mangiavacchi, N. (2000). Subgrid-scale interactions in a numerically simulated planar turbulent jet and implications for modelling. *J. Fluid Mech.* 408: 83–120.

Akselvoll, K., and Moin, P. (1996). Large-eddy simulation of turbulent confined coannular jets. *J. Fluid Mech.* 315: 387–411.

Andreopoulos, J. (1982). Measurements in a jet-pipe flow issuing perpendicularly into a cross stream. *J. Fluid Engigneering* 104: 493–499.

Andreopoulos, J., and Rodi, W. (1982). Experimental investigation of jets in a crossflow. *J. Fluid Mech.* 138: 93–127.

Bardina, J., Ferziger, J.H., and Reynolds, W.C. (1980). Improved subgrid scale models for large eddy simulation. *AIAA paper 80-1357.*

Boersma, B.J., Brethouwer, G., and Nieuwstadt, F.T.M. (1998). A numerical investigation on the effect of the inflow conditions on the self-similar region of a round jet. *Phys. Fluids* 10(4): 899–909.

Cabot, W., and Moin, P. (2000). Approximate wall boundary conditions in the large-eddy simulation of high Reynolds number flow. *Flow, Turbulence and Combustion* 63(1-4): 269–291.

Chen, K.S., and Hwang, J.Y. (1991). Experimental study on the mixing of one- and dual-line heated jets with a cold cross-flow in a confined channel. *AIAA J.* 29(3): 353–360.

Cziesla, T., Tandogan, E., and Mitra, N.K. (1997). Large-eddy simulation of heat transfer from impinging slot jets. *Num. Heat Transfer Part A: Applications* 32(1): 1-17.

Dai, Y., Kobayashi, T., and Taniguchi, N. (1995). Investigation of organized structures in plane turbulent jet flow by using large eddy simulation. In *Proc. of 10th Symposium on Turbulent Shear Flows*.

Degano, G.M., Campolo, M., Soldati, A., and Cortelezzi, L. (2001). Influence of inlet boundary conditions on the numerical calculation of a jet in crossflow. Manuscript submitted for publication.

Desjardin, P.E., and Frankel, S.H. (1998). Large eddy simulation of a nonpremixed reacting jet: application and assessment of subgrid-scale combustion models. *Phys. Fluids* 10(9): 2298–2314.

Domaradzki, J.A., and Saiki, E.M. (1997). A subgrid-scale model based on the estimation of unresolved scales of turbulence. *Phys. Fluids* 9(7): 2148–2164.

Ducros, F., Nicoud, F., and Schönfeld, T. (1997). Large eddy simulation of compressible flows on hybrid meshes. In *Proc. of 11th Symposium on Turbulent Shear Flows*: 28.1–28.6.

Erlebacher, G., Hussaini, M.Y., Speziale, C.G., and Zang, T.A. (1992). Toward the large-eddy simulation of compressible flows. *J. Fluid Mech.* 238: 155–185.

Fureby, C., and Grinstein, F.F. (1999). Monotonically integrated large eddy simulation of free shear flows. *AIAA J.* 37(5): 544–556.

Gamet, L., and Estivalezes, J.L. (1998). Application of Large-Eddy Simulations and Kirchoff Method to Jet Noise Prediction. *AIAA J.* 36(12): 2170–2178.

Gao, S., and Voke, P.R. (1995). Large-eddy simulation of turbulent heat transport in enclosed impinging jets. *Int. J. Heat and Fluid Flow* 16(5): 349–356.

Garnier, E., Sagaut, P., Comte, P., and Deville, M. (1999). On the use of shock-capturing schemes for large-eddy simulation. *J. Comp. Phys.* 16(5): 349–356.

Germano, M., Piomelli, U., Moin, P., and Cabot, W.H. (1991). A dynamic subgrid-scale eddy-viscosity model. *Phys. Fluids A* 153: 273–311.

Geurts, B. (1997). Inverse modeling for large-eddy simulation. *Phys. Fluids* 9(12): 3585–3587.

Ghosal, S., Lund, T.S., Moin, P., and Akselvoll K. (1995). A dynamic localization model for large-eddy simulation of turbulent flows. *J. Fluid Mech.* 286: 229–255.

Ghosal, S. (1996). An analysis of numerical errors in large-eddy simulations of turbulence. *J. Comp. Phys.* 125(1): 187–206.

Giles, M.B. (1990). Nonreflecting boundary conditions for Euler equation calculations. *AIAA J.* 28(12): 2050–2058.

Grinstein, F.F. (1994). Open boundary conditions in the simulation of subsonic turbulent shear flows. *J. Comp. Phys.* 115(1): 43–55.

Grinstein, F.F., Gutmark, E., and Parr, T. (1995). Near field dynamics of subsonic free square jets. A computational and experimental study. *Phys. Fluids* 7(6): 1483–1497.

Grinstein, F.F., and De Vore, C.R. (1996). Dynamics of coherent structures and transition to turbulence in free square jets. *Phys. Fluids* 8(5): 1237–1251.

Grinstein, F.F., Gutmark, E.J., Parr, T.P., Hanson-Parr, D.M., and Obeysekare, U. (1996). Streamwise and spanwise vortex interaction in an axisymmetric jet. A computational and experimental study. *Phys. Fluids* 8(6): 1515–1524.

Grinstein, F.F. (2001). Vortex dynamics and entrainment in rectangular free jets. *J. Fluid Mech.* 437: 69-101.

Grötzbach, G. (1993). Direct numerical and large-eddy simulations of turbulent channel flows. In *Encyclopedia of Fluid Mechanics* Cheremisinoff N. ed., Gulf Publishing, West Orange, NJ.

Hong, Y., and Mingde, S. (1999). Application and comparison of two SGS models in large eddy simulation of free turbulent jet flow. *Com. Nonlinear Science & Num. Simulation* 4(1): 12–19.

Horiuti, K. (1997). A new dynamic two-parameter mixed model for large-eddy simulation. *Phys. Fluids* 9(11): 3443–3464.

Hu, F.Q. (1996). On absorbing boundary conditions for linearized Euler equations by a perfectly matched layer. *J. Comput. Phys.* 129(1): 201–219.

Hughes, T.J.R., Mazzei, L., and Jansen, K.E. (2000). Large eddy simulation and the variational multiscale method. *Computing and Visualization in Science* 3(1-2): 47–59.

Hughes, T.J.R., Mazzei, L., Oberai, A.A., and Wray, A.A. (2001). The multiscale formulation of large eddy simulation: d ecay of homogeneous isotropic turbulence. *Phys. Fluids* 13(2): 505–512.

Jones, W.P., and Wille, M. (1996). Large eddy simulation of a plane jet in a cross-flow. *Int. J. Heat and Fluid Flow* 17(3): 296–306.

Kraichnan, R.H. (1976). Eddy viscosity in two and three dimensions. *J. Atmos. Sci.* 33(8): 1521–1536.

Kravchenko, A.G., and Moin, P. (2000). Numerical studies of flow over a circular cylinder at Re_D=3900. *Phys. Fluids* 12(2): 403–417.

Leonard, A. (1974). Energy cascade in large eddy simulations of turbulent fluid flows. *Adv. Geophys.* 18: 237.

Le Ribault, C., Sarkar, S., and Stanley, S.A. (1999). Large eddy simulation of a plane jet. *Phys. Fluids* 11(10): 3069–3083.

Lesieur, M., and Comte, P. (1997). Large-eddy simulations of compressible turbulent flows. In *AGARD-R-819*: A4.1–A4.39.

Liu, S., Meneveau, C., and Katz, J. (1994). On the properties of similarity subgrid-scale models as deduced from measurements in a turbulent jet. *J. Fluid Mech.* 275: 83–119.

McIlwain, S., and Pollard, A. (2002). Large eddy simulation of the effects of mild swirl on the near field of a round jet. *Phys. Fluids* 14(2): 653–661.

Meneveau, C., Lund, T.S., and Cabot, W.H. (1996). A Lagrangian dynamic subgrid-scale model of turbulence. *J. Fluid Mech.* 319: 353–385.

Mittal, R., and Moin, P. (1997). Suitability of upwind-biased finite difference schemes for large-eddy simulation of turbulent flows. *AIAA J.* 35(8): 1415–1417.

Nadaoka, K., Nihei, Y., and Yagi, H. (1999). Grid-averaged Lagrangian LES model for multiphase turbulent flow. *Int. J. Multiphase Flow* 25(8), 1619-1643.

Olsson, M., and Fuchs, L. (1996). Large eddy simulation of the proximal region of a spatially developing circular jet. *Phys. Fluids* 8(8): 2125–2137.

Olsson, M., and Fuchs, L. (1998). Large eddy simulations of a forced semiconfined circular impinging jet. *Phys. Fluids* 10(2): 476-486.

Poinsot, T.J., and Lele, S.K. (1992). Boundary conditions for direct simulations of compressible viscous flows. *J. Comp. Phys.* 101(1): 104–129.

Porté-Agel, F., Meneveau, C., and Parlange, M.B. (2000). A scale-dependent dynamic model for large-eddy simulation: application to a neutral atmospheric boundary layer. *J. Fluid Mech.* 415: 261-284.

Pruett, C.D. (2000). Eulerian Time-Domain Filtering for Spatial Large-Eddy Simulation. *AIAA J.* 38(9): 1634–1642.

Rizk, M.H., and Menon, S. (1988). Large-eddy simulations of axisymmetric excitation effects on a row of impinging jets. *Phys. Fluids* 31(7): 1892–1903.

Salvetti, M.V., and Banerjee, S. (1995). A priori tests of a new dynamic subgrid-scale model for finite-difference large-eddy simulations. *Phys. Fluids* 7(11): 2831–2847.

Salvetti, M.V., Orlandi, P., and Verzicco, R. (1996). Numerical simulations of transitional axisymmetric coaxial jets. *AIAA J.* 34(4): 736–743.

Salvetti, M.V., and Beux, F. (1998). The effect of the numerical scheme on the subgrid scale term in large-eddy simulation. *Phys. Fluids* 10(11): 3020–3022.

Sarghini, F., Piomelli, U., and Balaras, E. (1999). Scale-similar models for large-eddy simulations. *Phys. Fluids* 11(6): 1956–1607.

Schlüter, J.U., and Schönfeld, T. (2000). LES of jets in cross flow and its application to a gas turbine burner. *Flow, Turbulence and Combustion* 65: 177–203.

Schmidt, H. and Schumann, U. (1989). Coherent structure of the convective boundary layer derived from large-eddy simulations. *J. Fluid Mech.* 200: 511–562.

Schönfeld, T., and Rudgyard, M.A. (1999). Steady and unsteady flows simulations using the hybrid flow solver AVBP. *AIAA J.* 37(11): 1378–1385.

Sherif, S.A. and Pletcher, R. (1989). Measurements of the flow and turbulence characteristics of round jets in crossflow. *Trans. ASME: J. Fluids Engineering* 111: 165–171.

Smagorinsky, J. (1963). General circulation experiments with the primitive equations. I. The basic experiment *Mon. Weather Rev.* 91: 99–164.

Smith, S.H., and Mungal, M.G. (1998). Mixing, structure and scaling of the jet in crossflow. *J. Fluid Mech.* 357: 83–122.

Steiner, H., and Bushe, W.K. (2001). Large-eddy simulation of a turbulent reacting jet with conditional source-term estimation. *Phys. Fluids* 13(3): 754–769.

Stolz, S., and Adams, N.A. (1999). An approximate deconvolution procedure for large-eddy simulation. *Phys. Fluids* 11(7): 1699–1701.

Thompson, K.W. (1987). Time dependent boundary conditions for hyperbolic systems. *J. Comp. Phys.* 68(1): 1–24.

Toy, N., Savory, E., and McCusker, S. (1993). The interaction region associated with twin jets and a normal crossflow. In *AGARD-CP-534*.

Urbin, G., and Metais, O. (1997). Large-eddy simulations of three-dimensional spatially-developing jets In Voke, P.R., Kleiser, L., and Chollet, J.P. eds. *Direct and Large-Eddy Simulation II*. Dordrecht: Kluwer Academic Publishers.

Vasilyev, O.V., Lund T.S., and Moin, P. (1998). A general class of commutative filters for LES in complex geometries *J. Comp. Physics* 146(1): 82–104.

Voke, P.R., and Gao, S. (1998). Numerical study of heat transfer from an impinging jet. *Int. J. Heat and Mass Transfer* 41(4-5): 671-680.

Vreman, B., Geurts, B., and Kuerten, H. (1995). Subgrid-modelling in LES of compressible flow. *Appl. Sc. Res.* 54(3): 191–203.

Wall, C., Boersma, B.J., and Moin, P. (2000). An evaluation of the assumed beta probability density function subgrid-scale model for large eddy simulation of non premixed, turbulent combustion with heat release. *Phys. Fluids* 12(10): 2522–2529.

Weinberger, C., Rewerts, J., and Janicka, J. (1997). The influence of inlet conditions on a large eddy simulation of a turbulent plane jet. In *Proc. of 11th Symposium on Turbulent Shear Flows*:25.17–25.22.

Winckelmans, G.S., Wray, A.A., and Vasilyev, O.V. (1998). Testing of a new mixed model for LES: the Leonard model supplemented by a dynamic Smagorinsky term. In *Proc. Summer Program-1998* (ed. P. Moin and W. Reynolds), 367-388. Stanford University and NASA Ames Research Center (Center for Turbulence Research).

Yakhot, A., Orszag, S.A., Yakhot, V., and Israeli, M. (1989). Renormalization group formulation of large-eddy simulation. *J. Scient. Comp.* 4(2): 139–158.

Yuan, L.L., and Street, R.L. (1998). Trajectory and entrainment of a round jet in crossflow. *Phys. Fluids* 10(9): 2323–2335.

Yuan, L.L., Street, R.L., and Ferziger, J.H. (1999). Large-eddy simulations of a round jet in crossflow. *J. Fluid Mech.* 379: 71–104.

Zang, Y., Street, R.L., and Koseff, J. (1993). A dynamic mixed subgrid-scale model and its application to turbulent recirculating flows. *Phys. Fluids A* 5: 3186–.

Zhou, X., Luo, K.H., and Williams, J.J.R. (2001). Large-eddy simulation of a turbulent forced plume. *Eur. J. Mech. B-Fluids* 20: 233–254.

Reduced-order controllers for fluid mechanics applications

Luca Cortelezzi *

Department of Mechanical Engineering, McGill University, Montreal, H3A 2K6, Canada

Abstract. Controller complexity is a crucial parameter for control in engineering applications. Controllers with a large number of states are of no practical interest because of the amount of hardware and computer power necessary to compute a real-time control law. Consequently, it is crucial to reduce the order of the controller. This summary provides a background on some techniques useful for deriving reduced-order controllers for fluid mechanics applications, setting the stage for the specific topics described later in this text.

1 Introduction

Active fluid flow control is becoming one of the most promising and sophisticated cutting-edge research topics in fluid mechanics. Recent reviews (Gad-el-Hak, 2001; Lofdahl & Gad-el-Hak, 1999; Gad-el-Hak, 1994) summarize achievements and open questions in active fluid flow control.

Controller complexity is a crucial parameter in engineering applications because of the amount of hardware and computer power necessary to compute a real-time control law. Active control of an unsteady fluid flow in general requires nonlinear feedback laws. A general framework to obtain a desired controller is not yet available, although Cortelezzi *et al.* (1994) proposed a possible framework for deriving reduced-order, or low-order, controllers using a hierarchy of mathematical models of increasing complexity. A controller for a given problem could be derived in two steps: first by constructing a reduced-order model and then designing a controller for the reduced-order model. In recent years, this approach was successfully adopted by some researchers: Graham *et al.* (1999), Ravindran (2000), Tang & Aubry (2000), Dunstan *et al.* (2001), Raveh (2001), Lucia *et. al* (2002), Rediniotis *et. al* (2002), etc..

There are three main approaches to obtain a reduced-order model for fluid flow control: 1) physical insight and intuition of the scientist, 2) transformation of the governing (Navier-Stokes) equations into a set of ordinary differential equations using vortex methods, Galerkin's methods, proper orthogonal decomposition, etc., and 3) system identification. This article will discuss the first two approaches to obtain reduced-order controllers, while the third approach is discussed in M'Closkey *et. al* (2002) and M'Closkey *et. al* (2002). As a first step, one should start with a reduced-order model that captures the dynamic features of the flow that one wants to control. In general, the reduced-order model is a low-dimensional nonlinear system governed by a set of ordinary differential equations, for example, a point vortex model with few vortices, or a spectral decomposition with few "modes".

There are several advantages to working with a reduced-order model. First, it is often easier to derive the desired control strategy. Second, dynamical system analysis could be used to characterize the dynamics of the closed-loop system. Finally, the control strategy derived for the reduced-order model can produce fast numerical algorithms which are essential for controlling a real flow. In the following steps, one could apply the control strategy derived for the reduced-order model to higher or infinite dimensional models. As proposed by Cortelezzi (1996), the process of transferring the control strategy from a low-dimensional model to a higher or infinite dimensional model and eventually to a real flow should be guided and supported by a dynamical system and time series analysis. In general, the control strategy should be transferable from a reduced-order model to the next high-order model, provided that the phase space of the two models are topologically equivalent, i.e., they have the same dimensions and structures. However, the proof or even the verification of such equivalence is in general a complex task.

* Email: crtlz@cim.mcgill.ca

2 Reduced-order controller via vortex methods

This Section summarizes the results obtained in collaboration with Professors A. Leonard and J.C. Doyle of the California Institute of Technology, Professors J.S. Gibson and C.R. Anderson of the University of California at Los Angeles and two graduate students: Dr. H-L. Chang and Dr. Y-C. Chen.

The first example is the derivation of a reduced-order nonlinear controller for the two-dimensional unsteady separated flow past a plate perpendicular to the free-stream velocity. The flow past a plate is an unbounded flow dominated by large vortical structures. Physical insight and intuition suggest the use of Lagrangian schemes such as vortex methods to generate a hierarchy of models of increasing complexity. We introduce two inviscid vortex models. The first, a reduced-order point vortex model (PVM), simulates the unsteady separation from the tips of a plate by means of a pair of point vortices of time-dependent circulation (Cortelezzi, 1996). The circulation is predicted by an unsteady Kutta condition. The second, a high-order vortex blob model (VBM), simulates the unsteady separation with a collection of vortices with a finite core introduced in the flow at appropriate time intervals near the tips of the plate (Cortelezzi *et al.*, 1997). We consider as the control actuator a suction point placed on the downstream wall of the plate. The control objective is to confine the wake to a single vortex pair of constant circulation.

The simplicity of the reduced-order model allows the derivation, in closed form, of a nonlinear controller. The controller predicts, for any free-stream velocity, the suction necessary to inhibit the production of circulation when a vortex pair of sufficiently high circulation is present in the flow. We define the center of circulation of an ensemble of vortices. The center of circulation permits the treatment of an ensemble of vortex blobs as a single point vortex preserving certain properties of the ensemble. Consequently, this representation permits the verification of the topological equivalence of the phase space of the two models and it provides the means to apply the controller derived for the PVM to the VBM.

2.1 Point vortex model

In this subsection we introduce a low-dimensional model, also referred to as a reduced-order model, of two-dimensional unsteady separated flows past a finite plate. We assume that the regions of vorticity that separate from the boundary layer and are convected away are thin enough to justify a description by means of a vortex sheet. The consequent stretching and rolling up of the vortex sheet, due to the unsteadiness of the flow, suggests a more coarse description via point vortices (Brown & Michael, 1954; Rott, 1956; Cortelezzi & Leonard, 1993). The vortex sheet is not completely lost. It is assumed to be of negligible circulation and connects the tip of the plate to a point vortex of time-dependent circulation which is able to satisfy an unsteady Kutta condition. All the other vortices in the wake are represented by point vortices of fixed circulation.

There is experimental evidence that the near wake is nearly two dimensional and symmetric about the x-axis if the plate moves with a nonzero acceleration. Under these circumstances the problem can be simplified by imposing symmetry with respect to the real axis. In other words, one requires that the vortices shed from the top tip of the plate have equal and opposite circulation of the vortices shed from the bottom tip of the plate, $\Gamma_n^t = \Gamma_n$ and $\Gamma_n^b = -\Gamma_n$, and are located in complex conjugate positions, $z_n^t = z_n$ and $z_n^b = \bar{z_n}$, respectively.

The motion of the vortex pairs in the physical plane is described by the following set of equations:

$$\begin{cases} \dfrac{d\bar{z}_1}{dt} + (\bar{z}_1 + 2i)\dfrac{1}{\Gamma_1}\dfrac{d\Gamma_1}{dt} = w(z_1, t) \\ \dfrac{d\bar{z}_r}{dt} = w(z_r, t) \qquad r = 2, ..., N, \end{cases} \tag{1}$$

with the initial conditions:

$$\begin{cases} z_1(t_s) = 2i \\ z_r(t_s) = z_{r_s} \qquad r = 2, ..., N, \end{cases} \tag{2}$$

where w is the complex velocity. The top equation of motion predicts the position of the vortex pair of time-dependent circulation and the term containing $d\Gamma_1/dt$ is known as Brown and Michael's correction (Brown & Michael, 1954). The bottom $N-1$ equations of motion predict the position of the vortex pairs of constant circulation. t_s indicates the time when the rate of circulation production changes sign. Up to time t_s^- the circulation of the vortex pair 1 changes with time to satisfy the Kutta condition. At time $t = t_s$ this vortex pair has its circulation frozen and all the vortices are renumbered. At t_s^+ a new vortex pair of time-dependent circulation is introduced in the flow.

The reader is referred to Cortelezzi (1996) for a detailed implementation of this model.

2.2 Vortex blob model

In this subsection we introduce a high-dimensional model, a VBM (Chorin, 1973), of two-dimensional unsteady separated flows past a finite plate. We assume, as in the previous model, that the regions of vorticity that separate from the boundary layer and are convected away are thin enough to justify a description by means of a vortex sheet. The vortex sheet is represented by a collection of vortices with finite cores, also known as vortex blobs, introduced into the flow at appropriate time intervals near the tips of the plate. The evolution of the vortex sheet is then approximated by the motion of the vortex blobs. See Sarpkaya (1989) and Sarpkaya (1994) for references and discussion.

The motion of the blobs is predicted by the following equations:

$$\begin{cases} \dfrac{d\bar{z}_p^t}{dt} = w(z_p^t, t) \qquad p = 1, ..., N^t, \\ \dfrac{d\bar{z}_q^b}{dt} = w(z_q^b, t) \qquad q = 1, ..., N^b. \end{cases} \tag{3}$$

We integrate these equations in time using a 4th order Runge-Kutta scheme. A pair of new vortices is introduced in the flow domain at each time step. Note that the circulation of the new vortices, Γ_1^t and Γ_1^b, does not depend on time since it remains fixed once the vortices are introduced in the fluid domain.

The separation mechanism at the tips of the plate has to be implemented with care because it affects the final quality of the simulation. Different techniques have been proposed in the literature; see Sarpkaya (1989) and Sarpkaya (1994) for references and discussion. We approach the problem in a semi-classical fashion: Up to time $t = 0.1$ we use a pair of point vortices of time-dependent circulation. In other words, we use the PVM to start up the vortex blob simulation. At initial times, in fact, the separation is especially difficult to simulate with the VBM because there are only few vortex blobs in the flow and it is not clear where they should be placed. For $t > 0.1$ we use the first sub-step of the Runge-Kutta routine to place the new vortex blob in the flow while we are convecting all the other. The circulation of the new vortex blob is determined by imposing the Kutta condition. The new vortex blob is placed in the flow at a distance from the tip of the plate predicted by a two points velocity measurement taken around the tip of the plate. Then, the new vortex blob is convected for the remaining three sub-steps of the Runge-Kutta routine like all the other vortex blobs. Consequently, the position of the new vortex blob after one time step is improved and corrected during the three sub-steps. We use blobs with an algebraic profile for the vorticity and with a blob size $\sigma = 0.01$. This choice is a compromise between the need to reduce the high velocities generated by small vortex blobs and the need to avoid the overlapping of the vortex blobs with the plate.

2.3 Reduced-order nonlinear feedback control

In this section we introduce a nonlinear controller which is able to confine the wake behind the plate to a single vortex pair of constant circulation. The controller is derived for the PVM. In the following sections we will show that the controller can be successfully implemented also in a VBM.

There are different approaches one can use to control the wake past a plate by suction. One may try to control the position of the vortices or some features of the velocity field. We choose to control the amount of circulation injected in the flow since we believe that this is the most efficient way to control the wake. As shown by Cortelezzi (1996), suction and its rate of change contribute to the production of circulation in the same way that free-stream velocity and free-stream acceleration do. Consequently, suction is a powerful means to control the wake because it permits a balance of circulation generated by both the unsteady free-stream velocity and the time-evolution of the wake.

Our control objective is to inhibit the rate of circulation production after the starting vortex pair is shed in the flow. In other words, we want to predict the suction so that once the starting vortex pair is shed in the flow the Kutta condition remains satisfied without requiring a new vortex pair. The possibility of maintaining the wake so that it is confined to a controlled recirculating bubble has important implications for the general problem of drag reduction. Moreover, it can provide insight into vortex management techniques for the three-dimensional flow over a delta wing (Rao, 1987).

The controller closed form solution was obtained by Cortelezzi (1996). Let t_s be the time when the starting vortex pair is shed in the flow, i.e., the time when the rate of circulation production is zero. Then the suction s which implements our control strategy is a function of the position z_1 and circulation Γ_{1_s} of the vortex pair and of the free-stream velocity. The circulation Γ_{1_s} is maintained constant by the action of the controller.

2.4 Phase space of the vortex blob model

The application of a controller derived from a low-dimensional model to a model of higher or of infinite dimension is not trivial. Beside the old trial-and-error method a better and more scientific approach is to prove or to verify that the two models are dynamically equivalent; namely, that the phase spaces of the two models are topologically equivalent. If the two models are dynamically equivalent then the controller designed to govern the dynamics of the reduced-order model should, in principle, be able to govern the dynamics of the higher-order model. If, instead, the two models are not dynamically equivalent then the differences in the structure of their respective phase spaces and the knowledge of the controller for the lower-order model should provide the necessary information for designing a controller to govern the dynamics of the higher-order model.

The proof of the dynamic equivalence of the PVM and the VBM can be given by deriving the equation of motion for the center of circulation. The center of circulation is the liaison between PVM and VBM. It allows us to treat an ensemble of vortex blobs as a single point vortex and, consequently, to compare the results obtained from the VBM to the ones obtained from the PVM. Even though non trivial, the derivation of the equation of motion for the center of circulation is, in principle, possible. When the equation of motion for the center of circulation is available it can be analyzed as a dynamical system; i.e., we can compute its fixed points, limit cycles, etc. and study their structure and stability. The final step of this process is the construction of the phase space associated with the dynamics of the center of circulation and the comparison with the phase space associated with the dynamics of the point vortex. If the phase spaces of the two systems present the same types of fixed points, limit cycles, etc., then the two models are dynamically equivalent and the controller derived for the reduced-order model should, in principle, govern the dynamics of the higher-order system.

A less formal approach, the one taken in this study, is the numerical verification of the topological equivalence of the phase spaces. To verify the equivalence of the phase spaces, one should run a series of

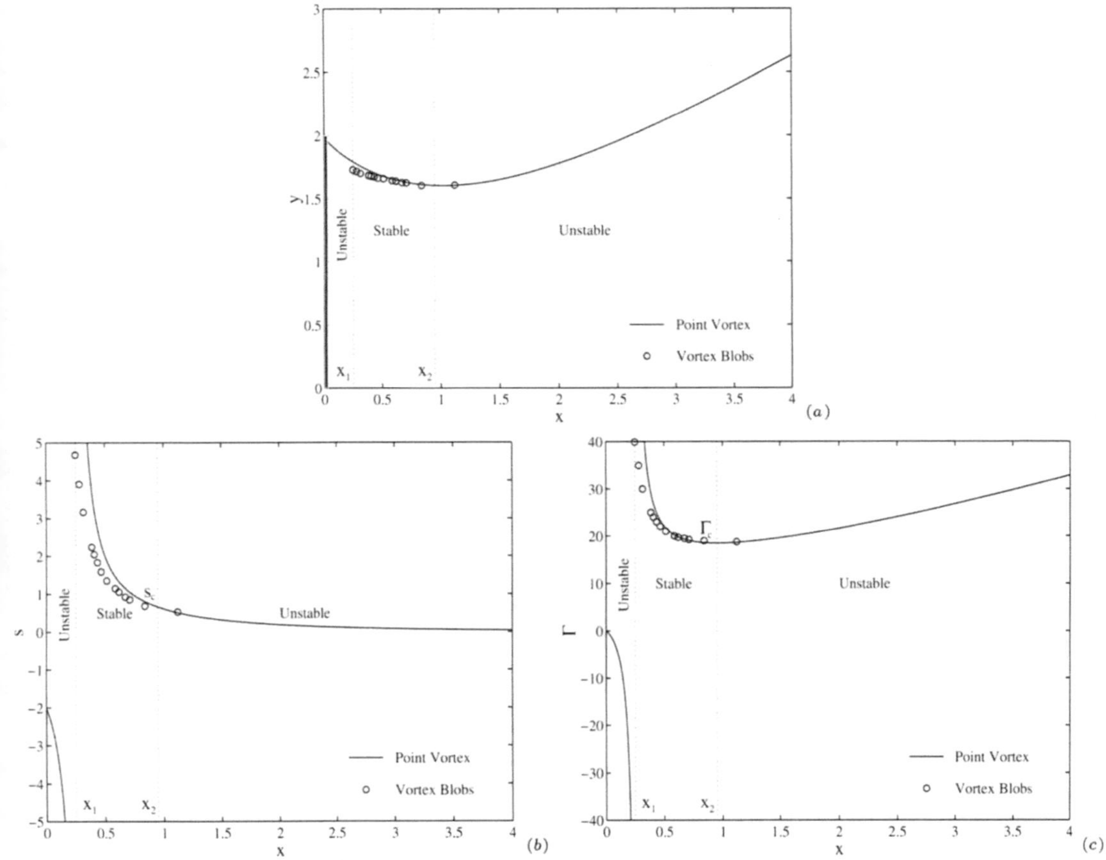

Figure 1. Comparison between PVM and VBM: Locus of the fixed points (a), suction (b), circulation of the top point vortex and circulation of the top center of circulation (c). Figure taken from Cortelezzi *et al.* (1997).

numerical experiments to confirm that the controller is able to drive the system, for example, near a stable fixed point of the reduced-order model. The numerical experiments are suggested, of course, on the basis of the knowledge of the phase space of the reduced-order model. A methodical reconstruction of the phase space of the higher-order model would provide crucial information about performance and limitation of the controller and possible suggestions in case the controller has to be redesigned.

The phase space structure of the PVM has been studied in detail by Cortelezzi (1996). The PVM presents fixed points only when the suction is nonzero (fixed points are those where the vortex is stationary when the free-stream velocity is unity and suction is constant). The locus of the fixed point is shown by the solid curve in Figure 1(a). Each point on this curve represents the position of a vortex which is stationary and satisfies the Kutta condition, and whose circulation is shown by the solid curve in Figure 1(c). The suction associated with each fixed point is shown by the solid curve in Figure 1(b). The reader should be aware that the discussion is restricted to the upper half of the domain due to the symmetry of the problem. Linear stability analysis shows that the fluid domain downstream the plate presents three stability regions. When $x < x_1$ the fixed points are unstable foci, when $x_1 < x < x_2$ they are stable nodes, and when $x > x_2$ they

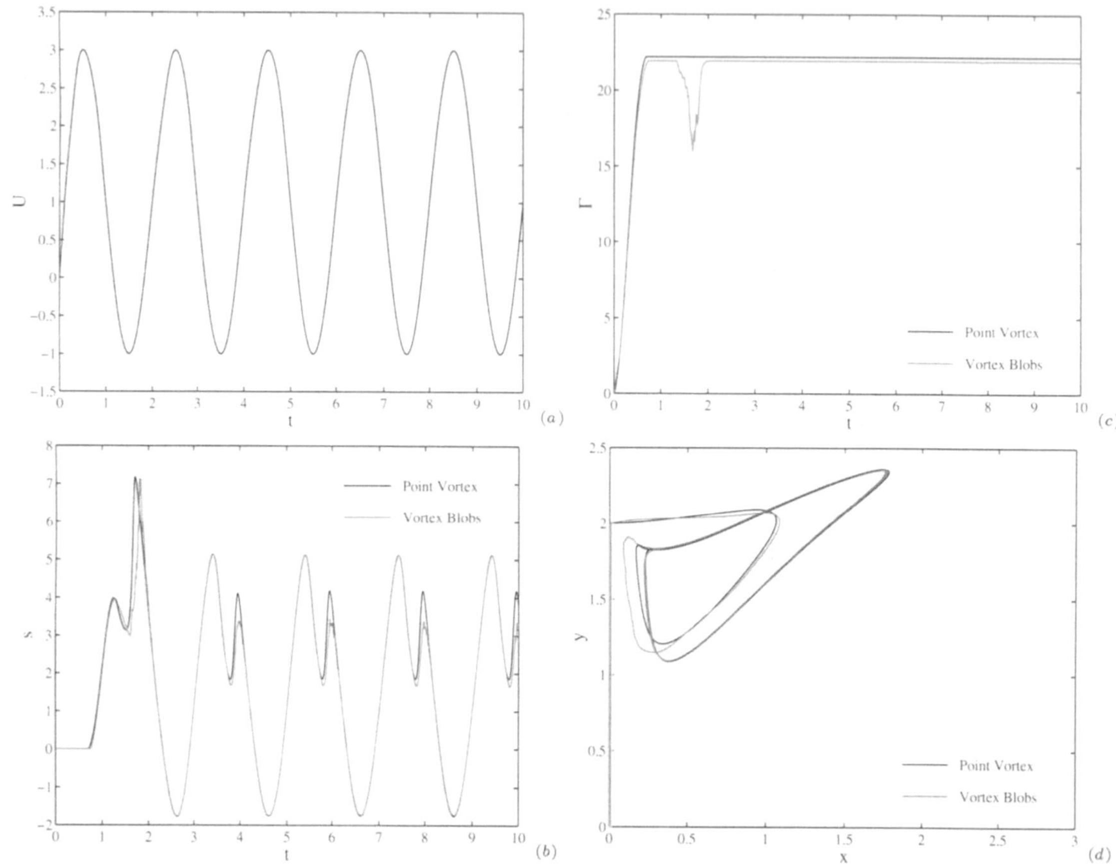

Figure 2. Comparison between PVM and VBM: Free-stream velocity (a), suction (b), circulation of the top point vortex and circulation of the top center of circulation (c), trajectory of the top vortex and trajectory of the top center of circulation (d). Figure taken from Cortelezzi *et al.* (1997).

are saddle points (see Figure 1(a)). We restrict our discussion to the sub-domain where $x > x_1$, $\Gamma_{1_s} > 0$, and $s > 0$. It is important to observe that in this sub-domain there are fixed points only when $\Gamma_{1_s} \geq \Gamma_c$. The circulation plays the role of a bifurcation parameter and the vector field undergoes a saddle-node bifurcation at $x = x_2$ when $\Gamma_{1_s} = \Gamma_c$.

We conducted a series of numerical experiments with the VBM to verify the existence of stable fixed points and their locus (fixed points now being those where the center of circulation is stationary when the free-stream velocity is unity and suction is constant). The complexity of the two system is entirely different: the PVM uses only 2 point vortices where the VBM uses more than 1000 blobs. Note also that although no symmetry was imposed on the VBM, the resulting simulations were symmetric with respect to the real axis.

The numerical experiments were designed using the information provided by the phase space of the PVM. Suction was set to the value predicted by Figure 1(b). The experiments were stopped when there was evidence that the system had reached a steady state. The outcome of the experiments is presented in Figure 1 using symbols. Figure 1(a) compares the stability regions of the two models. There is good agreement in the location of the fixed points although the stability region of the VBM appears to be slightly wider than that of the PVM. The symbol further downstream indicates the location of the fixed point when the circulation nearly equals Γ_c. Vortices with circulation lower than Γ_c drift irreversibly downstream. Figure 1(b) presents a comparison between the suction used by the two models to generate fixed points. Figure 1(c) presents a comparison between the circulation of the top point vortex and the circulation associated with the top center of circulation when the steady state is reached. Both figures show a good agreement between the two models. Figure 1(c) shows that the circulation associated with the center of circulation plays the role of a bifurcation parameter, and the bifurcation value is approximately Γ_c. This is crucial evidence of the dynamic equivalence of the two models. It guarantees, in fact, that both models present the same number and type of fixed points. Finally, Figures 1(b) and (c) substantiate the hypothesis that the region of stability of the VBM is slightly wider of the region of stability of the PVM.

The evidence provided by Figure 1 is sufficient to conclude that the phase space of the two models are indeed topologically equivalent when free-stream velocity and suction are constant. When the free-stream velocity is unsteady the dynamic equivalence of the two models should be verified using the Poincaré section. This step is not necessary in our case thanks to a theorem (see Guckenheimer & Holmes (1983), Ch. 4) which guarantees that the Poincaré section of a periodically perturbed system is topologically equivalent to the phase space of the unperturbed system, provided that the fixed points are hyperbolic and the perturbations are sufficiently small. Since both models present the same hyperbolic fixed points and since these fixed points nearly coincide, it follows that the Poincaré sections of the two models nearly coincide. Consequently, it is reasonable to expect that the controller derived for the PVM should control the dynamics of the higher dimensional VBM when the perturbations are small. In reality, as shown in the next Section, the controller derived for the PVM controls the dynamics of the VBM even when the perturbations of the mean free-stream velocity are large.

2.5 Results

Figure 2 provides evidence that the controller derived for the PVM controls the dynamics of the VBM even when the perturbations of the free-stream velocity are large. We ran the simulation twice: The first time we used the PVM with 2 point vortices and the second time the VBM with up to 1000 vortex blobs. In both cases we used exactly the same free-stream velocity and the same controller, the one derived for the PVM. In the first case we fed back to the controller the position and circulation of the point vortex while in the second case we fed back to the controller the position and circulation of the center of circulation. No symmetry was imposed on the vortex blob simulations while the point vortex simulations are strictly symmetric. Nevertheless, all simulations are nearly symmetric, consequently only the top half of the domain is shown.

Figure 2(a) shows the free-stream velocity, where the nondimensional amplitude of the oscillation is 2.0. Figure 2(d) shows the comparison between the trajectory of the top point vortex and the trajectory of the top center of circulation (flow is from left to right). The trajectories are driven almost to the same limit cycle trajectory. The comparison between the circulation of the top point vortex and the circulation of the top center of circulation is satisfactory (see Figure 2(c)). The difference in circulation amplitudes is remarkably small, about 2%. The circulation of the center of circulation presents a large drop (Figure 2(c)) because some vortex blobs are removed from the fluid through the suction point. However, the circulation quickly relaxes to the desired constant value thus providing some evidence of the robustness of the controller. Figure 2(b) shows the comparison between the two control signals. Overall, the suction required by the VBM follows quite closely the suction required by the PVM. On the one hand, there is an excellent agreement during the cyclic deceleration. On the other hand, the suction required by the VBM becomes noisy during the cyclic acceleration probably because of the variation in rotation and shape of the ensemble of vortex blobs.

Figure 3 presents a comparison between the two simulations (flow is from left to right). The comparison is made during one period of oscillation, $6 < t < 8$, as the vortices move clockwise on the limit cycle

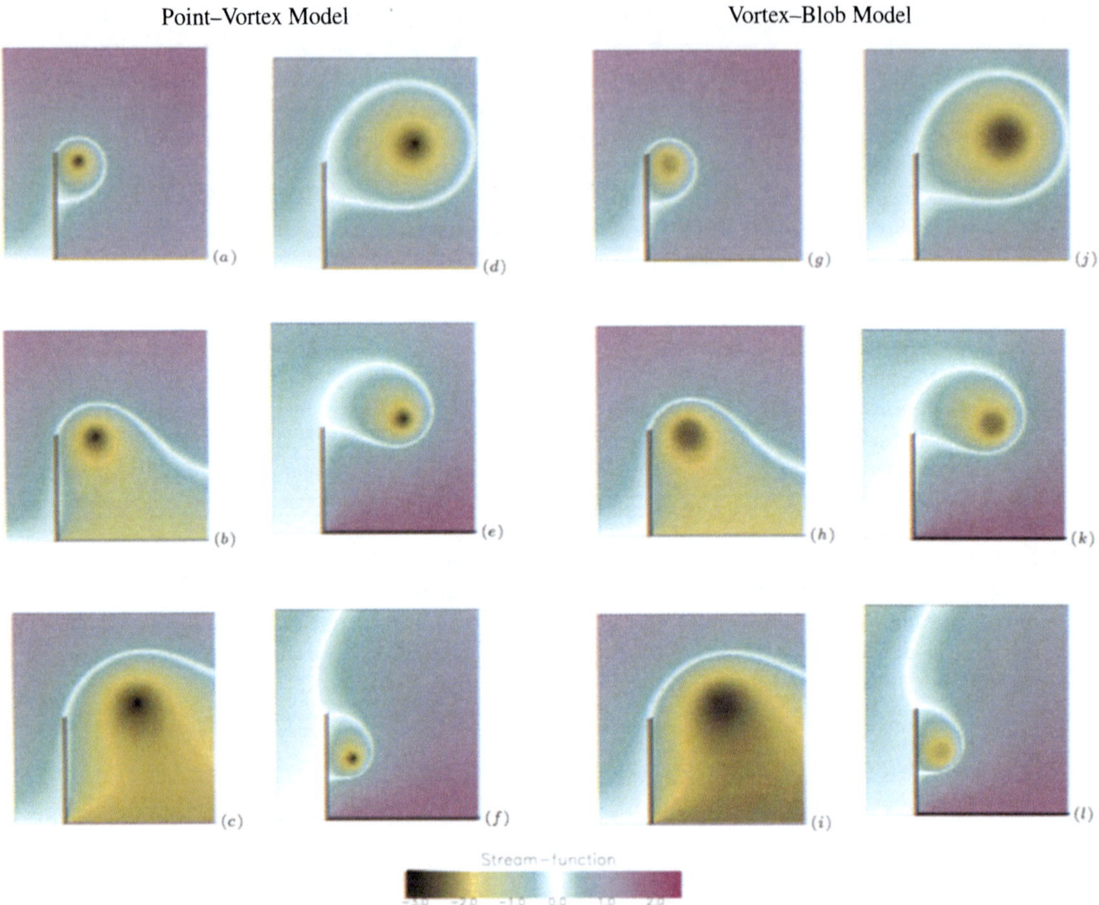

Figure 3. Comparison between PVM [(a),(b),(c),(d),(e),(f)] and VBM [(g),(h),(i),(j),(k),(l)]. Instantaneous stream function: t=6.2(a),(g); 6.4(b),(h); 6.7(c),(i); 7.0(d),(j); 7.2(e),(k); 7.6(f),(l). Figure taken from Cortelezzi *et al.* (1997).

trajectory. Note that the point vortex simulation involves only 2 vortices while the vortex blob simulation involves up to 1000 blobs. The instantaneous stream functions are remarkably similar, only the size of the vortex core, in black, is clearly different.

3 Reduced-order controller via spectral decomposition

This Section summarizes the results obtained in collaboration with Professors J.L. Speyer and J. Kim of the University of California at Los Angeles and three graduate students: Dr. K-H. Lee, Mr. S.M. Kang and Ms. V. Ryder.

The second example is the derivation of robust reduced-order controllers for channel flows using linear-quadratic-Gaussian (LQG) design, or, in modern terms, $\mathcal{H}_2$ design. The control objective is to reduce drag produced by skin friction, or, in other words, to reduce the wall-shear stresses generated by near-wall turbu-

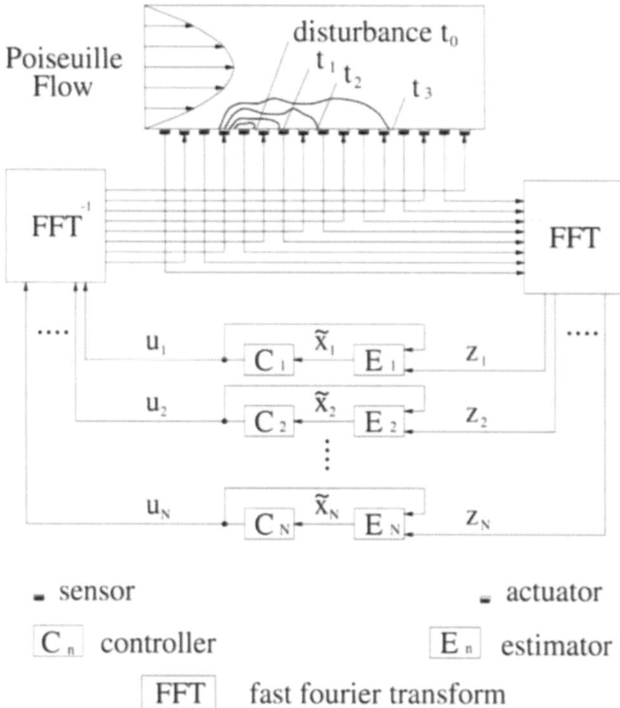

Figure 4. Controller architecture. Figure taken from Cortelezzi & Speyer (1998).

lence. To suppress perturbations evolving within the bottom boundary layer we apply blowing and suction at the bottom wall. For simplicity we assume that the actuators are uniformly distributed along the bottom wall. To detect and measure the deviations of the boundary layer from the laminar regime, we measure the gradient of the streamwise velocity component at given points along the bottom wall.

In this case, physical insight and intuition do not clearly suggest how to derive a reduced-order model. We derive a reduced-order model in a few steps (Cortelezzi & Speyer, 1998): We first linearize the Navier-Stokes equations with respect to a laminar parabolic profile, then we spectrally decompose the equations and use Galerkin's projection to obtain a set of ordinary differential equations, and finally these equations are cast by matrix transformation into state-space equations. In order to further reduce the order of the model, we transform the system matrices into Jordan Canonical form. This transformation decouples the modes of the system, in other word it reduces the system matrices to a block diagonal form. From the transformed system matrices the states that are equally well controllable and observable are determined and the reduced-order model is obtained by removing poorly controllable or observable states.

Figure 4 summarizes in a block diagram the control strategy and links with simplicity the mathematical formulation to its computational implementation. The controller can be programmed in a computer routine whose input is an array containing the gradients of the streamwise velocity component and whose output is an array containing the blowing and suction at the wall. The gradient of the streamwise velocity component is converted by a fast Fourier transform (FFT) into $\mathbf{z}_r$'s. Since the whole problem decouples with respect to the wave number, for each wave number r there is a pair of estimator $\mathbf{E}_r$ and control $\mathbf{C}_r$ blocks. Parallel computation produces $\mathbf{u}_r$'s. An inverse FFT converts $\mathbf{u}_r$'s into the actuating blowing and suction at the wall.

Figure 4 also provides the basic architecture for the potential implementation of the present controller in practical engineering applications. The gradient of the streamwise velocity component can be measured by micro-electro-mechanical-systems (MEMS) hot film sensors (Ho & Tai, 1996). Analog to digital converters (A/D) and digital signal processors (DSP) convert the measured gradients into $\mathbf{z}_r$'s. Each pair of estimator $\mathbf{E}_r$ and control $\mathbf{C}_r$ blocks is replaced by a microprocessor, and a parallel computation produces $\mathbf{u}_r$'s. A DSP and a digital to analog converter (D/A) produce the actuating signal. Finally, MEMS technology will provide the necessary hardware to implement the actuating signal. Note that a variety of actuators can mimic small amplitude blowing and suction at the wall: porous walls, micro pumps, deformable walls and thermal actuators (Ho & Tai, 1996).

3.1 State-space equations

We consider two-dimensional incompressible Poiseuille flow in a periodic channel of length L_x and height $2h$. The undisturbed velocity field has a parabolic profile with centerline velocity U_c. Since we are interested in controlling the transition of the boundary layer from laminar to turbulent regime, we consider small perturbation of flow quantities. The reader should be aware that with the term, "transition," we identify the disruption of the laminar regime in a fully developed boundary layer due to the growth of spatially localized near-wall perturbations. The linearized Navier-Stokes equations are written in terms of the perturbation stream-function ψ

$$(\partial_t + U\partial_x)\,\Delta\,\psi - U''\psi_x = Re^{-1}\,\Delta\,\Delta\psi, \tag{4}$$

to identically satisfy continuity. The problem is made dimensionless by using h as characteristic length and h/U_c as characteristic time. The Reynolds number is $Re = U_c h/\nu$.

To suppress perturbations evolving within the bottom boundary layer we apply blowing and suction at the bottom wall. For simplicity we assume that the actuators are uniformly distributed. Perturbations in the top boundary layer are left free to evolve. The corresponding boundary conditions are

$$\psi_x|_{y=-1} = -v_w(x,t), \qquad \psi_y|_{y=\pm1} = \psi|_{y=1} = 0, \tag{5}$$

where the control function $v_w(x)$ prescribes the amount of blowing and suction at the bottom wall. We impose that the mass of fluid injected equals the mass of fluid removed.

To detect and measure the deviations of the boundary layer from the laminar regime, we measure the gradient of the streamwise velocity component at given points $x = x_i$ along the bottom wall

$$z(x_i,t) = \psi_{yy}|_{y=-1}. \tag{6}$$

In other words, we measure the first term of the wall-shear stress, $\tau_{yx} = Re^{-1}(\psi_{yy} - \psi_{xx})|_{y=-1}$. The second term of the wall-shear stress is zero in the uncontrolled case and is known in the controlled case.

We define an optimal performance index J, or cost function, to design a controller for the LQG ($\mathcal{H}_2$) problem. Since we are interested in suppressing wall-shear stress, we define

$$J = \lim_{t_f\to\infty} \int_t^{t_f} \int_0^{L_x} (\psi_{yy}^2 + \psi_{xx}^2)|_{y=-1} dx dt. \tag{7}$$

The integrand represents the cost of the wall-shear stress being different from zero. Moreover, the integrand implicitly accounts for the cost of implementing the control itself. There are two reasons to minimize the cost of the controller: In any engineering application the energy available to drive the controller is limited; and a large control action may drive the system away from the region where the linear model is valid.

To reduce (4)-(6) to a set of first-order ordinary differential equations, we make a few transformations loosely based on Joshi *et al.* (1997). We write the stream-function as $\psi = \phi + \chi$ to embed the actuator into the evolution equation and to make the boundary conditions homogeneous. Substituting $\psi = \phi + \chi$ into equation (4), we obtain a forced equation for the Poiseuille flow

$$(\partial_t + U\partial_x) \triangle \phi - U''\phi_x = Re^{-1} \triangle \Delta\phi - (\partial_t + U\partial_x) \triangle \chi + U''\chi_x + Re^{-1} \triangle \Delta\chi, \tag{8}$$

with homogeneous boundary conditions $\phi|_{y=\pm1} = \phi_y|_{y=\pm1} = 0$. The forcing function χ satisfies the non-homogeneous boundary conditions (5), i.e., $\chi_x|_{y=-1} = -v_w(x,t)$, $\chi|_{y=1} = \chi_y|_{y=\pm1} = 0$. We also substitute $\psi = \phi + \chi$ into equations (6) and (7). The measurement equation (6) becomes

$$z(x_i, t) = (\phi_{yy} + \chi_{yy})|_{x=x_i, y=-1}, \tag{9}$$

while the cost function (7) takes the following form

$$J = \lim_{t_f \to \infty} \int_t^{t_f} \int_0^{L_x} [(\phi_{yy} + \chi_{yy})^2 + \chi_{xx}^2]_{y=-1} dx dt. \tag{10}$$

Subsequently, flow quantities are spectrally decomposed by using circular functions in the streamwise direction and Chebyshev polynomials in the vertical direction. We expand ϕ and χ as follows:

$$\phi = \sum_{n=1}^{N} \sum_{m=0}^{M} [a_{nm}(t) \cos(\alpha_n x) + b_{nm}(t) \sin(\alpha_n x)] C_m(y), \tag{11}$$

$$\chi = \sum_{n=1}^{N} [p_n(t) \cos(\alpha_n x) + q_n(t) \sin(\alpha_n x)] D(y), \tag{12}$$

where $\alpha_n = 2\pi n / L_x$. Functions C_m and D are combinations of Chebyshev polynomials constructed to satisfy the boundary conditions, i.e., $C_m(\pm1) = C'_m(\pm1) = D(1) = D'(\pm1) = 0$ and $D(-1) = 1$. We also expand the measurement function z as follows:

$$z = \sum_{n=1}^{N} [c_n(t) \cos(\alpha_n x) + d_n(t) \sin(\alpha_n x)]. \tag{13}$$

Substituting expansions (11), (12) and (13) into equations (8) and (9) and using Galerkin's projection, we obtain

$$\frac{d\mathbf{y}}{dt} = \mathbf{A}\mathbf{y} + \mathbf{B}_1\mathbf{u} + \mathbf{B}_2\frac{d\mathbf{u}}{dt}, \qquad \mathbf{z} = \mathbf{C}\mathbf{y} + \mathbf{D}_3\mathbf{u}. \tag{14}$$

To transform the above equations into standard state-space form, we define a new vector $\mathbf{x} = \mathbf{y} + \mathbf{B}_2\mathbf{u}$, and two new matrices $\mathbf{B} = \mathbf{B}_1 + \mathbf{A}\mathbf{B}_2, \mathbf{D} = \mathbf{D}_3 + \mathbf{C}\mathbf{B}_2$. Finally, we obtain the state-space equations

$$\frac{d\mathbf{x}}{dt} = \mathbf{A}\mathbf{x} + \mathbf{B}\mathbf{u}, \qquad \mathbf{z} = \mathbf{C}\mathbf{x} + \mathbf{D}\mathbf{u}, \tag{15}$$

with initial condition $\mathbf{x}(0) = \mathbf{x}_0$; where $\mathbf{x}$ is the internal state vector, $\mathbf{u}$ is the control vector and $\mathbf{z}$ is the measurement vector. Matrices $\mathbf{A}$, $\mathbf{B}$, $\mathbf{C}$ contain the dynamics of the Poiseuille flow, actuators and sensors, respectively. Matrix $\mathbf{D}$ contains the coupling between sensors and actuators. The cost function (10) becomes

$$J = \lim_{t_f \to \infty} \int_t^{t_f} [\mathbf{z}^T\mathbf{z} + \mathbf{u}^T\mathbf{W}^T\mathbf{W}\mathbf{u}] dt, \tag{16}$$

where the superscript ^{T} denotes transpose. The matrix $\mathbf{W}$ is obtained by spectrally decomposing the last term in the cost function (10).

3.2 Model reduction and controller design

The advantage of the present formulation is that the whole problem decouples with respect to the wave number. All matrices in (15) and (16) are block diagonal. The block diagonal structure of the matrix $\mathbf{A}$ was first recognized by Joshi *et al.* (1997). The above state-space system is consequently equivalent to N state-space sub-systems, one for each wave number. For a given wave number r the state-space equations are

$$\frac{d\mathbf{x}_r}{dt} = \mathbf{A}_r\mathbf{x}_r + \mathbf{B}_r\mathbf{u}_r, \qquad \mathbf{z}_r = \mathbf{C}_r\mathbf{x}_r + \mathbf{D}_r\mathbf{u}_r, \tag{17}$$

with initial condition $\mathbf{x}_r(0) = \mathbf{x}_{r0}$. The cost function also decouples with respect to the wave number and we obtain N optimal performance indexes. For a given wave number r the cost function is defined as follows

$$J_r = \lim_{t_f\to\infty}\int_t^{t_f}[\mathbf{z}_r^T\mathbf{z}_r + \mathbf{u}_r^T\mathbf{W}_r^T\mathbf{W}_r\mathbf{u}_r]dt. \tag{18}$$

Consequently, the design of an optimal and robust controller for the system (15) with (16) has been reduced to the independent design of N optimal and robust controllers, one for each wave number, for the sub-systems (17) with (18).

The challenge of the present study is to reduce the size of the controller. The controller of the full system would have $2N(M+1)$ states. A controller with thousands of states is of no interest in engineering applications because of the amount of hardware and computer power necessary to compute a real-time control law. We derive a lower order controller in two steps: First we construct a lower order model of (17), and subsequently, we design an optimal and robust controller for the reduced-order model. To obtain a lower order model, we transform (17) into Jordan Canonical form. The matrices $\hat{\mathbf{A}}_r, \hat{\mathbf{B}}_r, \hat{\mathbf{C}}_r, \mathbf{D}_r$ that describe the dynamics of the reduced-order model are obtained from the matrices in Jordan Canonical form by retaining rows and columns corresponding to equally well controllable or observable states. Hat denotes the quantities associated with the reduced-order model.

The design of an optimal and robust controller for the LQG ($\mathcal{H}_2$) problem is divided in two parts: the linear quadratic regulator (LQR) and the minimum variance estimator (Kalman filter) (Zhou *et al.*, 1996; Rhee & Speyer, 1996). LQR provides an optimal control law in terms of the internal state vector. In general, however, the internal state vector is not directly measurable. The Kalman filter provides an optimal estimate of the internal state vector in terms of the measurements vector $\mathbf{z}_r$. The result of the LQG ($\mathcal{H}_2$) design of an optimal and robust controller based on the reduced-order model of (17) is summarized by the following equations:

$$\mathbf{u}_r = -\hat{\mathbf{K}}_r\tilde{\mathbf{x}}_r, \tag{19}$$

$$\frac{d\tilde{\mathbf{x}}_r}{dt} = \hat{\mathbf{A}}_r\tilde{\mathbf{x}}_r + \hat{\mathbf{B}}_r\mathbf{u}_r + \hat{\mathbf{L}}_r[\mathbf{z}_r - \hat{\mathbf{C}}_r\tilde{\mathbf{x}}_r - \mathbf{D}_r\mathbf{u}_r], \tag{20}$$

with initial conditions $\tilde{\mathbf{x}}_r(0) = \mathbf{0}$. Equation (19) is the control law. The gains matrix $\hat{\mathbf{K}}_r$ is obtained by minimizing the following optimal performance index:

$$\hat{J}_r = \lim_{t_f\to\infty}\int_t^{t_f}[\hat{\mathbf{z}}_r^T\hat{\mathbf{z}}_r + \mathbf{u}_r^T\mathbf{W}_r^T\mathbf{W}_r\mathbf{u}_r]dt, \tag{21}$$

where $\hat{\mathbf{z}}_r = \hat{\mathbf{C}}_r\hat{\mathbf{x}}_r - \mathbf{D}_r\mathbf{u}_r$. Equation (20) is the minimum variance estimator. The matrix $\hat{\mathbf{L}}_r$ is obtained by minimizing the variance of the estimated state vector $\tilde{\mathbf{x}}_r$ with respect to the internal state vector $\hat{\mathbf{x}}_r$ assuming that the reduced-model of (17) is affected by additive Gaussian white noise. In this study the power spectral densities of the additive noise are used as design parameters to produce robust controllers. The initial condition $\tilde{\mathbf{x}}_r(0) = \mathbf{0}$ implies that the estimator starts with no information about $\hat{\mathbf{x}}_r$.

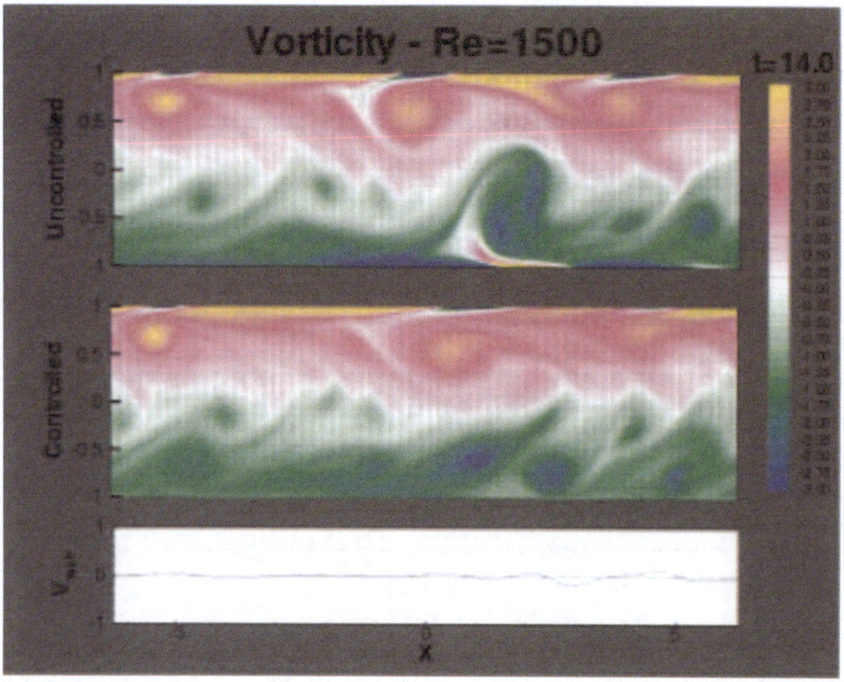

Figure 5. Vorticity field for the uncontrolled and controlled case and amplitude of the wall velocity, at time t=14. Figure taken from Cortelezzi *et al.* (1998).

3.3 Control of a two-dimensional channel flow

To design a controller for two-dimensional Poiseuille flow in a periodic channel at $Re = 1500$, we spectrally decompose the problem using $N = 32$ circular functions in the streamwise direction and $M = 60$ Chebyshev polynomials in the vertical direction. Hence, the order of the spectrally decomposed system is 3904. Using the model reduction technique previously described, we create a reduced-order model of order 384 representing a dramatic reduction of about 90%. Since the whole problem decouples with respect to the wave number, we design an estimator $\mathbf{E}_r$ and a controller $\mathbf{C}_r$ for each wave number r (see Figure 4). Wave number estimators and controllers operate in parallel and are tested on a finite-difference simulation of a two-dimensional turbulent channel flow at Re=1500 (Cortelezzi *et al.*, 1998).

This controller is tested on a two-dimensional channel flow at $Re = 1500$. Since there are no unstable modes at $Re = 1500$, we use a combination of channel and wall modes to create a worse scenario initial perturbation velocity field (Farrell, 1988). The maximum rms value of the initial perturbation velocity field is $u_{rms} = 0.3$, i.e. 30% of the centerline velocity of the undisturbed laminar flow. We construct the initial velocity field by superimposing a perturbation velocity field onto the laminar flow. The corresponding initial vorticity field produces pockets of positive and negative vorticity distributed along the center and walls of the channel. Although the initial disturbances in the two-dimensional channel flow will eventually decay, they permit testing the capability of the controller in suppressing near-wall disturbances that can increase drag in turbulent boundary layers. Note that this initial condition includes nonlinear effects that are not included in the linear model and also stimulates modes that are not included in the reduced-order model.

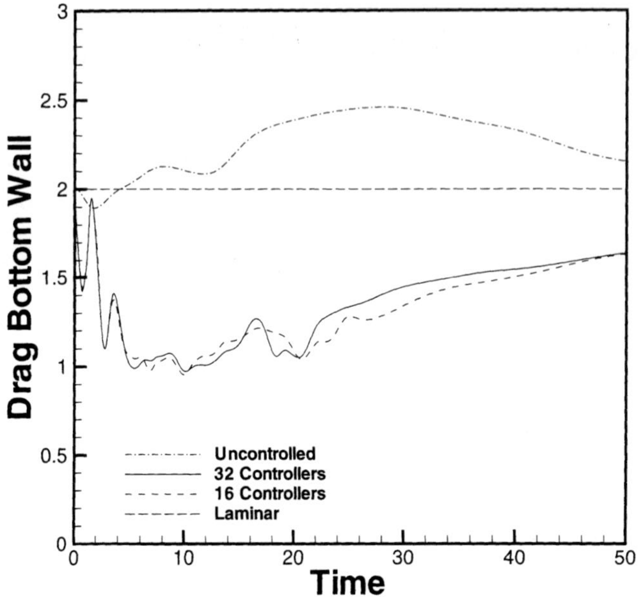

Figure 6. Comparison of the total drag measured at the bottom wall of the channel: laminar flow, uncontrolled turbulent flow, fully and partially controlled turbulent flow. Figure taken from Cortelezzi *et al.* (1998).

Figure 5 presents a comparison of the vorticity field in the controlled and uncontrolled cases at nondimensional time t=14 (flow is from left to right). Figure 5 also presents the amplitude of blowing and suction applied to the bottom wall of the channel. The uncontrolled case shows a major eruption along the bottom wall that affects nearly half of the bottom boundary layer as well as part of the top boundary layer. In particular, the pairing process of two large negative vortical structures interacts with a pocket of positive vorticity on the bottom wall. The vortex pairing entrains parts of the positive vorticity enhancing the separation of the boundary layer. In the controlled case almost the entire vorticity field has been strongly modified by the controller, high values of vorticity are confined to the top boundary layer. Near the bottom wall three negative vortical structures are kept rolling by the controller action although there is no significant vorticity along the wall. The controller injects and removes fluid from the rear and the front of the vortical structures, respectively. In this figure it is easy to see that wall velocity changes sign at the streamwise locations corresponding with the center of the vortical structures. The action of the controller has also reduced the interaction between top and bottom boundary layers resulting in an mild attenuation of the near-wall turbulence at the top wall.

Drag reduction is the dominant reason driving the effort of controlling near-wall turbulence. To estimate the performance of the controller, we compute the drag by integrating the wall-shear stresses along the bottom wall of the channel. Figure 6 compares the drag in the uncontrolled, laminar, fully controlled and partially controlled cases. The drag for the laminar case is exact and equals 2. In the uncontrolled and controlled cases the drag is initially 2 because the initial velocity field has been constructed by superimposing a zero-mean perturbation velocity field onto the laminar flow. In the uncontrolled case, the drag drops below

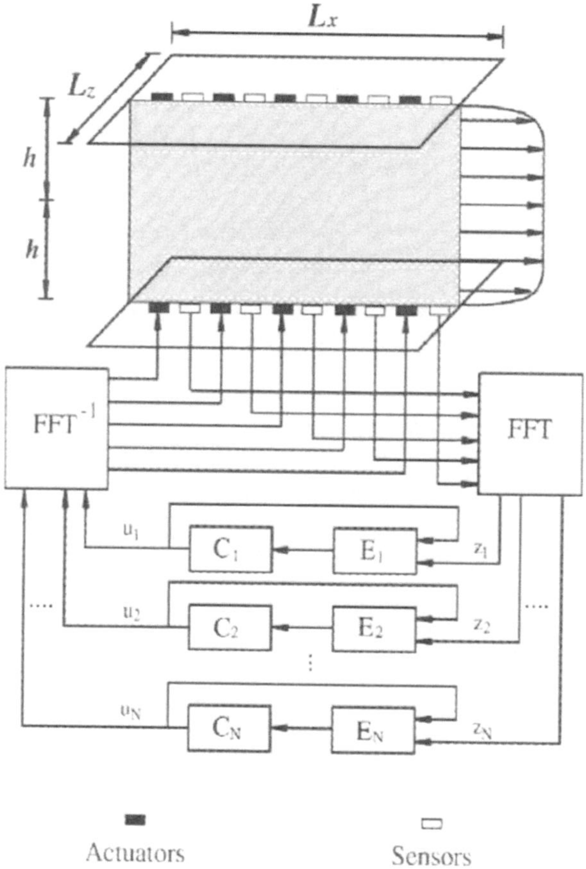

Figure 7. Schematic representation of turbulent channel flow equipped with sensors and actuators distributed in the streamwise direction in each $z - plane$. Figure taken from Lee *et al.* (2001).

the laminar value during the transient period. Subsequently, drag increases above the laminar value and with some fluctuation reaches its maximum value 2.46 around t=25. In the last half of the simulation the drag decreases toward the laminar value because disturbances are subdued by viscosity, since disturbances cannot sustain themselves in a two-dimensional channel flow at this Reynolds number. In the fully controlled case, the drag presents some large fluctuation below the laminar value during the first few time units at the onset of the flow. The amplitude of the fluctuations reduces with time while the value of the drag stays at about half of the laminar value. In the later part of the simulation the controller affects the entire computational domain and the flow is nearly relaminarized. The drag in the partially controlled case is remarkably similar to the fully controlled case. This results substantiates the notion that it is not necessary to control all the wave numbers.

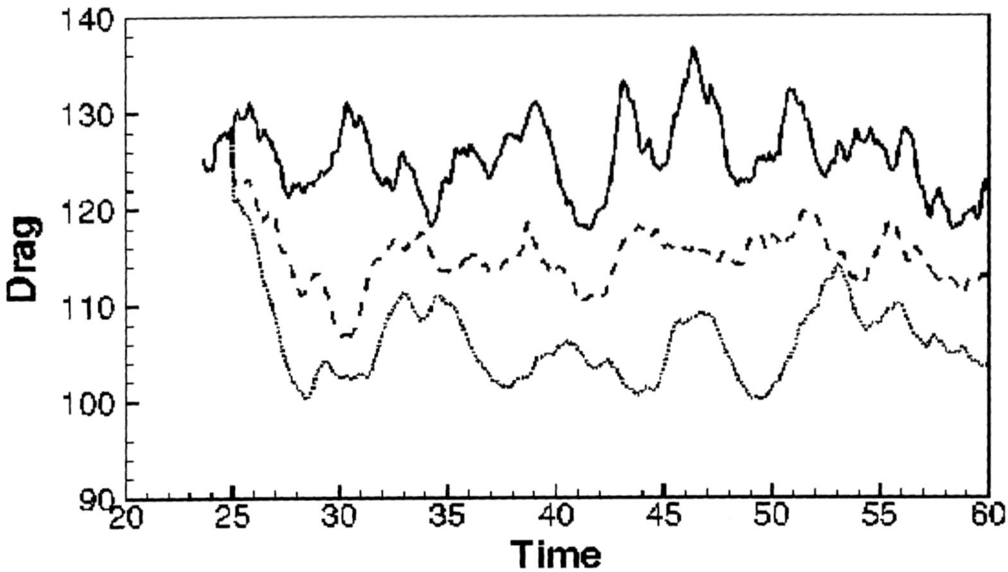

Figure 8. Time history of the drag for the *ad hoc*-controlled (dotted) 2D-controlled (dashed) and uncontrolled (solid) flows. Figure taken from Lee *et al.* (2001).

3.4 Distributed control of a three-dimensional fully developed channel flow

Figure 7 shows the configuration of the turbulent channel flow equipped with the controller tested for our study. We slice the channel with xy-planes equally spaced in the z-direction to reduce the order of the controller. We then construct the distributed two-dimensional controller by applying the controller developed from the linearized two-dimensional Navier-Stokes equations to each plane. See Lee *et. al* (2001) for more details.

To test the performance of a reduced-order distributed controller, we performed direct numerical simulation of a turbulent channel flow at $Re_\tau = 100$ with a computational domain of $(4\pi, 2, 4\pi/3)$ in the (x, y, z) directions. We designed a distributed two-dimensional controller in two steps. First, we designed reduced-order controllers for two-dimensional Poiseuille flow in a periodic channel of streamwise length $L_x = 4\pi$ at $Re = 5000$, which has the same mean wall-shear stress as turbulent channel flow at $Re_\tau = 100$. Subsequently, we fine-tuned the single-wave number reduced-order controllers in order to minimize the magnitude of the Fourier coefficients of the wall-shear stresses in turbulent channel flow at $Re_\tau = 100$. We used $N = 32$ and $M = 60$ in this linear model flow. Controllers operate at both top and bottom walls in parallel. If the two-dimensional controllers without model reduction were applied at each z-plane, then the order of the ensemble of controllers would be $64 \times 3904 = 249,856$. Using the model reduction technique previously described, we designed eight single-wave number controllers of order 12, corresponding to the eight lowest wave numbers. Consequently, the combined order of the controllers is $64 \times 96 = 6,144$. It represents a state-space reduction of about 97.5% with respect to the full-order system. See Lee *et. al* (2001) for more details.

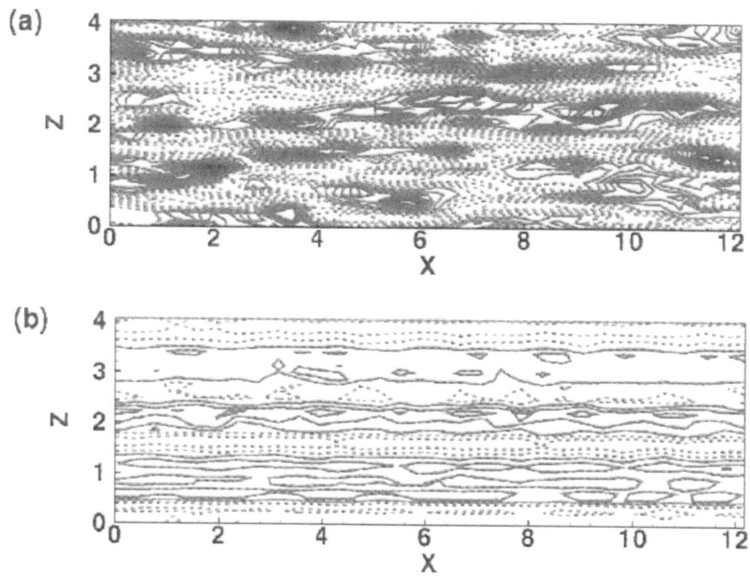

Figure 9. Contours of disturbance wall-shear stresses at the bottom wall; (a) uncontrolled flow; (b) 2D-controlled flow. Negative contours are dashed. Figure taken from Lee *et al.* (2001).

Figure 8 shows the time history of the drag in the uncontrolled and controlled flows. This two-dimensional control yields about a 10% drag reduction. Note that this observed drag reduction is a byproduct since our controller is designed to suppress the fluctuations of the streamwise wall-shear stress, not the mean wall-shear stress. Contours of the disturbance wall-shear stresses at the bottom wall in the controlled and uncontrolled flows at t=30 are shows in Figure 9. Contours for the controlled flow show the dramatic effect of the distributed two-dimensional controller. The long streaky wall-shear stress region spans almost the entire streamwise direction, indicating that the low wave number components (except the zero wave number that we do not control) are completely suppressed. The remaining spanwise variations, i.e., the alternating regions of high- and low-shear stress, are due to the fact that the two-dimensional controllers distributed along the streamwise direction are operated independently from one z-plane to another.

The above results demonstrate that our distributed two-dimensional controller designed from the linear model works remarkably well in suppressing near-wall disturbances in the fully developed turbulent flow. However, this controller does not account for the spanwise variations of fluid motion. In an attempt to reduce the remaining spanwise variations of the controlled flow, we introduce a simple *ad hoc* control augmentation scheme. This additional control, which generates blowing/suction to attenuate the spanwise variation of the wall-shear stress, is given as follows:

$$v_{ad} = C\left(\left.\frac{\partial u}{\partial y}\right|_w^{(x,z)} - \left.\frac{\partial u}{\partial y}\right|_w^{x} \right), \tag{22}$$

where $\partial u/\partial y|_w^{(x,z)}$ and $\partial u/\partial y|_w^{x}$ are the streamwise velocity gradients averaged over the xz-plane and the x-direction, respectively, and C is a constant to be adjusted for the best performance. The subscript *ad*

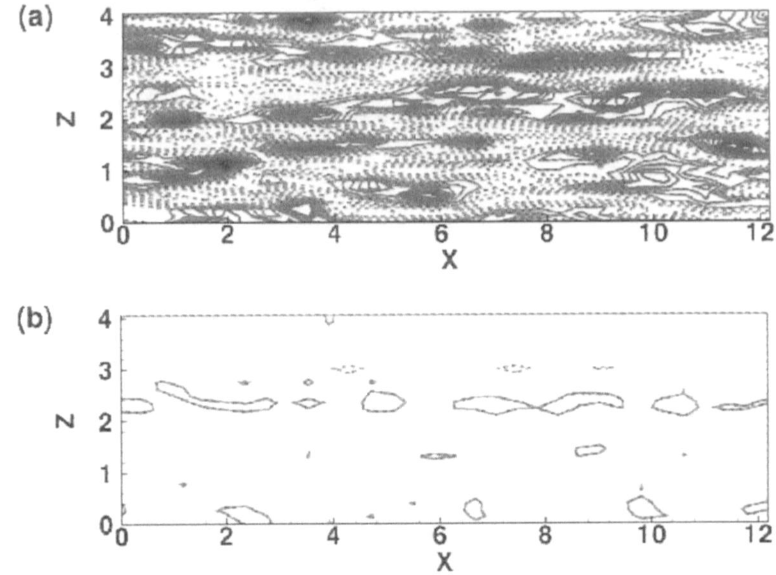

Figure 10. Contours of disturbance wall-shear stresses at the bottom wall; (a) uncontrolled flow; (b) *ad hoc*-controlled flow. Negative contours are dashed. Figure taken from Lee *et al.* (2001).

indicates the *ad hoc* control, and v_{ad} is a function of z only. Therefore, the new control input is defined by

$$v_w(x, z) = v_{2D} + v_{ad}, \tag{23}$$

where v_{2D} is the actuation velocity generated by the distributed two-dimensional controller.

Using the distributed two-dimensional controller augmented with this *ad hoc* control scheme, the control of the fully developed turbulent flow with $Re_\tau = 100$ increased drag reduction to about 17% as shown in Figure 8. The constant, C, in Equation (22) is adjusted such that the root-mean-square (rms) value of the actuation is maintained at $0.1u_\tau$, where u_τ is the wall shear velocity for the uncontrolled flow.

Figure 10 presents the comparison of the contours of the disturbance wall-shear stresses at the bottom wall between the *ad hoc* controlled flow and the uncontrolled flow at $t = 30$. Compared with Figure 9, the additional effort in the spanwise direction removes the pronounced peak-valley variation of the wall-shear stress that is observed in the controlled flow with the distributed two-dimensional controllers.

Although the present two-dimensional controller augmented by an *ad hoc* three-dimensional controller has shown a promising result, it is apparent that we need to develop a three-dimensional reduced-order controller using the same type of formulation presented in this Section. Extension of LQG/LTR design using three-dimensional channel flow models is in progress (Kang *et al.*, 1999; Kang *et al.*, 1999).

4 Concluding remarks

It is shown that there are several advantages to working with a reduced-order model. First, it is often easier to derive the desired control strategy. Second, dynamical system analysis could be used to characterize the dynamics of the closed-loop system. Finally, the control strategy derived for the reduced-order model might produce fast numerical algorithms which are essential for controlling a real flow. It is hoped that, in succeeding chapters of this text, specific techniques for the control of trasverse jets will be clear and will stimulate future researchers to investigate the control of this important flowfield.

References

BROWN, C. E., AND MICHAEL, W. H. 1954 Effect of leading-edge separation on the lift of a delta wing *J. Aero. Sci.* **21**, 690.

CHORIN, J. 1973 Numerical study of slightly viscous flows *J. Fluid Mech.* **57**, 785.

CORTELEZZI, L., AND LEONARD, A. 1993 Point vortex model for the unsteady separated flow past a semi-infinite plate with transverse motion *Fluid Dynamics Research* **11**, 263.

CORTELEZZI, L., LEONARD, A., AND DOYLE, J. C. 1994 An example of active circulation control of the unsteady separated flow past a semi-infinite plate *Journal of Fluid Mechanics* **260**, 127.

CORTELEZZI, L. 1996 Nonlinear Feedback Control of the Wake Past a Plate with a Suction Point on the Downstream Wall *Journal of Fluid Mechanics* **327**, 303.

CORTELEZZI, L., CHEN, Y-C., AND CHANG, H-L. 1997 Nonlinear Feedback Control of the Wake Past a Plate: from a Low-Order Model to a Higher-Order Model *Physics of Fluids* **9**, (7), 2009–2022.

CORTELEZZI, L., AND SPEYER, J. L. 1998 Robust Reduced-Order Controller of Laminar Boundary Layer Transitions *Physical Review E* **58**, (2), 1906.

CORTELEZZI, L., LEE, K-H., KIM, J., AND SPEYER, J. L. 1998 Skin-Friction Drag Reduction Via Robust Reduced-Order Linear Feedback Control *Physics of Fluids* **11**, (1-2), 79.

DUNSTAN, W.J., BITMEAD, R. R. AND SAVARESI, S. M. 2001 Fitting nonlinear low-order models for combustion instability control *Control Engineering Practice* **9**, (12), 1301.

FARRELL, B. F. 1988 Optimal excitation of perturbations in viscous shear flows *Phys. Fluids* **31**, (8), 2093.

GAD-EL-HAK, M. 2001 Flow control: The future *J. Aircraft*, **38**, (3), 402.

GAD-EL-HAK, M. 1999 Interactive control of turbulent boundary layers – A futuristic overview *AIAA Journal*, **32**, (9), 1753.

GRAHAM, W. R., PERAIRE, J. AND TANG, K. Y. 1999 Optimal control of vortex shedding using low-order models. Part II – Model-based control *Int. Journal for Numerical Methods in Fluids* **44**, (7), 973.

GUCKENHEIMER, J., AND HOLMES, P. 1983 Nonlinear oscillations, dynamical systems, and bifurcation of vector fields *Springer-Verlag, New York.*

HO, C. M. AND TAI, Y.C. 1996 Review - MEMS and its applications for flow control *J. Fluids Eng. - Trans. AMSE* **118**, 437.

JOSHI, S. S., SPEYER, J. L. & KIM, J. 1997 A systems theory approach to the feedback stabilization of infinitesimal and finite-amplitude disturbances in plane Poiseuille flow *J. Fluid Mech.* **332**, 157.

KANG, S. M., RYDER, V., CORTELEZZI, L. AND SPEYER, J. L. 1999 State-space formulation and control design for three-dimensional channel flows *American Control Conference*, San Diego, California, 2–4 June 1999.

KANG, S. M., CORTELEZZI, L. AND SPEYER, J. L. 1999 Performance of a linear controller for laminar boundary layer transition in three-dimensional channel flow *Proceedings of the 38th Conference on Decision and Control*, Phoenix, Arizona, 7–10 December, 1999.

LEE, K-H., CORTELEZZI, L., KIM, J. AND SPEYER, J. L. 2001 Application of reduced-order controller to turbulent flows for drag reduction *Physics of Fluids***13**, (5), 1321.

LOFDAHL, L. AND GAD-EL-HAK, M. 1999 MEMS applications in turbulence and flow control *Prog. in Aerospace Sciences*, **35**, (2), 101.

LUCIA, D. J., PACHTER, M. AND BERAN, P. S. 2002 Rocket nozzle flow control using a reduced-order computational fluid dynamics model *Journal of Guidance Control and Dynamics***25**, (3), 449.

M'CLOSKEY, R. T., KING, J., CORTELEZZI, L. AND KARAGOZIAN, A. R. 2002 The actively controlled jet in crossflow *J. Fluid Mech.* **452**, 325.

M'CLOSKEY, R. T., KING, J., CORTELEZZI, L. AND KARAGOZIAN, A. R. 2002 Active control of jets in crossflow In *Manipulation and Control of Transverse Jets,* Springer-Verlag.

RAO, D. M. 1987 Vortical flow management techniques *Prog. Aerospace Sci.* **24**, 173.

RAVEH, D. E. 2001 Reduced-order models for unsteady aerodynamics *AIAA Journal* **39**, (8), 1417–1429.

RAVINDRAN, S. S. 2000 A reduced-order approach for optimal control of fluids using proper orthogonal decomposition *Int. Journal for Numerical Methods in Fluids* **34**, (5), 425.

REDINIOTIS, O. K., KO, J. AND KURDILA, A. J. 2002 Reduced order nonlinear Navier-Stokes models for synthetic jets *Journal of Fluids Engineering – Transactions of th ASME* **124**, (2), 433–443.

RHEE, I. AND SPEYER, J. L. 1996 A game theoretic approach to a finite-time disturbance attenuation problem *IEEE Trans. Automatic Control* **36**, (9), 1021.

ROTT, N. 1956 Diffraction of a weak shock with vortex generation *J. Fluid Mech.* **1**, 111.

SARPKAYA 1989 Computational methods with vortices – The 1988 Freeman Scholar Lecture *ASME J. Fluids Eng.* **111**, 5.

SARPKAYA 1994 Vortex element methods for flow simulation *Advance in Applied Mechanics* **31**, 113.

TANG, S. AND AUBRY, N. 2000 Suppression of vortex shedding inspired by a low-dimensional model *J. Fluids and Structures* **14**, (4), 443.

ZHOU, K., DOYLE, J. C. AND GLOVER K. 1996 Robust and optimal control *Prentice Hall.*

Control of Micromixers, Jets, and Turbine Cooling using Evolution Strategies

Sibylle D. Müller[1] and Petros Koumoutsakos[12]

[1] Institute of Computational Sciences, Swiss Federal Institute of Technology (ETH) Zürich, Switzerland
[2] NASA Ames Research Center, Moffett Field, CA, USA

Abstract. We present a class of evolution strategies and we discuss three engineering applications in the fields of mixing control and turbomachinery: (i) flow in micromixers, (ii) jet flow, and (iii) turbine cooling. Evolution strategies are chosen as optimization method as they are capable of handling noisy and multimodal functions, inherent to these applications, in an automated fashion.

1 Introduction

Evolution strategies (ES) are parameter optimization techniques based on evolution principles such as mutation, selection, and adaptation. In the present article, we discuss ES as an optimization tool capable of handling diverse requirements for engineering applications such as the optimization of mixing (i) in micromixers and (ii) in jet flow, and (iii) the optimization of a cooling configuration in turbine blades. Numerical simulations are used to solve the governing equations of the flow or heat transfer problem. The control of mixing has important engineering applications ranging from micromixers for drug delivery to mixing in exhausts of aerodynamic configuration for noise reduction or to blade cooling in turbomachinery. The optimization of such processes has been subject of several theoretical and experimental works.

ES's offer several advantages for the optimization of such problems as these processes often involve noise, multimodality, or discontinuities of the objective function, which is confirmed by the results of our simulations. Moreover, often gradient information is not available which makes the use of gradient-free optimization techniques necessary. The inherently parallel character of ES's can be exploited in order to compensate for the high computational cost of the simulations.

The paper is organized as follows: In Section 2, we describe evolutionary algorithms in general and the evolution strategies implemented in this work. In Section 3, the micromixing device is described and the optimization results are presented. Jet mixing is presented in Section 4 while Section 5 describes the turbine blade cooling optimization, followed by a summary of our observations in Section 6.

2 Evolutionary Algorithms

Evolutionary algorithms are stochastic opimization algorithms which address the following problem: Minimize an objective function which is a mapping from a parameter vector $\boldsymbol{x} \in \mathcal{R}^n$ to $\mathcal{R}$.

They are based on biologically inspired principles such as reproduction, mutation, isolation, recombination, and selection applied on individuals in a population. The members of a population are capable of evolving over time, by adapting to their environment.

In contrast to many deterministic algorithms, evolutionary algorithms require only the value of the objective function for a given point in the parameter space, but do not need any sensitivity information. Certain evolutionary algorithms attempt to obtain this sensitivity information by exploring the search space and formulating prior information. They are comparatively easy to implement and work efficiently for global search. On the other hand, stochastic algorithms do not possess highly efficient convergence properties like e. g. the conjugate gradient method, but this disadvantage is compensated by the parallel character of evolutionary algorithms.

2.1 Evolution Strategies

Evolution strategies operate on a population with a number of individuals where each individual is represented by a real-valued vector.

The (1+1)-ES depicts the simplest evolutionary model with a population consisting of two individuals, one parent and one offspring. Here, each individual is identified by a parameter vector $\boldsymbol{x}$ with n components representing its genotype. As basic elements of biological evolution, principles of mutation and selection are applied in this scheme.

In an iterative process, the strategy generates an offspring represented by the vector $\boldsymbol{x}_N^{(g)}$ by adding a normally distributed random vector $\boldsymbol{z} \sim \mathcal{N}(0, {\sigma_i}^2)$ to the parent vector $\boldsymbol{x}_E^{(g)}$:

$$\boldsymbol{x}_N^{(g)} = \boldsymbol{x}_E^{(g)} + \boldsymbol{z} \tag{1}$$

In the following selection step, for both the parent and the offspring the capacity of survival is determined by calculating the values of the objective function. For minimization, the one with the smaller function value is then chosen to be the parent vector of the next generation.

The (1+1)-Evolution Strategy can be written as the following algorithm:

1. Generation of an offspring:

$$\boldsymbol{x}_N^{(g)} = \boldsymbol{x}_E^{(g)} + \delta^{(g)} \cdot \boldsymbol{z} \tag{2}$$

2. Selection:

$$\boldsymbol{x}_E^{(g+1)} = \begin{cases} \boldsymbol{x}_N^{(g)} & , \quad \text{if} \quad f(\boldsymbol{x}_N^{(g)}) \leq f(\boldsymbol{x}_E^{(g)}) \\ \boldsymbol{x}_E^{(g)} & , \quad \text{otherwise} \end{cases} \tag{3}$$

with

$\boldsymbol{x}$ parameter vector to be optimized, $\boldsymbol{x} \in \mathcal{R}^n$,
g generation counter,
E Index of the vector representing the parent (*E*lder),
N Index of the vector representing the offspring(*N*ewer),
δ Step size determining the length of the mutation step, $\delta \in \mathcal{R}_{>0}$,
$\boldsymbol{z}$ Realization of a $\mathcal{N}(\boldsymbol{0}, \boldsymbol{I})$ distributed random vector, $\boldsymbol{z} \in \mathcal{R}^n$.

3. Increase the generation counter, choose an appropriate value for $\delta^{(g)}$, and go to Step 1 if termination criteria do not hold.

The (1+1)-evolution strategy was developed by Rechenberg and Schwefel in 1963, initially for optimization of experimental fluid dynamics problems [2].

Generalizations of the (1+1) evolution scheme include plus and comma strategies, $(\mu+\lambda)$ and (μ, λ) evolution strategies where μ means the total number of parents and λ the number of offspring.

In "plus" selection schemes $(\mu+\lambda)$, both the parent and offspring population are considered for selection. In a $(\mu+\lambda)$ ES, one or more individuals of the parent generation may survive for as long of a period as no better offspring are produced. In "comma" selection schemes (μ, λ), μ out of λ individuals are selected to create the offspring of the next generation [3].

Another principle of biological evolution, recombination, can also be used in evolution strategies. The I in the $(\mu_{\mathrm{I}}, \lambda)$ scheme stands for *i*ntermediate recombination and means that the μ parents are averaged to create one individual parent.

2.2 Mechanisms of Step Size Adaptation

In evolution strategies the choice of strategy parameters such as population size, mutation step size, or selection scheme is critical to the behavior and the convergence properties of the strategy.

Particular importance is attached to the adaptation of the step size during the optimization, e.g. [3,6]. Step size adaptation is a crucial factor in the convergence rate as well as the global search properties of an evolution strategy. Figure 1 shows lines of equal probability to place children where the center of the circles and ellipses are the location of the parent individual. The different types of step size adaptation include the following cases:

- A single step size yields circles representing lines of equal probability, Figure 1 (left). This step size is a scalar and called *global* step size. It represents the standard deviation of the mutation.
- Step sizes, different in each coordinate direction of the parameter space, correspond to mutation ellipses which are parallel to the coordinate axes, Figure 1 (middle). These step sizes can be summarized in an n-dimensional vector and they are referred to as *individual* step sizes. The standard deviation of the mutation is different in each coordinate axis.

– Correlated mutations, generated by a linear transformation of a $\mathcal{N}(\mathbf{0}, \boldsymbol{I})$-distributed random vector, give rise to arbitrarily oriented mutation ellipses, Figure 1 (right). Step sizes can be presented by an $n \times n$ matrix, $\boldsymbol{C}$. The diagonal entries of $\boldsymbol{C}$ contain the variances of the mutation, and the off-diagonal entries are the covariances.

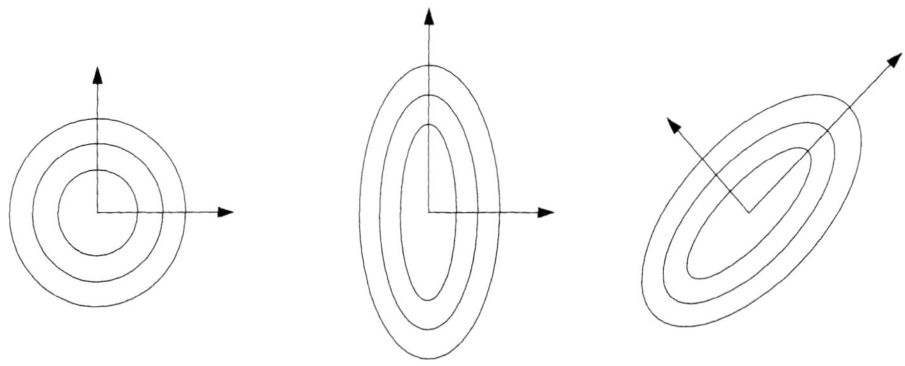

Fig. 1. Isoprobability contours in two dimensions for three probability densities used for the mutation step: Left: isotropic Gaussian density. Middle: scaled probability density. Right: scaled and rotated probability density.

2.3 Covariance Matrix Adaptation

The Covariance Matrix Adaptation (CMA) approximates a mutation distribution which is properly adapted to the local topology of the fitness landscape. The process performs a principle component analysis of the weighted mutation steps. In contrast to other approaches, the CMA shows invariance properties to general linear transformations of the parameter space [4].

Although rapid convergence and global convergence behavior are two contradictory requirements [3], it is indicated in [6] that the adaptation by the CMA-ES can even have positive effects on the global convergence behavior of the search process. Moreover, the CMA technique has been found to converge much faster than evolution strategies with only global and/or individual step size control for non-separable and badly scaled functions [6].

Following the notation in [1] for the $(\mu/_{\mathrm{I}}\mu, \lambda)$-CMA-ES, the offspring of generation $g+1$ are computed by

$$\boldsymbol{x}_k^{(g+1)} = \langle \boldsymbol{x} \rangle_\mu^{(g)} + \delta^{(g)} \boldsymbol{B}^{(g)} \boldsymbol{D}^{(g)} \boldsymbol{z}_k, \quad k = 1(1)\lambda, \tag{4}$$

where

$$\langle \boldsymbol{x} \rangle_\mu^{(g)} = \frac{1}{\mu} \sum_{i \in I_{sel}^{(g)}} \boldsymbol{x}_i^{(g)} \tag{5}$$

represents the center of mass of the selected individuals of generation g. $I_{sel}^{(g)}$ is the index set of selected individuals at generation g, $|I_{sel}^{(g)}| = \mu$. The random vectors $\boldsymbol{z}$ are $\mathcal{N}(\boldsymbol{0}, \boldsymbol{I})$ distributed and serve to generate offspring for generation $g+1$, as can be seen from equation (4). They can be used to calculate

$$\langle \boldsymbol{z} \rangle_{\mu}^{(g+1)} = \sum_{i \in I_{sel}^{(g+1)}} \boldsymbol{z}_{\mu}^{(g+1)}. \tag{6}$$

The columns of $\boldsymbol{B}^{(g)}$ represent eigenvectors of the covariance matrix $\boldsymbol{C}^{(g)}$. $\boldsymbol{D}^{(g)}$ is a diagonal matrix whose elements are identical to the square roots of the eigenvalues of $\boldsymbol{C}^{(g)}$. Hence, the relation of $\boldsymbol{B}^{(g)}$ and $\boldsymbol{D}^{(g)}$ to $\boldsymbol{C}^{(g)}$ can be expressed by

$$\boldsymbol{C}^{(g)} = \boldsymbol{B}^{(g)} \boldsymbol{D}^{(g)} \left(\boldsymbol{B}^{(g)} \boldsymbol{D}^{(g)} \right)^T, \tag{7}$$

and

$$\boldsymbol{C}^{(g)} \boldsymbol{b}_i^{(g)} = (d_{ii}^{(g)})^2 \cdot \boldsymbol{b}_i^{(g)}, \tag{8}$$

where $\boldsymbol{b}_i^{(g)}$ represents the i-th column of $\boldsymbol{B}^{(g)}$. Each covariance matrix corresponds to a hyperellipsoid which defines the surface of equal probability to place offspring. Here, the eigenvectors of the covariance matrix define the orientation of the hyperellipsoid and the eigenvalues define the squared ratio of the lengths of its axes.

Instead of using selection information of a single generation step, one could use a path that the population takes over a number of generations. Such a path is called an evolution path, $\boldsymbol{s}^{(g+1)}$, and is calculated by

$$\boldsymbol{s}^{(g+1)} = (1-c) \cdot \boldsymbol{s}^{(g)} + c_u \cdot \underbrace{\frac{\sqrt{\mu}}{\delta^{(g)}} \left(\langle \boldsymbol{x} \rangle_{\mu}^{(g+1)} - \langle \boldsymbol{x} \rangle_{\mu}^{(g)} \right)}_{= \sqrt{\mu}\, \boldsymbol{B}^{(g)} \boldsymbol{D}^{(g)} \langle \boldsymbol{z} \rangle_{\mu}^{(g+1)}}. \tag{9}$$

With the evolution path, the covariance matrix of generation $g+1$ is build by

$$\boldsymbol{C}^{(g+1)} = (1 - c_{cov}) \cdot \boldsymbol{C}^{(g)} + c_{cov} \cdot \boldsymbol{s}^{(g+1)} \left(\boldsymbol{s}^{(g+1)} \right)^T. \tag{10}$$

For the adaptation of the global step size, the evolution path $\boldsymbol{s}_{\delta}^{(g+1)}$, which is not scaled by $\boldsymbol{D}$, is calculated by

$$\boldsymbol{s}_{\delta}^{(g+1)} = (1 - c_{\delta}) \cdot \boldsymbol{s}_{\delta}^{(g)} + \\ + c_{\delta u} \cdot \underbrace{\boldsymbol{B}^{(g)} \left(\boldsymbol{D}^{(g)} \right)^{-1} \left(\boldsymbol{B}^{(g)} \right)^{-1} \frac{\sqrt{\mu}}{\delta^{(g)}} \left(\langle \boldsymbol{x} \rangle_{\mu}^{(g+1)} - \langle \boldsymbol{x} \rangle_{\mu}^{(g)} \right)}_{= \sqrt{\mu}\, \boldsymbol{B}^{(g)} \langle \boldsymbol{z} \rangle_{\mu}^{(g+1)}} \tag{11}$$

and its length is used to compute the overall step size for generation $g+1$

$$\delta^{(g+1)} = \delta^{(g)} \cdot \exp\left(\frac{\|\boldsymbol{s}_\delta^{(g+1)}\| - \hat{\chi}_n}{D\,\hat{\chi}_n}\right). \tag{12}$$

The underbraced terms in equations (9) and (11) emphasize that the evolution path for updating the covariance matrix does not depend on the global step size and that the global step size is not scaled by the lengths of the axes of the mutation ellipsoid.

Recommended strategy parameters are [1]:

$$\begin{array}{lll} c \ = \frac{1}{\sqrt{n}}, & c_{cov} = \frac{2}{n^2+n}, & c_\delta = \frac{1}{\sqrt{n}}, \\ c_u = \sqrt{c\cdot(2-c)}, & c_{\delta u} \ = \sqrt{c_\delta\cdot(2-c_\delta)}, & D = \sqrt{n}. \end{array} \tag{13}$$

Initial values are $\boldsymbol{s}^{(0)} = \mathbf{0}$ and $\boldsymbol{s}_\delta^{(0)} = \mathbf{0}$ and the covariance matrix $\boldsymbol{C}^{(0)}$ is the identity matrix, [1]. If $c_{cov} = 0$, then the mutation distribution is isotropic, and the strategy is called cumulative step size adaptation (CSA) evolution strategy.

3 Micromixer Optimization

3.1 Micromixing Design

The proposed mixer is actively controlled to enhance mixing in a straight channel. Flow in the main channel is manipulated by controlling time-dependent flow from six secondary channels. From these secondary channels, time-dependent cross-flow momentum is imparted on the main channel flow which alters the trajectories of flow-tracing particles. A micrograph of the mixing chip is shown in Figure 2 and the flow configuration is illustrated in Figure 3.

As seen from Figure 3, the main channel is $2h$ in height and $13.5h$ in length where h is a varying length scale. With a distance of $3h$ between secondary channels, they are $h/2$ in width and $5h$ in length. The inlet velocity $U(y)$ of the main channel

$$U(y, x=0) = U_m\left(1 - \left(\frac{y}{h}\right)^2\right), |y| \leq h \tag{14}$$

is parabolic, and the inlet velocity of each secondary channel set i

$$V_i = \hat{V}_i \sin(2\pi f_i t + \phi_i), i = 1, 2, 3 \tag{15}$$

is sinusoidal in time where f_i is the oscillation frequency, ϕ_i the phase shift relative to the first set of secondary channels, and $\hat{V}_i$ the velocity amplitude where $\hat{V}_i$ according to [9].

In order to study the performance of the mixer, we compute the mixing rate of the flow by numerical simulations of the governing Navier-Stokes equations at a Reynolds number $Re = 5$, with $U_m = 1\,\text{m/s}$ and $h = 1\,\text{m}$. In the following, we will skip the SI-units.

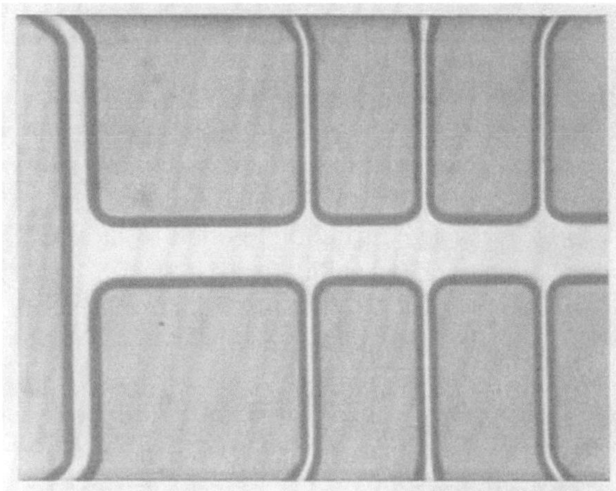

Fig. 2. Micrograph of the mixing chip

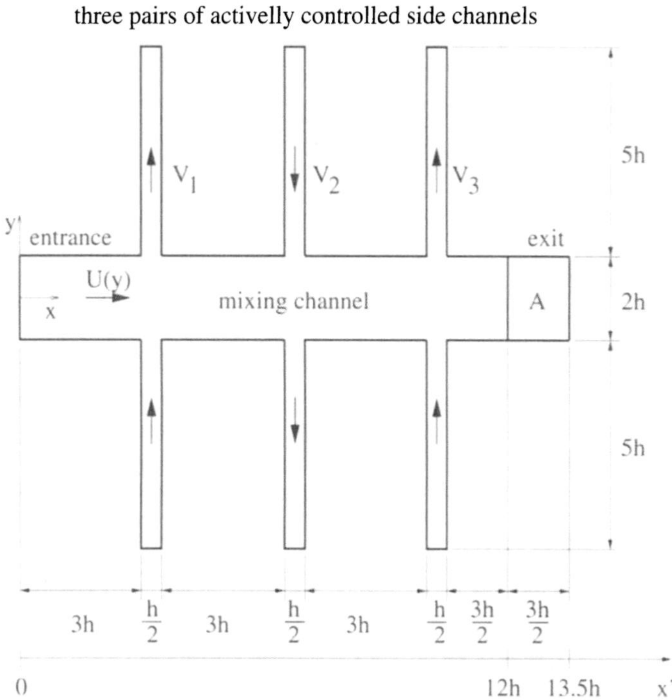

Fig. 3. Schematic of the flow configuration

3.2 A Theoretical Approach to Determine the Optimal Frequency

Dynamical system theory provides the means to compute the optimum frequency for a mixer with only one channel. In our assumption, optimum mixing occurs when a fluid particle at the lower wall of the main channel is moved to the upper wall and back to the lower wall during one actuation period, thus $l = 4h$ for the distance the particle has traveled during one period. v denotes the velocity of the fluid in the side channel. For the simulation, we set $U_m = 1$ and $h = 1$.

In a first approximation, we set v to the maximum velocity amplitude in the side channel, $v_1 = 2U_m$. This yields a frequency of

$$f = \frac{v_1}{l} = \frac{2U_m}{4h} = \frac{1}{2} \tag{16}$$

In a second approximation, we set v to the average velocity amplitude in the side channel, $v_2 = \frac{2}{3}v_1$, as found for two-dimensional parabolic velocity profiles. Note, however, that a parabolic profile is not used for the simulations. This yields a frequency of

$$f = \frac{v_2}{l} = \frac{\frac{2}{3} \cdot 2U_m}{4h} = \frac{1}{3} \tag{17}$$

In a third approximation, we set v to the average velocity in the side channel,

$$v_3 = v_2 \frac{1}{T/2} \int_0^{T/2} \sin\left(\frac{2\pi t}{T}\right) \mathrm{d}t = \frac{2}{\pi} v_2 \tag{18}$$

This yields a frequency of

$$f = \frac{v_3}{l} = \frac{2v_2}{\pi l} = \frac{2 \cdot \frac{2}{3} \cdot 2U_m}{\pi \cdot 4h} = \frac{2}{3\pi} \approx 0.21 \tag{19}$$

For a mixer with 3 side channels, we need to study the ratio between the 3 frequencies. One can learn that there is a difference if frequency ratios are 1, rational, or irrational. We distinguish the parameter vector, $\boldsymbol{x} = (f_1, f_2, f_3)$, with identical frequencies $\boldsymbol{x} = (1/2, 1/2, 1/2)$, "rational" frequencies $\boldsymbol{x} = (1/2, 1/3, 1/4)$, and "irrational" frequencies $\boldsymbol{x} = (1/(2\sqrt{5}), 1/(2\sqrt{2}), 1/2)$ as proposed in [9].

3.3 Optimization Parameters and Objective Function

The aim of the optimization is to obtain the parameter vector which leads to the most pronounced mixing rate in the micromixer. Actuation parameters are the frequency, the amplitude, and the phase shift for each pair of the secondary channels, yielding a total number of 9 actuation parameters. Within this work, we set amplitudes and phases to constant values, namely $\hat{V}_1 = \hat{V}_2 = \hat{V}_3 = 2U_m = 2$, and $\Phi_1 = \Phi_2 = 0$, $\Phi_3 = \pi$ as stated in [9]. As optimization parameters remain the three frequencies f_1, f_2, f_3 which vary within the limits $[0, 1]$, summarized in the parameter vector $\boldsymbol{x} = (f_1, f_2, f_3)$.

The Navier-Stokes and convection-diffusion equations are discretized using a second order finite volume technique and solved on a Cartesian grid using a standard computational fluid dynamics package [14]. The main flow channel contains 20×135 grid points in the span- and streamwise direction, respectively, while the secondary channels are discretized using 5×50 grid points each. The initial conditions involve a uniform concentration of zero and unity for the grid points below and above the symmetry line ($y = 0$), respectively. The mixing rate is estimated from the local variance of the concentration field,

$$m = \overline{c^2} - \bar{c}^2, \tag{20}$$

and $\bar{c}$ denotes the mean (spatial) concentration in the region A (Fig. 3) which extends between $x = 12h$ and $x = 13.5h$. If there is no mixing, $m = 0.25$, and if there is perfect mixing, $m = 0$.

The objective function is the mixing rate m, averaged between time $t = 90$ and time $t = 135$, a time regime in which the flow has reached steady-state. Time $t = 135$ corresponds to 10 flow through times which is defined as

$$T_f = \frac{L}{U_m} = \frac{13.5h}{U_m} = 13.5. \tag{21}$$

The CPU time for one function evaluation, i. e. one flow simulation until time $t = 135$, takes about 3 CPU hours on a Sun Sparc Ultra-2 processor.

The chosen optimization method is an evolution strategy with covariance matrix adaptation with intermediate recombination of the parents, called (μ_I, λ)-CMA-ES. The population consists of $\mu = 2$ parents and $\lambda \in [6, 8, 10]$ children. Using the explicit message passing (MPI), the optimization is performed in parallel on a cluster of $\lambda/2$ Sun workstations.

3.4 Optimization Results

For actuation of one, two, and three frequencies, the initial and the final parameters and mixing rates are given in Table 1.

Besides the three above-mentioned optimizations, we could not afford to run other optimizations with a different direct search technique and compare it with the ES due to the large computational cost. However, we compare the optimization results with dynamical system theory. For 3 actuated frequencies, we obtain the following mixing rates: The parameter vector $\boldsymbol{x} = (1/2, 1/2, 1/2)$ with identical frequencies yields the mixing rate $m = 0.1596$ while our initial "rational" frequency set $\boldsymbol{x} = (1/2, 1/3, 1/4)$ gives us $m = 0.0345$. "Irrational" frequencies as proposed in [9] $\boldsymbol{x} = (1/(2\sqrt{5}), 1/(2\sqrt{2}), 1/2)$ yield $m = 0.0285$ which is the best mixing found theoretically. As one can see, the evolutionary optimization yields a much better mixing rate of $m = 0.0213$ with the frequencies reported in Table 1. This result means a considerable improvement compared with numbers from a well-established theory. Figure 4 shows two snapshots of the flow in the micromixer at time $t = 45$ for the identical and optimal frequencies, respectively.

Number of actuated frequencies		3	2	1
Initial frequencies	f_1	0.25	0.5	0.5
	f_2	0.33	0.5	0
	f_3	0.5	0	0
Initial mixing rate	m	0.0345	0.17	0.18
Best frequencies	f_1	0.14	0.14	0.15
	f_2	0.32	0.32	0
	f_3	0.50	0	0
Best mixing rate	m	0.0213	0.037	0.06
Number of function evaluations		460	420	64

Table 1. Optimization results

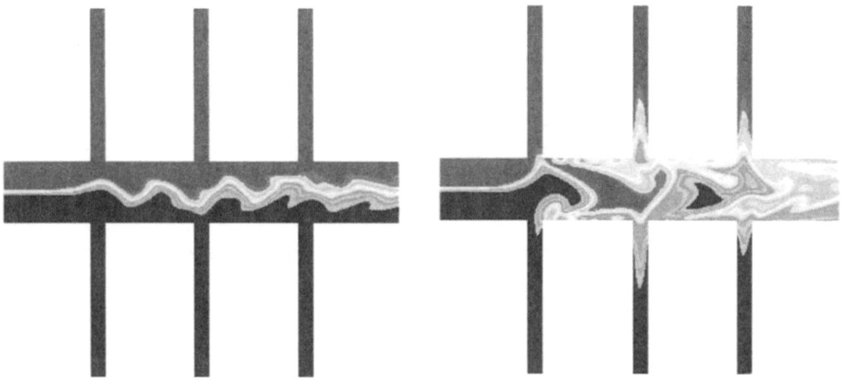

Fig. 4. Flow actuated by the frequencies $\boldsymbol{x} = (1/2, 1/2, 1/2)$ (left) and by the optimal frequencies $\boldsymbol{x} = (0.14, 0.32, 0.50)$ (right)

The difficulty for the optimization of two frequencies is the unsteadiness in the region where $f_1 \sim f_2$. We found that when the frequencies are similar, the dynamical behavior of the systems changes such that shedding of fluid packages occurs and averaging the mixing rate in the last third of the simulation time stops being a characterisic measure of the flow behavior. Actually, we did not learn this behavior before optimization. A direct search method without a capability of handling noise would not have been able to overcome the above difficulty and would have converged to the unsteady region. The recombination (= averaging) feature of the CMA-ES managed to escape this region. For the two-frequency actuation, the mixing rate as a function of the actuation parameters is reconstructed from the computed points during the optimization, see Fig. 5.

The optimal frequency of the first side channel appears to be independent of the number of actuated frequencies. In the three optimizations, the first parameter attains values of about $f_1 = 0.14 - 0.15$. The same phenomenon occurs

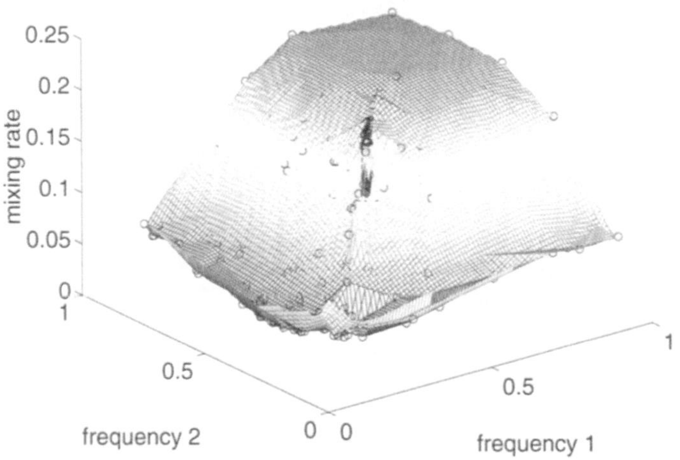

Fig. 5. Surface of the mixing rate as a function of the two actuation frequencies

for the second frequency for the optimization of two and three frequencies where $f_2 = 0.32$.

4 Jet Mixing

The control of jet flows has applications in various fields such as combustion, aerodynamic noise, and jet propulsion. The mixing rate of a jet can be altered by applying a suitable excitation at the jet orifice. We examine a prototypical configuration for jet mixing using a vortex filament method to simulate temporally varying jets. The vortex filaments are subject to azimuthal perturbations. Our objective is to maximize the length of all vortex filaments, i.e. to find the perturbation parameters (amplitudes and phase shifts) that cause the greatest perturbation. Optimization parameters are the phase shifts β_j, $(j = 2, ..., 5)$ and the amplitudes ϵ_j, $(j = 1, ..., 5)$ of the azimuthal perturbation. The phase shift β_1 is set to zero. The amplitude $\epsilon_j = 0.2$ given by [7] based on linear stability arguments corresponds to a perturbation of 4%. For the optimization, we implement a (1+1)-ES running for 300 generations with initial values given by [8]. With the phases constrained between 0 and 1, two optimizations are performed. The first optimization, for which the amplitudes are constrained between 0 and 1, yields an increase of 42 % for the filament length. In the second optimization with lower and upper amplitudes of 0 and 0.25, respectively, the filament length is increased by only 1 %. This indicates that the upper limit of the amplitudes has a high impact on the optimization result and therefore should be chosen carefully. Figures 6 and 7 show the filament structure of the upper half jet from an axial view. The structure resulting from the first optimization seen in Figure 7 has longer filaments than the one from the initial state seen in Figure 6.

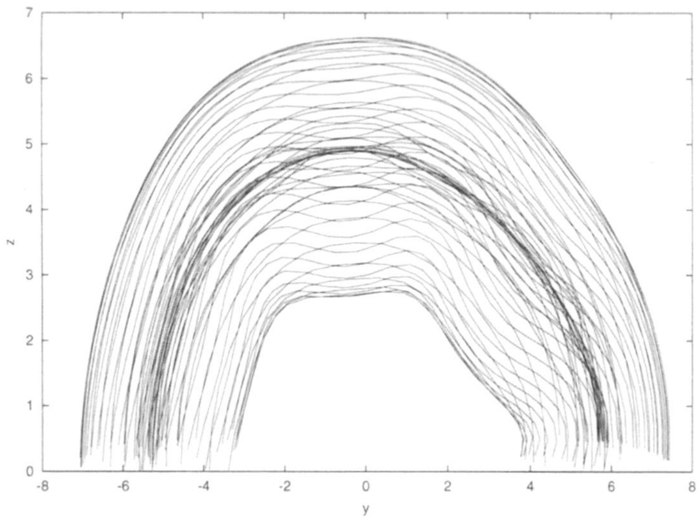

Fig. 6. Filament structure of the upper half of the jet, excitation with initial values

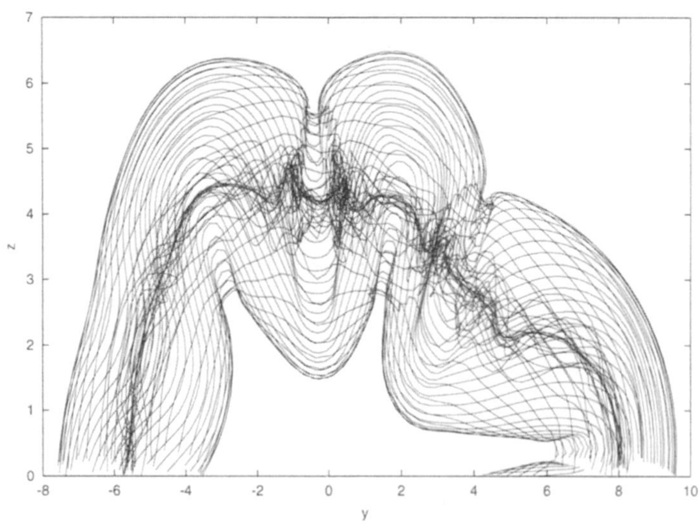

Fig. 7. Filament structure of the upper half of the jet, excitation with optimum values (first optimization)

5 Turbine Blade Film Cooling

An evolutionary algorithm is implemented in a realistic automated design cycle of turbine blade film cooling. This design cycle involves the use of empirical formulas for the flow calculation and the use of a commercial software package for the simulation of the heat transfer problem. The overall process consists of an engineering multiobjective optimization problem with constraints. In this work we consider the solution of this problem in the context of an automated optimization cycle using evolution strategies. In film cooling, the coolant leaves the blade through rows of holes, forming a protective film on the outer surface of the blade, separating the hot gas from the metal. The design of the cooling configuration aims at both minimizing the coolant supply and achieving optimum cooling, while obeying certain engineering constraints. For the simulation of the cooling configuration, the mass flow of the coolant and the surface temperature of the blade are computed, based on given external aerodynamic and thermal conditions for a two-dimensional blade geometry.

5.1 Description of a Blade Cooling Configuration

A typical vane cooling configuration is shown in Figure 8. It consists of three components: (i) an insert for impingement cooling; (ii) the trailing edge geometry for convection cooling; (iii) coolant air leaving the inside of the blade through holes and keeping attached as a film, thereby cooling the outer surface. The coolant flow is described by three different cooling mechanisms based on empirical correlations which compute the heat transfer coefficient and the temperature in the boundary layer [10], [11], [12]. The geometry as well as the external aerodynamic and thermal conditions are considered two-dimensional and the shape of the blade is fixed.

5.2 Optimization Algorithm and Parameters

Two optimization techniques [1] are chosen: The first method is an evolution strategy with covariance matrix adaptation (CMA-ES) while the second method is an evolution strategy with cumulative step size adaptation and isotropic mutation distribution (CSA-ES), see Section 2. Both methods work with a population size of $\mu = 3$ and $\lambda = 12$ and with intermediate recombination.

The optimization parameters are the number of film cooling rows (R_f), the position of film cooling rows ($s_{pos}(1 \text{ to } R_f)$), the injection angles of the holes, measured from the hole axis to the surface ($\alpha(1 \text{ to } R_f)$), the number of holes per row ($N_r(1 \text{ to } R_f)$), the number of impingement holes (N_j), and the loss parameter that is a function of the pressure loss in the trailing edge (A_{loss}).

The only integer parameter in the blade cooling optimization is the number of rows. Note that optimizing integer variables with an evolution strategy only makes sense if the number of possibilities to adjust the variables is large enough. Due to manufacturing requirements, the number of rows should not exceed about

10. With respect to this small number, this parameter is not included in the optimization process but rather changed manually for different optimization runs. First tries with $R_f = 4$ showed that the constraints for maximum and minimum temperature cannot be met. To homogenize the temperature distribution, we increased the number of rows to $R_f = 7$. This number was found to be sufficient for achieving the optimization goals and is used in this study.

All the other parameters are real valued. For each film row $j = 1, \ldots, R_f$, the position s_{pos}, the angle α, and the number of holes per row N_r can be chosen. The other two parameters are the number of impingement holes N_j and the loss parameter A_{loss}. There are $3 \cdot R_f + 2$ parameters altogether; in our case 23 parameters.

The limits of the optimization parameters are set to $0 <= s_{pos} <= 1$, $30^o <= \alpha <= 90^o$, $10 <= N_r <= 100$, $500 <= N_j <= 1000$, $1 <= A_{loss} <= 3$. The parameter range of s_{pos} covers the whole surface starting from $s_{pos} = 0$ at the trailing edge, continuing on the suction side to the stagnation point and on the pressure side back to the trailing edge. For the angle α, a special constraint is imposed by engineering considerations. In the leading edge of the blade, defined by values of s_{pos} between 0.49 and 0.53, holes with angles other than 90° are difficult to manufacture due to the high bend. Therefore, holes in the leading edge region should always be set to 90°. This is a constraint that can be neglected for the following reason: In all of the optimization runs, the objective function values are not sensitive to changes of the hole angles. This indicates that the influence of the injection angle is not sufficiently incorporated in the model. Instead of the constraint, we compare the outputs of the converged solution with the outputs computed with angles of 90° for rows of holes in the leading edge. For all of our optimization runs, the difference between both outputs was below 1%.

5.3 The Objective Function

The main goal of the optimization is to minimize the mass flow of the coolant, $\dot{m}_c$. At the same time, constraints have to be met. These constraints are expressed by one equality constraint $\varepsilon_{mean} = 0.5$ where ε_{mean} is the normalized mean surface temperature, and also by the two inequality constraints $\varepsilon_{max} <= 0.6$ and $\varepsilon_{min} >= 0.4$ where ε_{max} is a function of the minimum temperature at the outer surface, and where ε_{min} is a function of the maximum temperature. These constraints are incorporated in the objective function f using a distance method with Euclidian metric, $f = \left(\sum_{i=1}^{4} f_i^2\right)^{0.5}$ where $f_1 = \dot{m}_c$, $f_2 = 2|\varepsilon_{mean} - 0.5|$, and

$$f_3 = \begin{cases} 10000 \cdot (\varepsilon_{max} - 0.5)^4 & \text{if } \varepsilon_{max} >= 0.5, \\ 0 & \text{if } \varepsilon_{max} < 0.5. \end{cases}$$

$$f_4 = \begin{cases} 10000 \cdot (\varepsilon_{min} - 0.5)^4 & \text{if } \varepsilon_{min} <= 0.5, \\ 0 & \text{if } \varepsilon_{min} > 0.5. \end{cases}$$

The coefficients and the exponent can be considered weights and are chosen such that the function f_2 is in the same order of magnitude as f_3 and f_4 for the initial parameters.

The computation of the optimization objectives entails two steps: The mass flow of the coolant as well as the air temperatures and heat transfer coefficients in the boundary layer as a function of the geometry have to be computed first. In a second step, the steady heat equation is solved for the unknown temperature distribution using second order finite volume techniques subject to the aforementioned boundary conditions. The blade geometry is discretized using 660 x 20 mesh points and the solution is obtained applying a commercial package, STAR-CD [14]. One function evaluation is done in about 15 CPU seconds on a Sun Ultra 60/Creator 3D, 300 MHz.

5.4 Results

With the initial parameter configuration $s_{pos} = (0.2, 0.3, 0.4, 0.5, 0.6, 0.7, 0.8)$, $\alpha = 50^o$, $N_r = 50$, $N_j = 800$, $A_{loss} = 2$, we obtain the objectives: $\dot{m}_c = 0.367$, $\varepsilon_{mean} = 0.522$, $\varepsilon_{max} = 0.616$, $\varepsilon_{min} = 0.360$, $f = 4.22$.

A reduction of the initial objective function value by 60 % is achieved with the CMA-ES already after about 200 generations (= 2400 function evaluations). After about 1800 generations (= 21600 function evaluations), convergence is reached.

Optimum parameters $s_{pos} = (0.490, 0.500, 0.502, 0.508, 0.510, 0.706, 0.786)$, $\alpha = (65^o, 90^o, 89^o, 30^o, 30^o, 73^o, 30^o)$, $N_r = (100, 10, 13, 95, 100, 35, 41)$, $N_j = 501$, $A_{loss} = 2.97$ yield the outputs: $\dot{m}_c = 0.257$, $\varepsilon_{mean} = 0.502$, $\varepsilon_{max} = 0.595$, $\varepsilon_{min} = 0.400$, $f = 0.229$. Here, the reduction of the cost function equals 95 %.

The results of this study are summarized as follows:

- The original row configuration (rows equidistantly placed) was drastically modified. The final configuration suggests to place five rows in the leading edge and two rows on the pressure side, as seen in Figure 5.4.
  The number of holes per row attains small values for the rows on the pressure side whereas two rows in the leading edge are equal to the upper limit, $N_r = 100$. This result makes sense when considering that the hot gas is approaching the leading edge of the turbine, thus requiring efficient cooling in this area.
- The mass flow can be reduced significantly and the mean temperature distribution comes closer to the desired one. At the same time, the maximum and minimum temperatures are not exceeded, as seen from the constraints $\varepsilon_{max} < 0.6$ and $\varepsilon_{min} > 0.4$ that are met.
- As one can see from Figure 9 where the improvement of the multi-objective function is shown, convergence is reached after about 1800 generations (= 21600 function evaluations in a population with 12 offspring) for the evolution strategy with covariance matrix adaptation. Convergence is defined to be reached when the difference of the cost function for the best individual is less than 10^{-3} (in absolute values) over 10 generations. On the other hand,

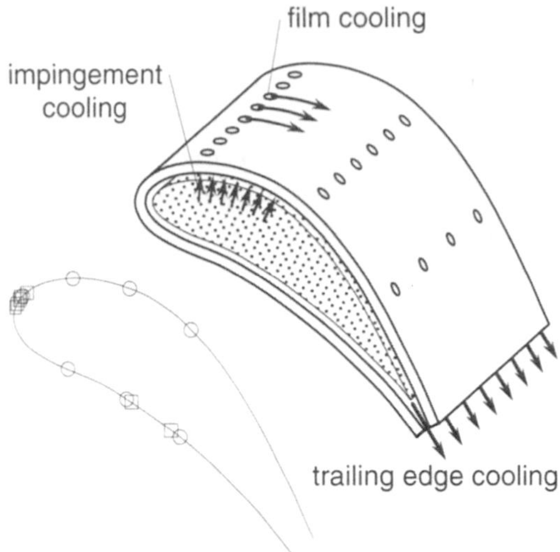

Fig. 8. The three cooling mechanisms and the initial (circles) and final (squares) row positions from the optimization with the CMA-ES.

the CSA-ES converges slower and cannot reach the same good value as the CMA-ES within the same number of iterations. The difference in convergence properties of these two types of evolution strategies agrees well with results from the optimization of several test functions [1].

- Although evolution strategies are often considered good candidates for global optimization, they may also converge to local optima. Therefore, to find a good optimum it is recommended to perform several optimization runs with different start configurations. Since the computations in our case are relatively expensive, we optimized with only two different initial settings of the strategy parameters. The converged function values in both optimization runs did not significantly $(1 - 2\%)$ differ from each other. The final model parameters were not the same but they represented a very similar configuration with five rows close to each other in the leading edge region and two rows of holes on the pressure side.

Interpreting the results, one should keep in mind that we use an empirical model that may not take into full account the properties of the flow and heat problem. Moreover, the results could be improved by replacing the combination of the several objectives into one single objective function by searching the Pareto-front of the multi-objective problem.

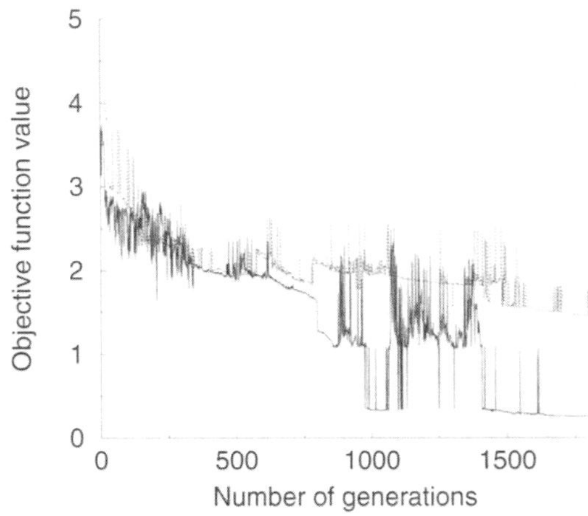

Fig. 9. Convergence of the objective function f (— CMA-ES, ··· CSA-ES)

6 Summary and Conclusions

Evolution strategies were applied to three engineering problems, (i) flow in micromixers, (ii) jet flow, and (iii) turbine blade cooling. In case of the micromixing example, the evolutionary optimization results outperformed results from a well-established theory. Moreover, the chosen evolution strategy was able to handle noise during the optimization process and did not get trapped by one of the suboptimal solutions. For jet flow mixing, even a simple evolution strategy, the (1+1)-ES can yield a great improvement over solutions found by engineering intuition. Turbine blade cooling is an example for the use of evolution strategies to optimize efficiently a complex engineering problem in an automated fashion. In summary, the results show that these optimization methods are highly suitable for optimization in applications that are characterized by noise, multimodality, and no availability of gradient information.

References

1. Hansen, N., Ostermeier, A., "Convergence Properties of Evolution Strategies with the Derandomized Covariance Matrix Adaptation: The $(\mu/\mu_I, \lambda)$-CMA-ES," *Proceedings of the 5th European Congress on Intelligent Techniques and Soft Computing (EUFIT'97)*, pp. 650-654, 1997.
2. Rechenberg, I., "Evolutionsstrategie: Optimierung technischer System nach Prinzipien der biologischen Evolution," Fromann-Holzboog, Stuttgart, 1973.
3. Schwefel, H.-P., "Evolution and Optimum Seeking," John Wiley and Sons, New York, 1995.
4. Hansen, N., Ostermeier, A., "Adapting Arbitrary Normal Mutation Distributions in Evolution Strategies: The Covariance Matrix Adaptation," *Proceedings of the IEEE*

International Conference on Evolutionary Computation (ICEC'96), pp. 312-317, 1996.
5. Ostermeier, A., "Schrittweitenadaptation in der Evolutionsstrategie mit einem entstochastisierten Ansatz." Dissertation, Fachbereich 6 der Technische Universität Berlin, 1997.
6. Hansen, N., "Verallgemeinerte individuelle Schrittweitenregelung in der Evolutionsstrategie: Eine Untersuchung zur entstochastisierten, koordinatensystemunabhängigen Adaptation der Mutationsverteilung," Mensch & Buch Verlag, Berlin, 1998.
7. J.E. Martin and E. Meiburg, "Numerical investigation of three-dimensionally evolving jets under helical perturbations," *Journal of Fluid Mechanics,* Vol. 243, pp. 457-487, 1992.
8. J.E. Martin, *Personal communication,* 2001.
9. Volpert, M., Mezic, I., Meinhart, C.D., Dahleh, M., "Modeling and Analysis of Mixing in an Actively Controlled Micromixer," *Unpublished report, University of Santa Barbara, CA*, 2000.
10. McGreehan, W.F., Schotsch, M.J. "Flow characteristics of long orifices with rotation and corner radiusing," ASME paper 87-GT-162, 1987.
11. Goldstein, R.J., "Film cooling," *Advances in Heat Transfer*, Vol. 7, pp. 321-379, 1971.
12. Seller, J.P., "Gaseous film cooling with multiple ejection stations," *AIAA Journal*, Vol. 1, No. 9, pp. 2154-2156, 1963.
13. Bestle, D., "Analyse und Optimierung von Mehrkörpersystemen," Springer, Berlin, 1993.
14. *Star-CD User Manual,* Computational Dynamics Ltd., London, 1997.

Multi-Objective Evolutionary Algorithm for Optimization of Combustion Processes

Dirk Büche[1], Peter Stoll[2], and Petros Koumoutsakos[13]

[1] Institute of Computational Science, Swiss Federal Institute of Technology (ETH), CH - 8092 Zürich, Switzerland
E-mail: *{bueche, petros}@inf.ethz.ch*
http://www.icos.ethz.ch

[2] Alstom (Switzerland) AG, Segelhof, CH - 5405 Dättwil, Switzerland
E-mail *peter.stoll@power.alstom.com*
http://www.alstom.ch

[3] NASA Ames Research Center, Moffett Field, CA, USA
http://www.arc.nasa.gov

Abstract. This work introduces a multi-objective evolutionary algorithm capable of handling noisy problems like experimental setups with a particular emphasis on robustness against unexpected measurements (outliers). The algorithm is based on the Strength Pareto Evolutionary Algorithm (SPEA) of Zitzler and Thiele and includes the new concepts of domination dependent lifetime, re-evaluation of solutions and modifications in the update of the archive. Several tests on prototypical functions underline the improvements in convergence speed and robustness of the extended algorithm. The proposed algorithm is implemented to the Pareto optimization of the combustion process of a stationary gas turbine in an industrial setup. The free parameters of the optimization are the fuel injection rates through transverse jets. The Pareto front is constructed for the objectives of minimization of NO_x emissions and reduction of the pressure fluctuations (pulsation) of the flame. Both objectives are conflicting affecting the environment and the lifetime of the turbine, respectively. The optimization leads a Pareto front corresponding to reduced emissions and pulsation of the burner. The physical implications of the solutions are discussed and the algorithm is evaluated.

1 Introduction

A central component in the design of a gas turbine is the design of the burners in the combustion chamber. The burners mix air and fuel and combust them continuously. This is different from Diesel engines, which combust in a cyclic manner. For complex problems such as the combustion process, numerical simulations are not widely used as a predictive tool due to the complexity of the physical phenomena under investigation. Thus an experimental setup is used for the optimization of the combustion process.
The design of a burner addresses two main objectives: First, the burner should

mix air and fuel uniformly to achieve lean premixed combustion, since the presence of less mixed and/or stoichiometric combustion results in increased emissions and a non-homogeneous temperature distribution may damage the turbine blade. Second, the burner should produce a stable combustion flame, avoiding undesired pulsations. Pulsations are thermoacoustic waves, which occur in particular for lean combustion when operating under part load condition. They reduce the lifetime of the turbine by fatigue and by destroying the film cooling along the blades surface. These two objectives are conflicting, thus motivating the requirement for a variety of designs as manifested on a Pareto front. The Pareto front is defined as the set of solutions, which represent the best compromises between the objectives, i.e. no other solution can be found that is superior in all objectives.

A classical and still widely employed approach to handle multiple objectives is the aggregation [1] of all objectives into a single, a priori defined figure of merit. Objectives are usually aggregated by a weighted-sum or a constraint approach. This weighting behavior implies prior knowledge about the problem and is dependent on the a priori unknown shape of the Pareto front. While point-to-point search methods like gradient descent methods require an aggregation and converge to one Pareto solution at a time, evolutionary algorithms can converge to the Pareto-set in a single optimization run.

Evolutionary Algorithms (EAs) are biologically inspired optimization algorithms, incorporating operators such as mutation, recombination and fitness based selection of parameters. EAs use a set of solutions (population), to converge to the optimal design(s). The population-based search allows to converge towards the Pareto front in a single optimization run. Therefore much effort has been spent over the past twenty years on the development and application of evolutionary algorithms for Pareto optimization. Promising methods have been proposed and compared by several researchers [2] [3] [4]. An exhaustive list of references can be found on the web page of Coello [5]. Although the number of applications in the field of multi-objective (Pareto) optimization is increasing, problems with noisy objective functions are rarely considered, even though noise is present in almost every real-world application. As evolutionary algorithms do not require gradient information, they are already inherently robust to small amounts of noise, a feature which is sufficient for many problems. In the considered experiments however, large-amplitude noise is induced from various sources, such as unsteady operating conditions, limited measurement precision, and time averaging in restricted sampling time. In addition, measurements may fail, leading to erroneous outlier, characterized by nonphysical objective values. Standard multi-objective evolutionary algorithms cannot handle these difficulties, and there is a need to extend their basic components to overcome these difficulties.

Two recent publications [6] [7] adapt the Pareto ranking scheme [8] to noisy solutions by defining probabilities of dominance between them. Both methods assume either a uniform or normal distribution of the noise and can benefit from a priori knowledge of its magnitude. In this paper we introduce three new principles in order to improve robustness against noise.

This paper is organized as follows: First, the principles of multi-objective optimization are described and a reference evolutionary algorithm is presented. Then we present extensions for the method in order to handle noise. These extensions are analyzed on noisy test functions, which reflect the noise characteristics of the intended application. Finally the application of the method to the optimization of a gas turbine burner is presented, showing the capabilities of the extended algorithm. The optimization leads a Pareto front corresponding to reduced emissions and pulsation of the burner. The physical implications of the solutions are discussed and the algorithm is evaluated.

2 Multi-Objective Evolutionary Algorithms for Noisy Applications

2.1 Definition of Multi-Objective Optimization

A multi-objective optimization problem can be described by an objective vector f and a corresponding set of design variables x. Without loss of generality we can consider the minimization of f. Formally:

$$\begin{aligned} &\min f(x) = (f_1(x), f_2(x), ..., f_m(x)) \in F, \\ &\text{where } x = (x_1, x_2, ..., x_n) \in X, \end{aligned} \tag{1}$$

where $X \in \mathcal{R}^n$ is the n-dimensional design space, $F \in \mathcal{R}^m$ is the m-dimensional objective space. A partial ordering can be applied to solutions in the objective space F by the dominance criterion. A solution a in X is said to dominate a solution b in X ($a \succ b$), if it is superior or equal in all objectives and at least superior in one objective. This is expressed as:

$$\begin{aligned} a \succ b, \text{ if } \ \forall\, i \in \{1, 2, ..., m\}: &\ \ f_i(a) \leq f_i(b) \ \wedge \\ \exists\, j \in \{1, 2, ..., m\}: &\ \ f_j(a) < f_j(b) \end{aligned} \tag{2}$$

The solution a is said to be indifferent to a solution c, if neither solution is dominating the other one (Fig. 1). When no a priori preference is defined among the objectives, dominance is the only way to determine, if one solution performs better than the other [9]. The complete set of nondominated solutions is referred to the Pareto set of solutions, after the work of the engineer and economist Vilfredo Pareto [10] and represents the best solutions to the problem.
In other words, starting from a Pareto solution, one objective can only be improved at the expense of at least one other objective. From the Pareto definition, two targets have to be considered by the formulation of an evolutionary algorithm for Pareto optimization. On the one hand, the algorithm must be able to converge sufficiently fast towards the Pareto front, while on the other hand, it must preserve diversity among its population in order to be able to spread over the whole Pareto front. A common difficulty is the focusing of the population on a certain part of the Pareto front, which is known as genetic drift [11]. In single objective optimization the latter issue is unimportant, since convergence to a single (global) optimum is desired.

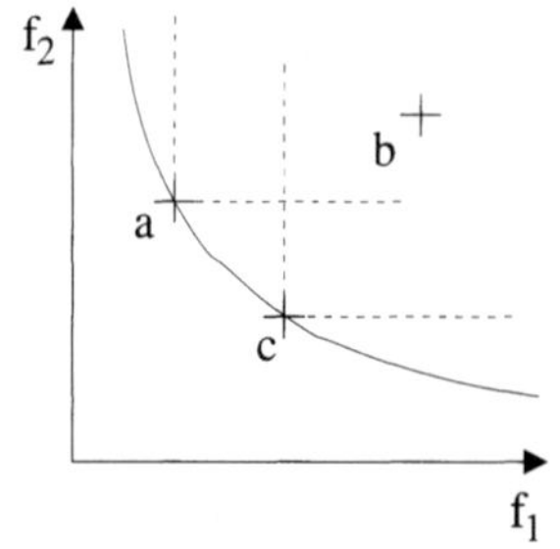

Fig. 1. Illustration of the dominance principle. For this two-objective minimization problem, the solution a is dominating b, since both objective values of a are smaller than the ones of b. a is indifferent to c, since each solution is superior in one objective. In addition, the Pareto front [curved line] is given.

2.2 Multi-objective Evolutionary Algorithm

Evolutionary Algorithms are optimization algorithms, incorporating concepts such as fitness based selection, recombination and mutation. EAs start with a set of λ randomly generated solutions, which are referred to as the population P_λ. For a multi-objective problem, a *selection operator* selects in average the less dominated solutions from P_λ and places them in a parent population P_μ of size μ. A selection operator is described in Section 2.3.
The *recombination operator* chooses randomly always two individuals $a, b \in X$ from P_μ and recombines them into an individual $c \in X$ by a linear interpolation:

$$c_i = a_i \; p + b_i \; (1 - p), \quad \text{with } p \in \mathrm{U}(0,1), \tag{3}$$

and p is equal for all design variables. For the *mutation operator*, a simple method with constant normal distributed mutation is implemented. Each design variable c_i is mutated by:

$$c_i = \begin{cases} c_i + N(0, \sigma_M^2), & \text{if } p_i < p_M, \;\; p_i \in \mathrm{U}(0,1) \\ c_i, & \text{otherwise,} \end{cases} \tag{4}$$

where $N(0, \sigma_M^2)$ is a normally distributed random number with zero mean and standard deviation σ_M. p_M is the probability of mutating a design variable and p_i is evaluated for every design variable c_i separately. This normally distributed mutation reflects the natural principle that small mutations occur more often than large ones.

2.3 Strength Pareto Evolutionary Algorithm

The Strength Pareto Evolutionary Algorithm (SPEA) of Zitzler and Thiele [2] is a well-established Pareto-optimization algorithm. The advantages and drawbacks of the method have been extensively discussed [12] [2]. SPEA describes a selection operator, while the recombination and mutation operator can be used from a single objective algorithm.

The algorithm entails a fitness assignment and selection mechanism. SPEA uses the nondominated solutions for the fitness assignment. First, the fitness of each nondominated solution is computed as the fraction of the population, which it dominates. Then the fitness of a dominated individual is computed as one plus the fitness of each nondominated solution by which it is dominated. This fitness assignment guarantees that the fitness of nondominated solutions is always lower than the fitness of the dominated.

Elitism is a technique of preserving always the best solutions obtained so far. In multi-objective optimization, elitism is performed by storing the nondominated solutions in an archive P_a [13]. In the selection process individuals of the current population P_λ and of the archive P_a are competing in a binary tournament where contrary to the standard tournament selection the solution with the lower fitness wins. In order to preserve diversity in the archive and to keep its size limited, a clustering algorithm is used. Clustering removes solutions in areas of high density as measured in the objective space.

The studies of Zitzler and Thiele [2] have illustrated that elitism improves the performance of multi-objective evolutionary algorithms on noise-free test problems. Some researchers state elitism as a necessity for multi-objective optimization [13], since information may be lost by the stochastic selection operator. However, this advantage is debatable for noisy objective functions.

2.4 Definition of Noise in Applications

In experiments and industrial configurations we can always detect different results for repeated measurements of the same operating point. The differences are attributed to noise and unobserved factors in the setup.

We define the noise, which is present in all measured experiments, as **experimental noise**. It is often modeled by a normal distribution with defined mean and standard deviation.

In addition, during an automated optimization cycle, an experimental measurement may fail completely, producing **outliers**, i.e. arbitrary nonphysical results. This occurs very rarely, but may have large impact on the automated process optimization if not recognized by a supervisor or captured by some penalty function. Outliers cannot be described by a statistical model with given mean and deviation, but are best modeled by a probability of occurrence.

2.5 Improvements for Noisy Objectives

The presence of noise affects the fitness assigned to an individual. This may cause inferior solutions to occasionally win in the selection process. Multi-objective evolutionary algorithms, which implement elitism, would then select these solutions into the archive, thus misleading the entire optimization run by participating in the selection process. More important, these solutions may dominate other

solutions in the archive and in the worst case all other solutions in the archive are then removed.
On the other hand, noise reduces the selection pressure [14], suggesting that elitism, which is increasing the selection pressure, should be used to compensate. To successfully use elitism in a noisy environment, modifications are needed to ensure fast convergence while maintaining a robust algorithm.
We propose three modifications for an extended multi-objective algorithm for noisy environments:

1. *Domination dependent lifetime:* In contrast to elitism, which may preserve elite (nondominated) solutions for an infinite time, a maximal lifetime κ is assigned to each individual. For evolution strategies, algorithms with implemented lifetime κ are referred as (μ, κ, λ) algorithms [15]. In this work this concept is extended to multiple objectives so that lifetime is variable and related to the dominance of a solution. The lifetime is shortened, if the solution dominates a major part of the archive. This limits the impact of a solution and safeguards against outliers.
2. *Re-evaluation of solutions:* Solutions with expired lifetime are usually deleted. We propose to re-evaluate the nondominated among these solutions and add them to the population. This enables good solutions to stay in the evolutionary process, but re-evaluated outliers will not re-enter the archive.
3. *Extended update of the archive:* The SPEA algorithm updates the archive always with the current population. We extend the update to *all* solutions with non-expired lifetime. This hinders loss of information.

A dominance dependent lifetime, which is measured in generations, is assigned to each individual. For dominating less than a user defined fraction c_1 of the archive P_a, the maximal lifetime $\kappa = \kappa_{max}$ is assigned to the individual. For dominating more than a fraction c_2 of P_a, the minimal lifetime of $\kappa = 1$ is assigned. In-between these two fractions, the lifetime is interpolated in discrete steps of one generation.
With these features the extended SPEA uses the advantage of an archive as convergence accelerator, but it reduces the risk induced by outliers. The extended algorithm is given by:

Algorithm 1

1. **begin**
2. Generate an initial population P_λ of random individuals and an empty archive P_a.
3. Define a maximal lifetime κ_{max} for individuals (in generations).
4. Evaluate the objectives of the individuals in P_λ.
5. **while** termination criterion is not fulfilled **do**
6.   Assign lifetime to the individuals in P_λ dependent on dominance.
7.   Update P_a: Find the nondominated within all solutions with non-expired lifetime.
   Reduce their number by clustering, if necessary.

8. Assign fitness to the individuals in P_a and P_λ.
9. Select the parent population P_μ from P_λ and P_a.
10. Recombine P_μ to a new population P_λ.
11. Mutate the individuals in P_λ.
12. Select the solutions from P_a with expiring lifetime and add a copy for re-evaluation to the population P_λ
13. Evaluate the objectives of the individuals in P_λ.
14. **end while**
15. **end**

3 Performance Comparison

3.1 Generation of Test Functions

From Deb [16], a two-objective minimization problem for an arbitrary number of design variables $x_{1\ldots n}$ is chosen and implemented as the first noise free **test function 1**:

$$f_1 = x_1,$$
$$f_2 = \frac{1}{x_1}\left(1 + \sum_{i=2}^{n} x_i^2\right), \tag{5}$$

with $x_1 \in [0.5;2]$, $x_{2\ldots n} \in [-2.0;2]$. For the *experimental noise*, we assume a normal distribution with zero mean and standard deviation σ_N. A noisy test function is generated by adding this noise to test function 1, leading to **test function 2**:

$$f_1 = x_1 + N(0,\sigma_N^2),$$
$$f_2 = \frac{1}{x_1}\left(1 + \sum_{i=2}^{n} x_i^2\right) + N(0,\sigma_N^2). \tag{6}$$

A second type of noise was introduced in Section 2.4 as the random occurrence of *outliers*. For modeling the occurrence of *outliers* in a test function, we define a probability p_o for their occurrence. Since we consider a minimization problem, reducing the objective value has a stronger influence on the optimization process. Therefore we divide the objective value by a factor of 10, if an outlier occurs. The large factor is chosen in order to produce a significant change in the objective value. In mathematical form, we define **test function 3** as:

$$f_1 = \begin{cases} \frac{1}{10}x_1, & \text{if } p < p_o,\ p \in \mathrm{U}(0,1) \\ x_1, & \text{otherwise} \end{cases}$$
$$f_2 = \begin{cases} \frac{1}{10\,x_1}\left(1 + \sum_{i=2}^{n} x_i^2\right), & \text{if } p < p_o,\ p \in \mathrm{U}(0,1) \\ \frac{1}{x_1}\left(1 + \sum_{i=2}^{n} x_i^2\right), & \text{otherwise.} \end{cases} \tag{7}$$

The probability p_o of an outliner is small.

3.2 Performance Measures

In order to compare different optimization methods on the three test functions, performance measures are needed. In multi-objective optimization, the definition of the quality of an optimization depends on the convergence speed of the optimization as well as on the uniformity of the approximation of the Pareto front. Since the test functions contain noise in their objective functions, measuring the performance in objective space is difficult. Instead the performance of the optimizer is measured in design space. The performance measure is set as the distance in design space of evaluated solutions to the actual (analytically defined) Pareto front.
To evaluate this distance, 10 points are distributed uniformly in design space along the analytical Pareto front. To each of these points the closest evaluated solution of an optimization is searched and the distance is computed. The mean of the resulting 10 distances is taken as performance measure. For the three test functions the exact Pareto front in the genotypic space is given by $x_1 \in [0.5; 2]$ and $x_{2\dots n} = 0$. The 10 uniformly distributed points are:

$$x_1(k) = \frac{1}{2} + \frac{1}{6}k, \quad x_{2,\dots,n}(k) = 0, \tag{8}$$

with $k = 0, \dots, 9$.

3.3 Comparison of Standard and Extended SPEA Algorithm

In the following, the performance of the *extended SPEA* from Section 2.5 is numerically analyzed on the test functions. The extended algorithm is compared with the *standard SPEA* of Zitzler and Thiele [2] and with a *non-elitistic SPEA*. The non-elitistic SPEA is obtained by setting the lifetime of all individuals of standard SPEA to one. It is used in order to analyze the influence of elitism.
For all optimization algorithms, a parent and child population of $\mu = \lambda = 60$ is used, with an archive size of 15. The recombination and mutation operators of Section 2.2 are used with a mutation probability of $p_M = 0.2$ per design variable and a relative standard deviation of $\sigma_M = 0.1$. For the extended algorithm, the fractions c_1 and c_2 are set to 0.1 and 0.3, respectively and a maximal lifetime $\kappa_{max} = 4$ is used. Since evolutionary algorithms are semi-stochastic algorithms, the result of 100 optimization runs is averaged for each test function. The number of design variables is set to 7. This number is equal to the number of design variables of the burner optimization problem, which is addressed in the next section. For test function 2 and 3, the standard deviation is set to $\sigma_N = 0.8$ and an error probability of $p_o = 1\%$ per objective is defined, respectively.
The optimization results are given in Fig. 2. The performance measure P is plotted in a logarithmic scale over the number of evaluated solutions N. The measure P reflects distance of the optimization to uniformly distributed points along the analytic Pareto front. For all 3 test functions, the extended SPEA shows comparable or improved performance compared to the two other implementations. The differences for the noise free test function 1 are small. While

for test function 2, which includes normally distributed noise, the non-elitistic SPEA converges worst, the elitistic SPEA converges worst for test function 3, which includes outliers.

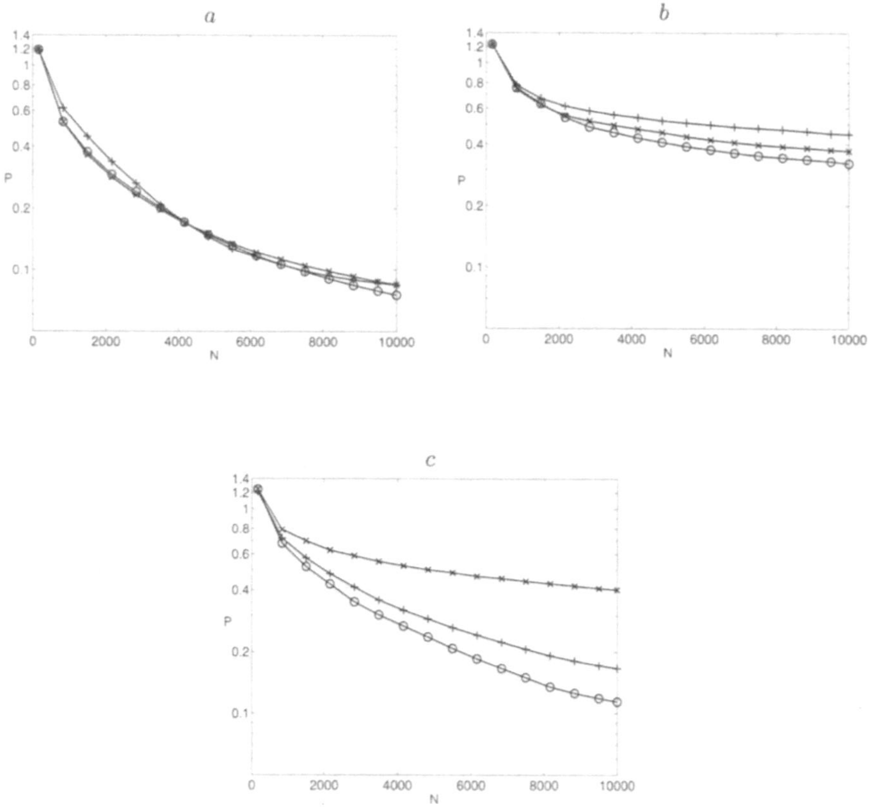

Fig. 2. Convergence of the extended SPEA [circular symbol] on the test function 1, 2 and 3 compared with the standard SPEA [cross symbol] and a non-elitistic SPEA [plus symbol]. The test function are either noise free (a), contain normally distributed noise (b) or outliers with a small probability (c).

4 Optimization of a Burner in a Gas Turbine Combustion Test-Rig

4.1 Atmospheric combustor test-rig

We consider the optimization of a single burner in an atmospheric test-rig as illustrated in Fig. 3. Preheated air enters the test-rig from the plenum chamber and is premixed with fuel in the low-emission burner by swirl. The burner

stabilizes the combustion flame in a predefined combustion area by a controlled vortex breakdown. The burned air leaves the test-rig through an exhaust. The fuel is natural gas or oil and is injected transversely though injection holes, which are uniformly distributed along the burner.
The fuel mass flows through the injection holes are the design variables of the setup. The mass flow distribution is controlled by 8 continuous valves. Each valve controls the mass flow through a set of adjacent injection holes along the burner axis. In order to keep the operating conditions constant, the total fuel mass flow is fixed, reducing the number of free design variables for the optimization from 8 to 7. The NO_x emissions and the pulsation of the burner are the two objectives to be minimized.

4.2 Optimization results

An optimization run is performed using the extended SPEA algorithm and evaluating a total of 326 different burner settings. All solutions are plotted in Fig. 4. The initial solution is marked in the figure and represents a setting with equal mass flow through all valves. The solutions found by the optimization process dominate the initial solution, i.e. are superior in both objectives. The occurrence of a wide Pareto front underlines the conflict in minimizing both objectives and just (Pareto) compromise solutions can be found.
In the Figure, boxes mark five different areas along the Pareto front. For the solutions within the boxes, the valve settings are printed. For better illustration, the settings are connected with a line.
Box 1 and 5 are at the extreme ends of the Pareto front. Box 1 represents Pareto solutions with high NO_x emissions, but low pulsation. The corresponding valve settings show an increased fuel mass flow at valves 1, 2 and 4, while the flow at valves 5 and 6 is reduced. The fundamental mechanism corresponding to these settings is the fact that the increased mass flow through valves 1 and 2 leads to rich combustion in the center of the burner. Fig. 3 shows the arrangement of the valves in the combustor. The rich combustion zone stabilizes the flame, but increases the NO_x emissions. The lean zones are in the middle of the row of injection holes at valves 5 and 6.
Box 5 contains solutions with minimal NO_x emissions, but high pulsation. The mass flow through each valve is about equal, generating no rich combustion zones. Compared to the initial solution, the small mass flow increase at valves 5 and 8 leads to lower NO_x emissions, while the pulsation is unchanged.

5 Conclusions

A novel multi-objective evolutionary algorithm is introduced with increased robustness for applications prone to noise and outliers and based on the SPEA algorithm. The scheme introduces the concepts of domination-dependent lifetime, the re-evaluation of nondominated solutions and an extended update mechanism

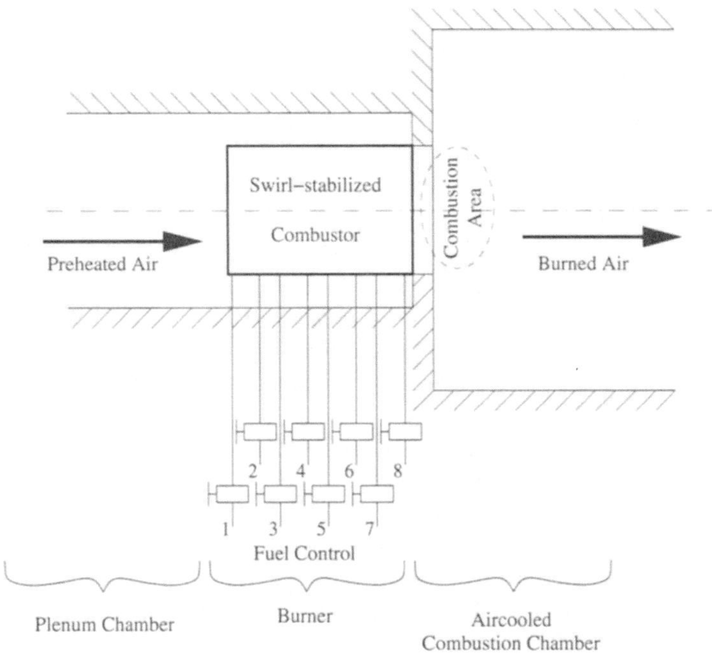

Fig. 3. Sketch of the atmospheric combustion test-rig with a low-emission swirl stabilized burner. The rates of fuel mass flow through the injection holes are the design variables of the setup. The NO_x emissions and the pulsation of the burner are to be minimized.

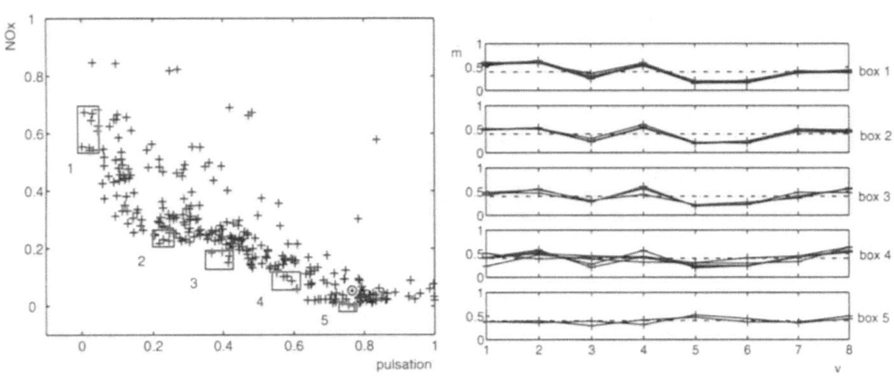

Fig. 4. All measured solutions of the burner optimization run [plus symbol] and initial solution [circular symbol] (left figure). Mass flow $\dot{m}$ through the valves $V_{i,i=1\ldots8}$ for solutions within the 5 marked boxes along the Pareto front (right figure).

for the archive.
Noisy and noise-free test functions are presented to evaluate the extended algorithm. A decrease in convergence is observed when noise is introduced. The concept of elitism is analyzed in the presence of noise. In the absence of noise, elitism can be used as a convergence accelerator. However, for different types of noise, elitism can imply a significant disadvantage, since the optimization can get misled by outliers.
The new extended algorithm overcomes the problem of premature convergence by introducing dominance dependent lifetime and accelerates the convergence by using an archive. The archive is modified by the re-evaluation of nondominated solutions and the extended update of the archive. For all test problems, the proposed algorithm shows equal or superior convergence compared to the standard and non-elitistic SPEA.
A further advantage is, that the extended SPEA can handle moving optima over time. The algorithm re-evaluates solutions after a limited lifetime, therefore minimizing the chance of convergence to local minima.
The algorithm is successfully applied to an automated optimization in the design process of gas turbine burners. The process produces in an automated fashion an experimental Pareto front for minimizing pulsation and emissions of an industrial burner. Automated optimization can be considered a supporting tool in a design process, complementing physical understanding as well as trial-and-error design. Future work will focus on using larger numbers of simpler (on/off) valves. This will allow more flexibility in the fuel distribution and also allows axially asymmetric distribution. The present algorithm is under modification to account for these configurations.

6 Acknowledgments

The results were obtained at the atmospheric test-rig of Alstom Power in Baden-Dättwil, Switzerland. For the realization of the experiments the authors wish to thank Bruno Schuermans. Special thanks to Rolf Dornberger and Christian Oliver Paschereit, whose initiatives led to this work.

References

1. Horn, J.: Multicriterion decision making. In Bäck, T., Fogel, B., D., Michalewicz, Z., eds.: Handbook of Evolutionary Computation, Sec. F1.9: pp. 1-15. (1997)
2. Zitzler, E., Thiele, L.: Multiobjective evolutionary algorithms: A comparative case study and the strength Pareto approach. IEEE Trans. on Evolutionary Computation **3** (1999) 257–271
3. Van Veldhuizen, D.A., Lamont, G.B.: On measuring multiobjective evolutionary algorithm performance. In: Proceedings of the 2000 Congress on Evolutionary Computation. (2000) 204–211
4. Coello Coello, C.A.: An updated survey of evolutionary multiobjective optimization techniques: State of the art and future trends. In: Congress on Evolutionary Computation. (1999) 3–13

5. Coello Coello, C.A.: List of references on evolutionary multi-objective optimization, (http://www.lania.mx/~ccoello/EMOO/EMOObib.html, Last accessed July 2002)
6. Teich, J.: Pareto-front exploration with uncertain objectives. In et al., Z., ed.: Proceedings of the First Conference on Evolutionary Multi-Criterion Optimization. (2001)
7. Hughes, E.J.: Evolutionary multi-objective ranking with uncertainty and noise. In et al., Z., ed.: Proceedings of the First Conference on Evolutionary Multi-Criterion Optimization. (2001)
8. Goldberg, D.E.: Genetic Algorithms in Search, Optimization, and Machine Learning. Addison-Wesley Publishing Company (1989)
9. Fonseca, M.C., Fleming, P.J.: Multi-objective genetic algorithms made easy: Selection, sharing and mating restrictions. In: Proceedings of the 1st International Conference on Genetic Algorithms in Engineering Systems: Innovations and Application. (1995) 45–52
10. Pareto, V.: Manuale die Economia Politica. Societa Editrice Libraria, Milano, Italy (1906)
11. Goldberg, D.E., Segrest, P.: Finite Markov chain analysis of genetic algorithms. In Grafenstette, ed.: Proceedings of the Second International Conference on Genetic Algorithms. (1987)
12. Zitzler, E., Deb, K., Thiele, L., Coello Coello, C.A., Corne, D.W.: Proceedings of the First Conference on Evolutionary Multi-Criterion Optimization. Springer-Verlag (2001)
13. Van Veldhuizen, D.A., Lamont, G.B.: Multiobjective evolutionary algorithms: Analyzing state-of-the-art. Evolutionary Computation **8** (2000) 125–14
14. Miller, B.L., Goldberg, D.E.: Genetic algorithms, selection schemes, and the varying effects of noise. Illigal report no. 95005, University of Illinois at Urbana-Champaign, Illinois Genetic Algorithm Laboratory (1995)
15. Bäck, T., Hoffmeister, F., Schwefel, H.P.: A survey of evolution strategies. In Belew, R.K., ed.: Proceedings of the Fourth International Conference on Genetic Algorithms and their Applications. (1991)
16. Deb, K.: Multi-objective genetic algorithms: Problem difficulties and construction of test problems. Evolutionary Computation **7** (1999) 205–230

Part III

Reactive Jets in Crossflow and Multiphase Jets

Jets in Crossflow – Effects of Heat Release

M. G. Mungal and E. F. Hasselbrink, Jr.

Mechanical Engineering Department, Stanford University, Stanford, CA 94305-3032, USA

Abstract. Detailed measurements of the velocity field in the symmetry plane of two jets and two jet flames in crossflow are obtained using Particle Image Velocimetry. The jets issue into a wind tunnel at density-weighted jet-to-cross flow velocity ratios $r = 10$ and $r = 21$, with corresponding Reynolds numbers 6,000 and 12,800. The changes in the flow due to the effects of heat release are examined and ensemble statistics of the velocity field are derived from these datasets. Simultaneous Planar Laser-Induced Fluorescence imaging of the OH radical is performed in selected regions of the flames. Results suggests that flame/flow interaction is strong near the lifted flame base, but increasingly weaker further downstream.

1 Introduction

The scope of the present experiments is to compare jets at two blowing ratios, $r = 10$ and $r = 21$, in both non-reacting and burning (unpiloted) cases. These cases were chosen for comparison to the work Smith and Mungal (1998). The selection also reflects their observation that when $r \geq 10$, the jet is not strongly influenced by the presence of the wall. In all cases, the jet was comprised of 99.0% methane. Flame stability considerations determine the crossflow velocity, which was chosen to achieve the highest possible Reynolds number, yet support stable flames which do not blow out (Kalghatgi 1981). This limits the wind tunnel speed and the maximum Reynolds number to 6,300 and 12,600 for the two cases.

The velocity field in the symmetry *(x-y)* plane of each jet and flame is obtained using PIV as described in the previous paper (Mungal and Su, this volume). In the flame cases, simultaneous PLIF imaging of the OH radical and PIV are also performed at five key locations in order to investigate the interaction between combustion reaction and turbulent flow.

2 Experimental Considerations

The flow facility is the same crossflow tunnel used by Smith and Mungal (1998) with minor improvements for PLIF imaging consisting of the addition of quartz windows for UV laser access. Methane (99.0% commercial grade) is seeded with 0.3μ (nominal) alumina particles, using a stirred fluidized-bed seeder and a conical cyclone to remove large agglomerates of particles. The jet is injected normal to the crossflow through a 25cm long tube (6.35mm OD, 4.72mm ID, 316 stainless steel), protruding 7cm into the wind tunnel. The jet tube is located within an annular tube (9.5mm OD) which can be used as a hydrogen pilot, but no pilot flames is used since this is problematic for the jet in crossflow as discussed by Hasselbrink and Mungal (1998).

Becker and Yamazaki (1978) found that buoyancy in turbulent jet flames increases the entrainment rate and modifies the velocity profiles. They also note that it is difficult to reach conditions in the laboratory in which buoyancy has negligible effect over the entire flame. Defining a Richardson number $Ri = gd/uj^2$, they found that buoyancy begins to affect the flow at a scaled distance along the jet axis, $\xi = Ri^{1/3}x/d \approx 1.5$. Buoyancy effects are moderate for $1.5 < \xi < 3$, and severe above this value. Their findings are based on experiments on free jet flames, however; transverse jet flames are expected to scale differently from free jets, and the buoyancy forces in our experiment are perpendicular to the jet momentum flux. Their findings are therefore used only as an estimate of the buoyancy's magnitude. For the two jet flames studied here, L_f/d^* is 79 and 86 (where d^* is the jet momentum diameter) while ξ_L (ξ measured at the flame length, L) is 2.7 and 1.8 for the $r = 10$ and 21 cases respectively.

2.1 OH PLIF imaging

OH PLIF imaging is one of several possible techniques for detecting reaction zones in turbulent flames. The excitation/collection scheme in the present work is excitation of the $A^2\Sigma^+ \leftarrow X^2(1,0)$ band of OH near 283nm, combined with detection of the (1,1) and (0,0) bands near 310nm. The $Q_1(6)$ transition is selected in order to minimize signal dependence on temperature (i.e., ground state population) and local composition. Barlow and Collignon (1991) have found that, although the quenching cross-section for OH fluorescence varies by about a factor of two across a typical flame, population fraction in the $Q_1(6)$ line varies oppositely, so that the signal should be proportional to OH concentration, to within 20% of peak OH concentration. An additional 15% uncertainty in this proportionality is introduced by laser sheet extinction as it propagates through a nonuniform density of particles before reaching the imaged region of the flow.

2.2 PIV image processing and uncertainty

PIV image processing is performed with custom software, StanPIV, written as part of the present work (Hasselbrink, 1999). StanPIV utilizes some of the more recently published box-offset advances in PIV image processing (Raffel et al., 1998) to achieve significantly higher probability of detection, as well as a factor of six reduction of systematic measurement bias error. Data yield is better than 98.5% in the flame cases (99.5% in nonreacting cases), improved from less than 90% using commercial single-pass PIV with 32^2 interrogation windows.

In flows with high temperature gradients, uncertainty is dominated by thermophoresis (Muñiz et al., 1996), a phenomenon in which a particle experiences a force opposite the direction of the fluid temperature gradient. Thermophoretic velocities in flames are (worst-case: assuming 2000K/mm temperature gradient at 1300K, and using air properties) about 15cm/sec. Although this may not be a large relative error in the measured velocity, it can introduce significant error for derivative quantities such as strain rates. The smallest grid spacing used in the present work is about 0.43mm, which introduces a maximum uncertainty in central-difference first derivatives of about $240s^{-1}$. This value is quite large compared to extinction strain rates for methane diffusion flames, which are on the order of $400s^{-1}$. Uncertainty in vorticity, dilatation, and strain rates is larger by $\sqrt{2}$ because the calculation involves 4 velocity measurements for central differences. Hence the maximum uncertainty due to thermophoresis is approximately $340s^{-1}$.

3 Results: Flow Structure

3.1 Near-field flow structure

Figure 1 shows instantaneous and ensemble-averaged velocity profiles in the potential core region of the $r = 10$ jet. The meandering of the instantaneous profile is to be contrasted with the smooth mean profile. Streamtraces, integrated from the velocity field, highlight the asymmetry, and often reveal vortical "sinks" (in two dimensions) on the windward side, as well as a persistent "source" on the lee side of the jet. In contrast, free and coflowing jet flowfields have no points in the instantaneous flow where velocity magnitude is zero, in the lab frame of reference (Muñiz and Mungal, 2001}. The differences from average profiles observed for free jets are subtle, and are more pronounced in the $r = 10$ jet by $x/d = 5$, the velocity profile is skewed towards the lee side, and the u velocity does not decay to zero rapidly on the lee side.

Figure 2 shows streamtraces in the near-field, overlaying color contours of the two-dimensional divergence, defined as $\nabla_{2D}\cdot\mathbf{u} = \partial u/\partial x + \partial v/\partial y$. This quantity is calculated using central differences from the velocity data. The lee-side source observed in the instantaneous field is a dominant feature of the near-field symmetry plane; in fact, in this plane, all time-averaged flow on the lee side of the jet is apparently sourced from this node. The existence of this node was also found by Kelso et al. 1996) for lower r jets, using a flying hotwire technique, and by Sykes et al. (1986), in a computational study. In both of these studies, the jet issued from a wall. The results of Kelso et al. were of much lower resolution than the present work, and the computation by Sykes et al. had a slip boundary condition at the wall; nonetheless, the present findings corroborate their discoveries, and demonstrate that the node exists without the presence of a wall boundary.

The out-of-plane flow pattern can be better understood by considering the out-of-plane extensional strain rate, $\partial w/\partial z$. Since the flame base is only present at the upper right corner of the $r = 10$ image, elsewhere the flow is divergence-free. Therefore the out-of-plane extensional strain rate is $\partial w/\partial z = -\ \nabla_{2D}\cdot\mathbf{u}$. From the in-plane streamline topology and $\nabla_{2D}\cdot\mathbf{u}$, it is possible to infer the out-of-plane streamline structure. Low values of $\nabla_{2D}\cdot\mathbf{u}$ in the jet result from the entrainment into the jet, and the out-of-plane growth of the jet. The high $\nabla_{2D}\cdot\mathbf{u}$ behind the jet suggests high compressive out-of-plane strains, $\partial w/\partial z$ has large negative values in this regions. Since $w = 0$ on the symmetry plane, this suggests opposed flow on the lee side, as illustrated in Fig. 2. The apparent lee-side source is labeled "S". From this cartoon, it is also clear that another critical point must exist in the streamline pattern: a re-attachment point, labeled "A", must exist somewhere on either the jet tube or the tunnel wall - its location in the cartoon was chosen arbitrarily. We note that a corresponding region of negative $\nabla_{2D}\cdot\mathbf{u}$ in front of the jet does not exist, suggesting that as crossflow fluid approaches the jet, it is entrained into the jet flow rather than passing around it. This underscores yet another important difference between transverse jets and cylinders in crossflow, as emphasized in numerous other ways by Fric and Roshko (1994) and others.

3.2 Far-field flow structure

Figures 3, 4 each show color maps of six normalized flowfield variables (left-to-right, then top-to-bottom): u, v, u', v', ω_z, $\nabla_{2D}\cdot\mathbf{u}$ for the $r = 10$ non reacting and reacting case. The $r = 20$ case is not shown for brevity. The calculated jet-exit centerline streamtrace is shown in each figure for reference. As expected, more noise is evident in the fluctuation data than in the averaged velocities; the fluctuation statistics are also much more sensitive to occasional spurious outliers.

Starting with the map for the nonreacting jet (Fig. 3) the u contours (upper left corner) show the somewhat surprising result that the locus of points with maximum u in the jet cross-section do not fall on the jet-center streamline. This behavior is consistent with the finding by Kamotani and Greber (1972) that the passive scalar trajectory does not penetrate as far into the crosswind as the central streamline. The map of v/v_∞ (upper right) shows that the v component of velocity can exceed the crossflow velocity, by a factor as large as about 1.8, and that this maximum value is usually found along the center streamline. However, on the lee side, the effect of the jet on the v component is most pronounced: reverse flows with velocity of about $-1.2v_\infty$ can be found just behind the jet and this region of low v/v_∞ persists downstream.

RMS fluctuations are shown in the second row of color maps. The turbulence fluctuations are as high as $u'_{rms}/u_j = 0.3$, and $v'_{rms}/u_j = 0.2$, with the largest values found (not surprisingly) in the jet shear layer. Unlike u, the locus of u'_{rms} and v'_{rms} maxima correspond very closely to the location of the center streamline.

The effects of igniting this methane jet are illustrated by the color maps of Fig. 4. Not surprisingly, heat release increases both u and v; perhaps somewhat surprising is the width of the region over which the u field is affected. The effect of heat release in the jet affects the surrounding flow significantly, over a distance which appears to scale with the size of the flame. Correspondingly, the u contours also become noticeably broader near $y/d = 10$; this becomes more evident in profiles (not shown). The influence of heat release is perhaps most striking in the v color map. Interestingly, the base of the flame outlined in this figure, and in flow visualization photographs, show that the flamebase is not orthogonal to the center streamline; instead, it is nearly orthogonal to the crossflow. Furthermore, no longer does $v/v_\infty \rightarrow 1$ in the far field. With the flame, $v/v_\infty \rightarrow 2.6$ at the flame tip. It is noted that this value is close to $(T_f/T_o)^{1/2}$, where T_f is the maximum flame temperature (probably very close to the adiabatic flame temperature of a stoichiometric methane/air flame, 2200K), and T_o is the unburned gas temperature. This result suggests that the quantity ρv^2 is approximately conserved through the flame, as can be inferred from the Rankine-Hugoniot relations for a deflagration.

The flame appears to influence the structure of the v'_{rms} to a greater degree than the u'_{rms} field. In particular, the contours of v'_{rms} appear to be highest in regions corresponding to the location of the flame envelope. We note that the bifurcation in the v'_{rms} contours does not become pronounced until near the end of the data window, suggesting that this may be a buoyancy effect.

3.3 Results: OH PLIF Imaging

Simultaneous OH images were obtained at 5 imaging locations: the lee and windward flamebase, the lee and windward mid-flame, and the flame tip for both flames. A full discussion of results can be found in Hasselbrink (1999) and Hasselbrink and Mungal (2001). Detailed measurements on the stabilization velocities measured at the flame base can be found in Hasselbrink and Mungal (1998). Sample results are shown in Figs 5 for the flame side, and in Fig. 6 for the flame base.

Figure 5 shows examples of measurements taken further along the flame surface, away from the stabilization region as shown by the inset figure. The left image shows instantaneous streamlines and the flame as marked by OH contours, superposed onto the vorticity, ω_z, computed from the velocity data. As was first reported by Rehm and Clemens (1998), there is a good correspondence between the OH region and the vorticity. The right image of Fig. 5 also shows that compressive strain, S_{xx}, is a good marker of the location of the flame. Additional details of the vorticity and strain rates and their correspondence can be found in Han and Mungal (2000) who used simultaneous PIV and PLIF of CH to uniquely determine the flame location.

Based on the analysis of Ruetsch et al. (1995), the acceleration through the flame base is expected to be such that the burned fluid velocity exceeds the unburned fluid velocity by a factor proportional to the square root of the temperature ratio. For methane flames, this is a factor of 2.7, presuming the adiabatic flame temperature is reached. Further insights can be provided by ensemble averaging the OH and velocity fields at the lee-side stabilization point of the jet flame in crossflow, Fig. 6. The line contours show the OH concentration with highest values marking the mean flamebase, while the color contours mark the magnitude of the velocity. The mean streamlines are also shown to mark the path of fluid through the flamebase. The plots on the right show the velocity magnitude and OH signal along the central streamline. A very good correspondence is observed between the two so that the flow accelerates as more OH is present, which can be interpreted as an increase in heat release. Most interestingly, the average velocity magnitude change is seen to be about 2.6 which is quite close to the square root of the density ratio of unburned to burned gas, $\rho_{ub}/\rho_b \sim 7$, for the present fuel.

4 Summary and Conclusions

Turbulent transverse jets at high jet-to-crossflow blowing ratios have been investigated using both analytical and experimental techniques, with the primary goal of understanding the overall structure of the velocity field, and how it is influenced by combustion heat release. Major results of the present work are as follows:

An apparent "source" exists in the time-averaged two-dimensional flow in the symmetry plane. A similar result was found by Kelso et al. (1996) for jets issuing from a wall. The source is located in the lee of the jet, centered at about $(x/d, y/d) = (1,1)$ in the laboratory coordinate system. The source corresponds to a region of large $\partial w/\partial z$, indicating that there is compressive strain, i.e. opposedflow, in the out-of-plane direction on the lee side. In contrast, on the leading edge, $\partial w/\partial z$ is not nearly as large, indicating that much of the crossflow approaching the leading edge is entrained into the jet, rather than diverted around it.

The center streamline from the jet nozzle follows a power-law trajectory when normalized by the outer length scale, *rd*. The power-law exponent is 1/3, as predicted by scaling laws, and the leading coefficient measured is 2.1 to within experimental uncertainty. Combustion heat release causes a deviation in the trajectory, such that the jet penetrates slightly further into the crossflow.

Combustion heat release accelerates the flow within the heated region. As the flame tip is approached, maximum $v/v_\infty \rightarrow 2.7$. This factor may be a simple result of momentum conservation. Therefore, the momentum flux per unit area, J, must be approximately the same between the two flows, and since $J \sim \rho v^2 A$, it follows that ρv^2 is nearly equivalent in the two cases. Therefore if density drops by a factor of T_f/T_o, flow speed must increase by $(T_f/T_o)^{1/2}$. An intriguing implication of this result is that the mass entrainment, $m \sim \rho u A$, is reduced by the factor $(T_f/T_o)^{1/2}$. An investigation of the reduction of entrainment due to heat release, via direct measurement of entrainment into straight jets confirms this result (Han and Mungal, 2001).

References

Barlow, R.S. and Collignon, A. (1991). Linear LIF measurements of OH in nonpremixed methane-air flames: When are quenching corrections unnecessary? AIAA Paper 91-0179, *29th Aerospace Sciences Meeting*, Reno, NV.

Becker, H.A. and Yamazaki, S. (1978). Entrainment, momentum flux and temperature in vertical free turbulent diffusion flames. *Combustion and Flame* 33:123 -149.

Fric, T.F., and Roshko, A. (1994). Vortical structure in the wake of a transverse jet. *Journal of Fluid Mechanics* 279:1-47.

Han, Donghee and M. G. Mungal (2000), Simultaneous Measurement of Velocity and CH Layer Distribution in Turbulent Non-premixed Flames, *Twenty-Eighth Symposium (International) on Combustion*, 261-267.

Han, Donghee and M. G. Mungal (2001), Direct Measurement of Entrainment in Reacting/Non-Reacting Turbulent Jets, *Combustion and Flame*, 124: 370-386.

Hasselbrink, E. F. Jr. (1999) Transverse Jets and Jet Flames: Structure, Scaling, and Effects of Heat Release. Ph.D. Thesis, Stanford University, 1999.

Hasselbrink, E.F. Jr. and Mungal, M.G. (1998). Observations on the stabilization region of lifted nonpremixed methane transverse jet flames. *Twenty-Seventh Symposium (International) on Combustion,* 1167-1173.

Hasselbrink, E.F. Jr. and Mungal, M.G. (2001) Transverse Jets and Jet Flames. Part 2. Velocity and OH Field Imaging, *Journal of Fluid Mechanics* 443:27-68.

Kalghatgi, G.T. (1981). Blowout stability of gaseous jet diffusion flames Part II: Effect of cross-wind. *Combustion Science and Technology* 26:241-244.

Kelso, R.M., Lim, T.T., and Perry, A.E. (1996). An experimental study of round jets in cross-flow. *Journal of Fluid Mechanics* 306:111-144.

Muñiz, L., Martinez, R.M., and Mungal, M.G. (1996). Application of PIV to turbulent reacting flows. *Eighth International Symposium on Applications of Laser Techniques to Fluid Mechanics* , Lisbon, Portugal.

Muñiz, L., and Mungal, M.G. (1997). Instantaneous flame-stabilization velocities in lifted jet diffusion flames. *Combustion and Flame* 111 (1/2), 16-31.

Muñiz, L., and Mungal, M.G. (2001), Effects of Heat Release and Buoyancy on Flow Structure and Entrainment in Turbulent Nonpremixed Flames, *Combustion and Flame*, 126:1402-1420.

Raffel, M., Willert, C., and Kompenhans, J. (1998). *Particle Image Velocimetry, A Practical Guide.* Berlin: Springer-Verlag.

Rehm, J.E., and Clemens, N.T. (1998). The relationship between vorticity/strain and reaction zone structure in turbulent non-premixed jet flames. *Twenty-Seventh Symposium (International) on Combustion* . The Combustion Institute, Pittsburg, PA, USA.

Ruetsch, G., Vervisch, L., and Liñan, A. (1995). Effects of heat release on triple flames. *Physics of Fluids* 7 (6):1447-1454.

Smith, S.H., and Mungal, M.G. (1998). Mixing, structure and scaling of the jet in cross flow. *Journal of Fluid Mechanics* 357:83-122.

Sykes, W., Lewellen, W., and Parker, S. (1986). On the vorticity dynamics of a turbulent jet in a crossflow. *Journal of Fluid Mechanics* 168:393-413.

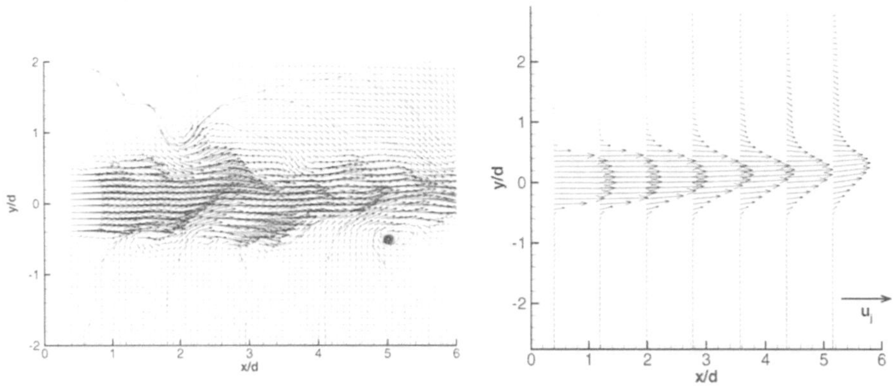

Fig. 1 (a) instantaneous velocity and (b) average velocity fields in the potential core region for $r = 10$ jet.

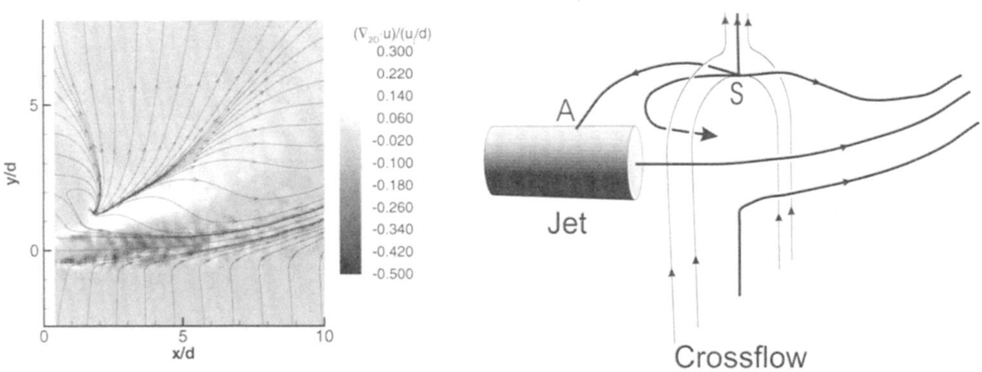

Fig. 2 Near-field streamtraces and two-dimensional divergence. Left: Two-dimensional divergence, overlaid with two-dimensional streamtraces, in the near field. Streamtraces are everywhere parallel to measured 2-D velocity field, but are not contours of constant streamfunction. Right: interpretation of three-dimensional near-field streamlines. Area of high 2-D divergence measured on lee side of the jet indicates highly compressive out-of-plane strain, implying opposing out-of-plane streamlines which meet at the stagnation point, S. However, the near-zero 2-D divergence on the windward side implies that streamlines do not diverge strongly out-of-plane on the windward side; instead, they are entrained into the jet.

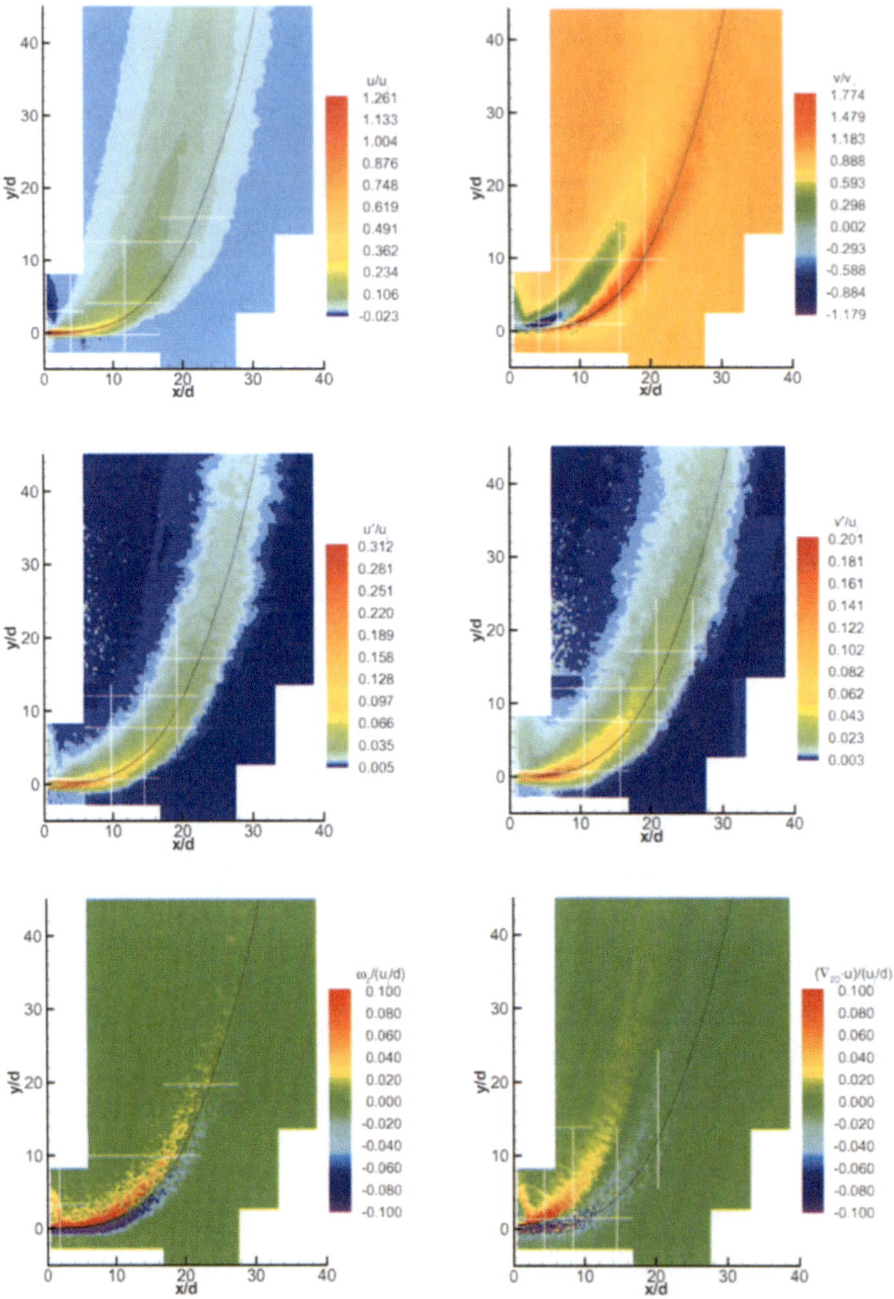

Fig. 3. $r = 10$ nonreacting jet: color maps of normalized mean velocities, RMS fluctuations, mean vorticity and mean divergence. Quantities are labeled in individual legends. Contour levels span maximum to minimum values.

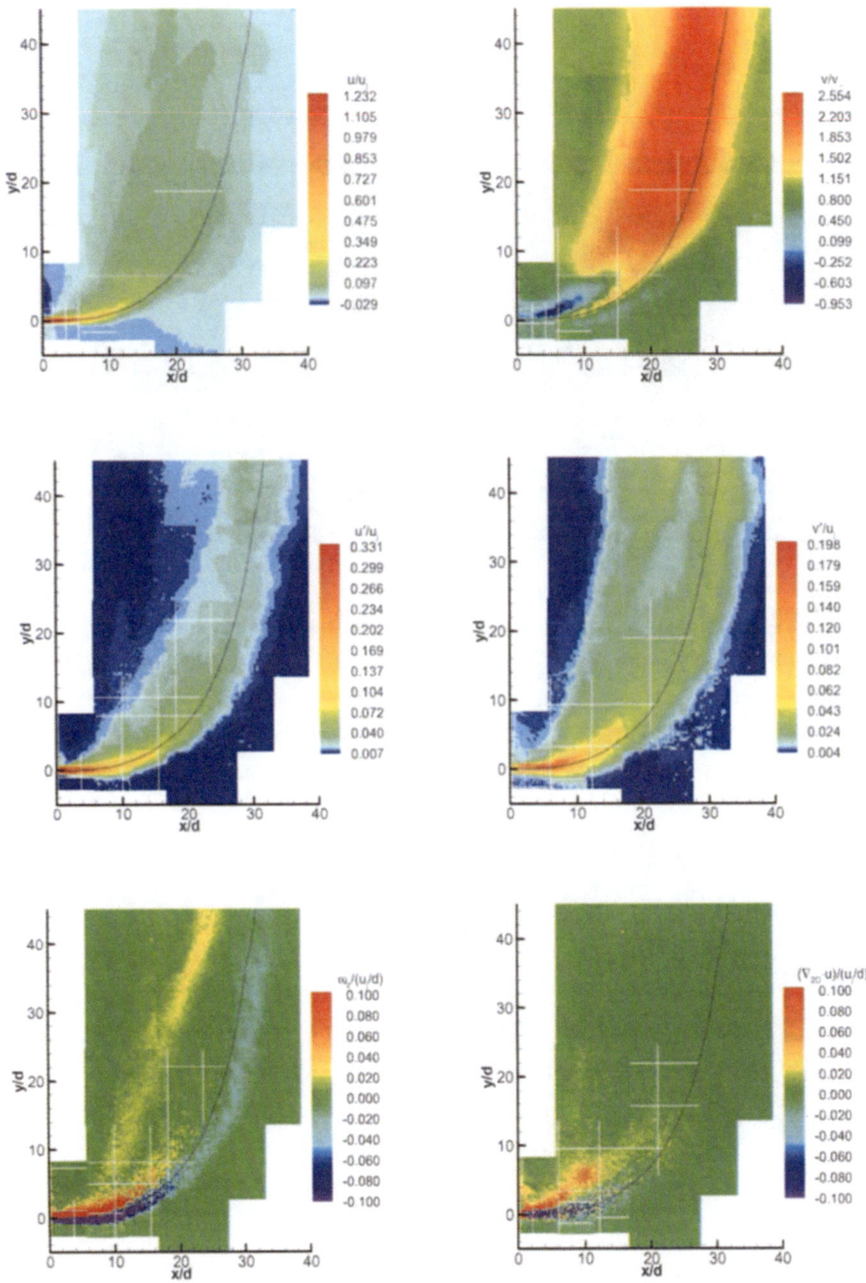

Fig. 4. $r = 10$ reacting jet: color maps of normalized mean velocities, RMS fluctuations, mean vorticity and mean divergence. Quantities are labeled in individual legends. Contour levels span maximum to minimum values.

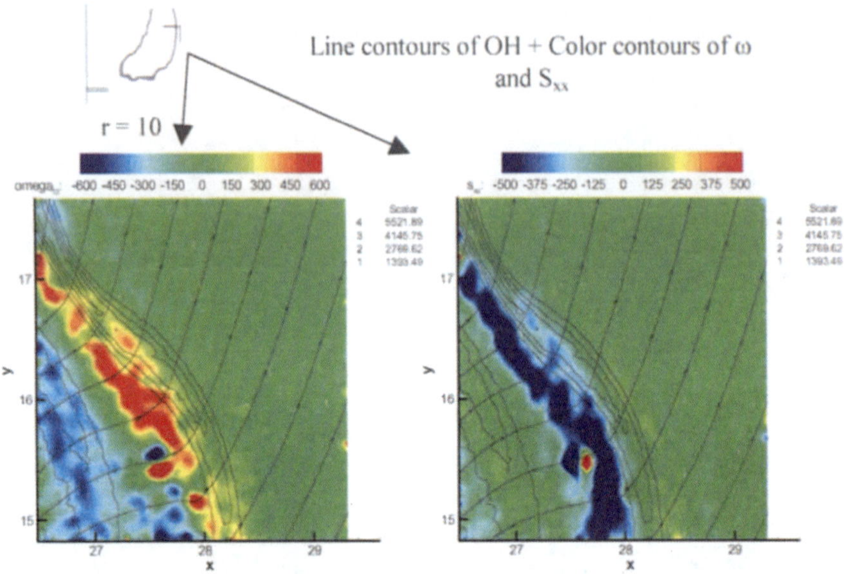

Fig. 5. Vorticity and S_{xx} strain rate (color contours) overlaid with OH contours (black line contours) overlaid. Streamlines are also shown in the lab reference frame (Hasselbrink, 1999, Mungal 2001).

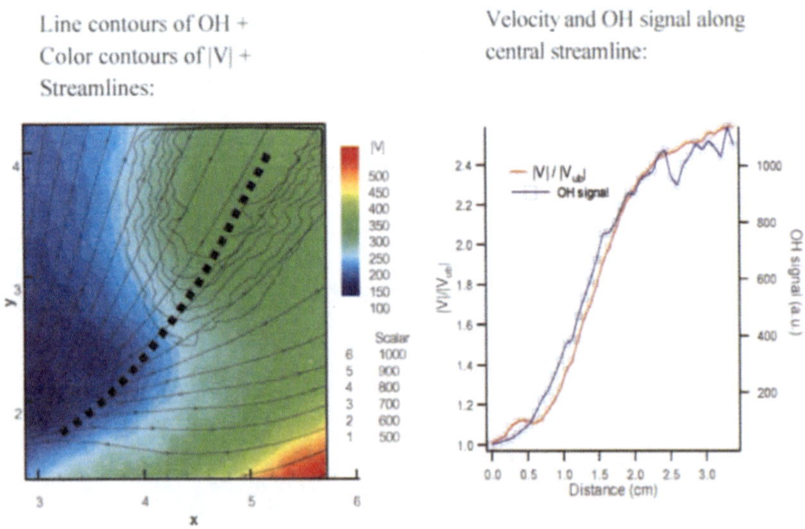

Fig. 6. Acceleration through lee-side lifted flame base, r = 10: time-averaged simultaneous OH PLIF/PIV data. Left: Black line contours show region near flamebase where OH signal is high. Color contours show velocity magnitude in cm/s. Streamlines from velocity field overlaid. Right: Velocity magnitude (normalized by unburned fluid velocity) and OH signal compared along dotted streamline, showing acceleration of fluid due to heat release (Hasselbrink and Mungal, 2001)

Jets in Crossflow - NO_X Control Using the Two-Stage Lagrangian Model

M. G. Mungal and Donghee Han

Mechanical Engineering Department, Stanford University, Stanford, CA 94305-3032, USA

Abstract. This paper presents results of the application of the Two-Stage Lagrangian (TSL) model of Broadwell and Lutz to the basic and advanced reburning processes in a 300 kW natural gas fired Boiler Simulator Facility (BSF). The injection of the reburning fuel and overfire air is modeled as independent deflected jets in crossflow and the TSL model is applied while each jet completely mixes with the main flue gas stream of the boiler. The entrainment rate to the jet, which is required as a model input, is derived from control volume analysis using the experimentally determined jet-trajectory. The comparison with the experimental data shows good agreement for relatively high reburn zone stoichiometric ratios (SR~0.99) while the removal of NO is overestimated for richer conditions (SR~0.95 or lower). The model in general follows the trend observed in the experiment and is able to quantitatively and rapidly predict the NO removal in the gas reburning process.

1 Introduction

It is known that the emission of oxides of nitrogen (NO_X) from combustion processes can be reduced by the gas reburning process Zamansky et al. (1997) which is a promising method for controlling NO_X in practical sized facilities. Advanced gas reburning technology is composed of three main processes: (1) the injection of reburning fuel, which is typically CH_4 or natural gas, to the products of a main burner, (2) the addition of N-agent, typically ammonia or urea, to the reburn zone, and (3) the injection of overfire air which completes the process by burning out remaining combustibles. The basic reburning process is identical to that of the advanced reburning except for the absence of the N-agent injection process. Zamansky et al. (1997) have performed a series of experiments on various reburning conditions in a test facility, the Boiler Simulator Facility (BSF), which can simulate a practical size boiler.

In this study, the Two-Stage Lagrangian (TSL) model which was developed by Broadwell and Lutz (1998) is applied to rapidly predict the performance of NO_X emission control using gas reburning processes while avoiding full, detailed numerical simulations. Earlier related efforts have been made by Zamansky et al. (1998) to apply kinetic codes to simulate the reburning process while the mixing processes were modeled as instantaneous, but distributed in time. Although their study gave some qualitative understanding of the reburning process, the result did not estimate the NO_X level quantitatively or predict proper trends. The TSL model retains the full chemistry, but important characteristics of jets are included using experimental inputs for a description of the mixing, including jet entrainment and a flame sheet residence time. This allows the model to deal with both the flow and chemistry during the turbulent mixing process.

2 Experimental Detail

The facility in which the experimental data were obtained is called the Boiler Simulator Facility (BSF) which is a 300kW facility, 7m tall, with inner diameter of 56cm. It has a vertically down-fired radiant furnace and a horizontal convective path. The main burner is operated in a fuel lean condition with stoichiometric ratio (SR) of 1.1, and initial NO concentration set to 600ppm by adding a relevant amount of NH_3. The reburning fuel is injected into the product of the main burner at a temperature of 1710K in the form of 8 transverse jets originating from a central pipe with the jets oriented 27° upstream. Natural gas is used as the reburn fuel and its composition is given as 90.7% CH_4, 7.5% C_2H_6, 1.5% N_2, 0.25% CO_2 and 0.05% O_2. Reburn fuel is carried in a gaseous transport medium (bottled nitrogen) so that the jet exit velocity is sufficient for good mixing. The injection of the reburning fuel makes the main mixture stream slightly rich, SR=0.89~0.99, and forms the reburn zone where the NO can be reduced. For the N-agent, urea is sprayed into the main stream in the form of an aqueous mixture according to Nitrogen Stoichiometric Ratio of 1.5 (NSR = $[N]_{N\text{-agent}}$ / [NO], where [NO] is measured at 600K for the basic reburning case). Overfire air is added in the form of 24 transverse jets making the final mixture lean, SR=1.15.

3 Two-Stage Lagrangian (TSL) model

Broadwell et al. (1982), Broadwell and Breidenthal (1982) and Tyson et al. (1981) have developed the Two-Stage Lagrangian model based on observations of various experiments on jets and mixing layers (Broadwell and Mungal, 1991). Experimental evidence for organized structure in jets and jets in crossflow from low to high Reynolds numbers can be found in Mungal et al. (1991, 1992). Broadwell and Lutz (1998) applied the model to jet diffusion flames to estimate NO_X production. The TSL model separates the jet diffusion flame into two reaction zones, which are the flame sheet reactor and the homogeneous core reactor. The homogeneous core reactor is modeled as a perfectly stirred reactor while the flame sheet reactor can be modeled either by a diffusion flame or a perfectly-stirred reactor. Using the latter for the flame sheet introduces the flame sheet residence time (τ) as one additional parameter, but significantly reduces the computing time which makes the model more practical. The reduction in computational burden due to using the perfectly-stirred reactor as the flame sheet enables the addition of the buoyancy and radiation to the model formulation.

The basic structure of the TSL model is explained in Fig. 1. First, the entrainment to a jet, obtained from experimental data, is provided as a model input. The entrained ambient fluid is mixed with a stoichiometric amount of the jet fluid in the flame sheet reactor and reacts for a specified flame sheet residence time. Then, the products of the flame sheet reactor are homogeneously mixed with the jet fluid and form the homogenous core reactor. This process is repeated up until the flame tip when all the fuel is consumed. The homogeneous core acts as an internal reburn zone in a flame where NO_X formed in the flame sheet may be consumed. Therefore, a main benefit of the TSL to other flamelet models is to have a homogeneous core reactor where the products from the flame sheet can be dumped and allowed to react with the mixture composed of the jet and previously formed product. In this work, the flame sheet reactor is consistently modeled as a perfectly stirred reactor.

4 Entrainment rate of a deflected jet

As explained in the previous section, the TSL model requires the entrainment rate as a model input. Hasselbrink and Mungal (1996) have proposed a control volume analysis to estimate the entrainment into the transverse jets using the experimental trajectory of such jets. In their analysis, the entrainment is estimated by calculating the momentum of the crossflow fluid required to turn the flow to the experimentally determined trajectory. It is known that the trajectory of transverse jets is given as

$$\frac{y}{rd} = A\left(\frac{x}{rd}\right)^B \quad , \text{ where } \quad r = \sqrt{\frac{\rho_j V_j^2}{\rho_\infty V_\infty^2}} \tag{1}$$

The constants A and B are determined from previous works (Margason, 1993, Storms, 1965, Pratte and Baines 1967) where the constant A ranges from 1.5 to 2.0 and B ranges from 0.25 to 0.38. Using (1) with the momentum balance, the expression for the entrainment can be obtained as

$$\frac{\dot{m}}{\dot{m}_j} = 1 + \frac{V_j}{V_\infty}\left(\frac{1}{A}\frac{y}{rd}\right)^{(1-B)/B} \tag{2}$$

The previous work concerned jets at normal injection, however, in this study jets are injected obliquely so here we extended the approach to deflected jets heading at various angles with respect to the main flow. Wu (1973) has proposed a coordinate transformation that makes the trajectory of deflected jets scale as given in Eqn. (1) given by

$$Y = y/\cos\theta, \quad X = x - y\tan\theta \tag{3}$$

Using the new coordinate system and the experimental result of Platten and Keffer (1971), the trajectory was obtained by Wu (1973) as

$$\frac{Y}{rd} = A\left(\frac{X}{rd}\right)^B \tag{4}$$

where A and B were determined as 1.75 and 0.38 respectively. The jet velocity ratios that Platten and Keffer (1971) used ranged from 4 to 8, which are relatively small for the jets used in the reburn fuel injection. Han et al. (2000) performed Mie scattering experiments in deflected jets with higher velocity ratios ranging from 5 to 20. They used the same coordinate transformation as (3) and obtained A=2.1 and B=0.29. Considering all these results, A=2.0 and B=1/3 were consistently used in the present calculations. The entrainment rate is obtained by solving mass and momentum conservation equations in the control volume containing the jet,

mass $$\dot{m} = \dot{m}_j + \dot{m}_e \tag{5a}$$

x-momentum $$\dot{m}_j V_j \sin\theta + \dot{m}_e V_\infty = \dot{m}u \tag{5b}$$

$$y\text{-momentum} \qquad \dot{m}_j V_j \cos\theta = \dot{m} v \tag{5c}$$

The resulting relation for the entrainment is given as the following equation.

$$\frac{\dot{m}}{\dot{m}_j} = 1 + \frac{V_j}{V_\infty}\left(\frac{dx}{dy}\cos\theta - \sin\theta\right) \tag{6}$$

The slope (*dx/dy*) in (6) can be obtained by differentiating equation (4) with the relations (3) in it. Fig. 2 shows the difference in entrainment for jets injected at different angles. It can be observed that the jet heading upstream is a more compact mixer in space compared to the normally-injected transverse jet, which results in a longer effective reburning zone – the distance from the complete mixing of the reburn fuel to the overfire air injection.

5 Model Formulation

Figure 3 shows the schematic of the facility with the modeling schemes used in each stage. Required inputs for each stage are also indicated on the left hand side. First, the main burner region is modeled as a plug flow reactor (PFR) and solved using CHEMKIN-II subroutines (Kee at al., 1992) with the known temperature profile. Then, the injection of the reburning fuel is modeled with the TSL model assuming that the 8 separate jets do not interact with each other and the jet injection temperature is kept low (300K) due to the water-cooling of the ejectors. The TSL model is used up to the point where the amount of entrainment equals the main flux and complete mixing is assumed from that point onward. The spraying of N-agent is assumed to lead to instantaneous mixing since the droplets are probably well distributed. Chemically, urea is modeled as its two decomposition products NH_3 and HNCO. The overfire air injection is modeled similarly as the injection of the reburning fuel, but assuming normal injection.

The reaction mechanism is based on the GRI-Mech-2.11 (Bowman et al., 1995). In addition, thermal $DeNO_X$ mechanisms which were proposed by Zamansky et al. (1998) are implemented to model the advanced reburning process. Since it is known that the HCCO + NO reactions are important for the reburning process (Miller at al., 1998), the reaction mechanisms and its rate coefficients are revised for these reactions. Miller et al. (1998) explained the importance of considering the branching ratios of HCCO + NO reactions, since HCNO is fated to be recycled to NO by reacting with O and OH radicals. The rate coefficients they suggested are:

$$HCCO + NO \leftrightarrow HCNO + CO \qquad k_1 = 1.17\times10^{11}\, T^{0.65}\ \text{cm}^3\text{/mole-sec} \tag{7a}$$

$$HCCO + NO \leftrightarrow HCN + CO_2 \qquad k_2 = 1.45\times10^{16}\, T^{-0.968}\exp(-648/RT)\ \text{cm}^3\text{/mole-sec} \tag{7b}$$

The branching ratio of these reactions is comparable to the value Glarborg et al. (1998) suggested for the temperature range (600K ~ 1710K) used in this modeling. In our final mechanism, these reactions are added with the rate coefficients shown in (7), and also the reactions converting HCNO to NO (HCNO + O and HCNO + OH) are included with the same rate coefficients as Miller et al. (1998). The final mechanism includes 50 species and 315 reactions. The effect of using these reaction mechanisms is compared with the original GRI-Mech-2.11 result for the basic reburning case.

6 Results and Discussion

6.1 NO Prediction

For comparison with TSL predictions, Fig. 3(a) shows the results obtained by Zamansky et al. (1998) using the time-distributed mixing model (ODF). Quantitative results and trends are lacking. Figure 3(b) shows the prediction of the NO remaining after the advanced reburning processes with different N-agent injection temperatures using the TSL model, when the reburning zone was kept just slightly rich, or SR of 0.99. The result for the basic reburning is also provided. For basic reburning, *i.e.* reburning without the addition of N-agent, the current model estimates 51% remaining NO in terms of the ratio of the final to initial NO mole fraction. The experimental result, in which water was sprayed at various temperatures instead of N-agent, show the remaining NO to be between 49 to 53%. For the advanced reburning process which injects the N-agent at various positions, and thus at different temperatures (1380K~1650K), the model predicts 30 to 40% remaining NO depending upon the N-agent injection temperatures. We also note that a significant portion of the NO reduction occurs during the mixing of the reburning fuel and overfire air. This observation and the improvement of the prediction over previous modeling (Zamansky et al., 1998) suggest that the appropriate modeling of the mixing process in addition to the correct chemical mechanism is essential for NO_X estimation.

Figure 4(a) shows the result of the basic reburning performance for the different reburn heat inputs. Higher heat input means a richer condition in the reburning zone, or lower SR. The estimations for the richer side are not as good as the cases with higher SR's. The result of using the original GRI-Mech-2.11 is provided for comparison. As we can expect, the implementation of the branching ratios of HCCO + NO reactions with the reactions converting HCNO to NO significantly reduces the overestimation. The difference with the experimental data at richer conditions could be partly attributed to the complexity in chemistry at lower SR's, but more investigation into both the chemistry and mixing issues is warranted. Fig. 4(b) shows the result for the dependence of NO reduction on the initial NO concentration for basic reburning with a reburning zone SR of 0.89. The model overestimates the reduction as observed in Fig. 4(b), but follows the trend of the experimental results closely. To confirm that the model result follows the trend of the experiment, the SR of the reburn zone is slightly changed by adjusting the amount of reburn fuel injection, and results are shown in the same plot.

6.2 Sensitivity to model parameters

The sensitivities of the model to C value ($\tau/(d_0/u_0)$) and entrainment coefficient are also investigated, but not shown for brevity. The model shows relatively high sensitivity to the flame sheet residence time, C, for advanced reburning, and moderate sensitivity for basic reburning. Glarborg et al. (1998) have performed modeling of basic reburning with a mixing model that is similar to the TSL except that it does not have the flame sheet reactor and uses an (arbitrary) exponential entrainment rate. They concluded that the effect of mixing is not important in natural gas (mixture of CH_4 and C_2H_6) reburning, which is contrary to this work. Thus, the flame sheet and determination of its residence time have a significant effect on the modeling and indeed improves the result.

The sensitivity to entrainment coefficient is not as significant as the reburn zone residence time for both basic and advanced reburning. In the basic and advanced reburning cases, the general trend is that smaller B tends to have lower reduction capability, but the dependence is weak. Insensitivity to entrainment rate was also found by Cha et al. (1998). Increase in NO level is predicted with higher jet temperatures for both basic and advanced reburning, which implies that the cooling of the jet nozzle system is essential for efficient reduction of NO level. The effect of the deflection angle of reburn fuel jet on NO level is also examined with the most significant effect of using deflected jets is the spatial compactness in mixing.

Although these results discussed in this section have not been verified by any experiment yet, even the qualitative understanding of the performance in NO_X reduction due to these parameters can be important. In this aspect, the TSL model has significant benefit over other models by containing all this information in a compact nature. Of course, the effectiveness of the TSL model in predicting the trends should be verified by the experiments in future.

7 Conclusions

The modeling of basic and advanced reburning processes using the TSL model demonstrates its effectiveness in predicting NO_X reduction. The benefit of using the TSL model comes from its implementation of the mixing process which allows continuous reaction while entrainment and other important aspects of the flow are retained. The reasonable applicability of the model, despite the uncertainties, suggests the importance of adequate modeling of the mixing process for a correct prediction of the NO reduction.

The model has a relatively high sensitivity to the flame sheet residence time, which implies that the role of the flame sheet is important in this model. The choice of the flame sheet residence time is consistent with earlier usage and with experimental measurements of the strain rate in a jet flame.

The uncertainties in the modeling can be summarized as uncertainties in entrainment rate, chemistry and radiation. Therefore, the model can be improved by more fundamental study in these aspects. These include more accurate determination of the trajectory or entrainment of the jet flames with the quantitative effect of the important factors, such as buoyancy, heat release, confinement and angle of injection. The improvement in the reaction mechanisms, rate coefficients and branching ratios can directly affect the result.

In spite of all the uncertainties mentioned, the TSL model is a significant improvement to other mixing models used in gas reburning predictions. It contains the important characteristics of the jet such as a realistic entrainment rate, the flame sheet where fast reactions occur, and the homogeneous reaction zone where internal reburn can occur. Along with the small computational burden and the connectivity to other reactor models such as PSR's and PFR's, the TSL model shows great potential as a practical design tool for realistic devices using turbulent reacting jets.

Further details on all aspects of this work can be found in Han et al. (1999).

References

Becker, H. A. and Yamazaki, S. (1978). Entrainment, momentum flux and temperature in vertical free turbulent diffusion flames, *Combust. Flame* 33:123-149.

Bowman, C. T., Hanson, R. K., Davidson, D. F., Gardiner, W. C., Lissianski, V., Smith, G.P., Frenklach, M. and Goldenberg, M., (1995). "GRIMECH-2.11, http://www.me.berkeley.edu/ gri_mech/."

Broadwell, J. E. (1982). A Model of Turbulent Diffusion Flames and Nitric Oxide Generation, Part I. TRW Document No. 38515-6001-UT-OG.

Broadwell, J. E. and Breidenthal, R. (1982). A simple model of mixing and chemical reaction in a turbulent mixing layer. *J. Fluid Mech.* 125:397-410.

Broadwell, J. E. and Lutz, A. E. (1998). A turbulent jet chemical reaction model: NOx production in jet flames, *Combust. Flame* 114:319-335.

Broadwell, J. E. and Mungal, M. G. (1991). Large-Scale Structures and Molecular Mixing *Phys. Fluids A* 3 (5):1193-1206.

Cha, C. M., Kramlich, J. C. and Kosály, G. (1998). *Twenty-seventh Symposium (International) on Combustion*, The Combustion Institute, Pittsburgh, PA.

Donbar, J. M., Driscoll, J. F. and Carter, C. D. (1998). *AIAA Thirty-sixth Aerospace Sciences Meeting and Exhibit*, Reno, NV, Paper No. 98-0151.

Glarborg, P., Alzueta, M. U., Dam-Johansen, K. and Miller, J. A. (1998). Kinetic modeling of hydrocarbon nitric oxide interactions in a flow reactor. Combust. Flame 115:1-27.

Han, D., M. G. Mungal, V. M. Zamansky & T. J. Tyson (1999). Prediction of NOx Control by Basic and Advanced Gas Reburning Using the Two-Stage Lagrangian Model, *Combust. Flame*, 119, 483-493.

Han, D., Orozco, V. and Mungal, M. G. (2000). Gross-Entrainment Behavior of Turbulent Jets Injected Obliquely into a Uniform Crossflow, *AIAA Jr.*, 38 (9), 1643-1649.

Hasselbrink, E. F. and Mungal, M. G. (1996). An analysis of the time-averaged properties of the far field of the transverse jet, *AIAA Thirty-fourth Aerospace Sciences Meeting and Exhibit*, Reno, NV, Paper No. 96-0201.

Kee, R. J., Rupley, F. M. and Miller, J. A. (1992). Chemkin II : A Fortran Chemical Kinetics Package for the Analysis of Gas Phase Chemical Kinetics, *Sandia National Laboratories* Report No. SAND89-8009.

Margason, R. J. (1968). The Path of a Jet Directed at Large Angles to a Subsonic Free Stream, NASA TN D-4919.

Miller, J. A., Durant, J. L. and Glarborg, P. (1998). *Western States Section / The Combustion Institute* 1998 Spring Meeting, Berkeley, CA, Paper No. 98S-15.

Mungal, M. G., P. S. Karasso and A. Lozano (1991). The Visible Structure of Turbulent Jet Diffusion Flames - Large-Scale Organization and Flame Tip Oscillation, *Comb. Sci. & Tech.*, 76, 165-185.

Mungal, M. G., A. Lozano & I. van Cruyningen (1992). Large-Scale Dynamics in High Reynolds Number Jets and Jet Flames, *Expts. Fluids*, 12, 141-150.

Platten, J. L. and Keffer, J. F. (1971). *J. Applied Mech*, Trans ASME v 38 Ser E n 4:756-758.

Pratte, B. D. and Baines, W. D. (1967). Profiles of a round turbulent jet in a cross flow, *J. Hyd. Div.*, Amer. Soc. Civ. Eng., HY6:53-64.

Storms, K. R., (1965). Low-Speed Wind Tunnel Investigation of a Jet Directed Normal to the Wind, Rep. 885, Aeronautics Lab., University of Washington.

Takagi, T., Shin, H. D. and Ishio, A. (1981). Properties of turbulence in turbulent diffusion flames, *Combust. Flame* 40:121-140.

Tyson, T. J., Kau, C. J. and Broadwell, J. E. (1981). "A Model of Turbulent Diffusion Flames and Nitric Oxide Generation, Part II," Energy and Environmental Research Corporation.

Wu, J. (1973). Near-field trajectory of turbulent jets discharged at various inclinations into a uniform crossflow. *AIAA J.* v 11, n 11:1579-1581.

Zamansky, V. M., Maly, P. M., Ho, L., Lissianski, V., Rusli, D. and Gardiner, W. C., (1998). *Twenty-seventh Symposium (International) on Combustion*, The Combustion Institute, Pittsburgh, PA.

Zamansky, V. M., Maly, P. M. and Ho, L. (1997). *Proceedings of the 1997 ASME International Joint Power Generation Conference*, Denver, CO.

Zamansky, V. M., Sheldon, M. S. and Maly P. M. (1998). *Western States Section / The Combustion Institute 1998 Spring Meeting*, Berkeley, CA, Paper No. 98S-30.

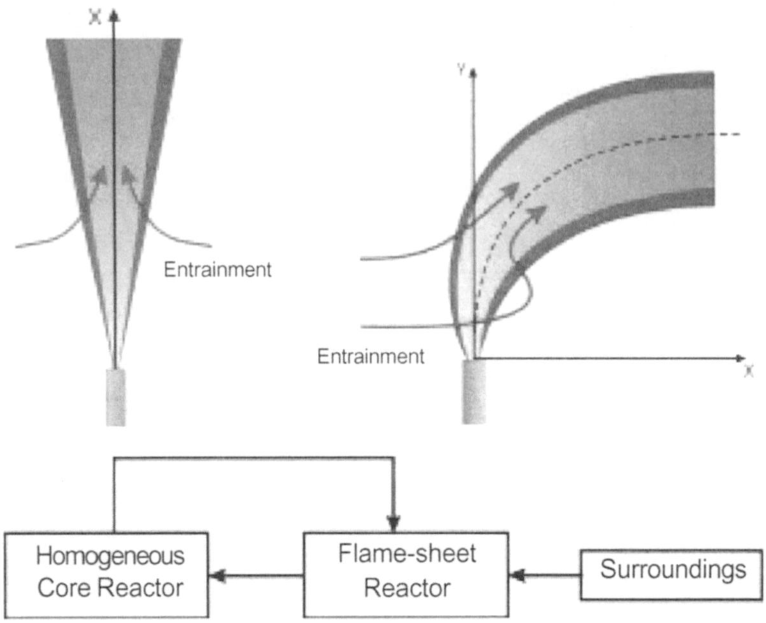

Fig. 1 Basic structure of TSL model applied to free jet flame and transverse jet flame

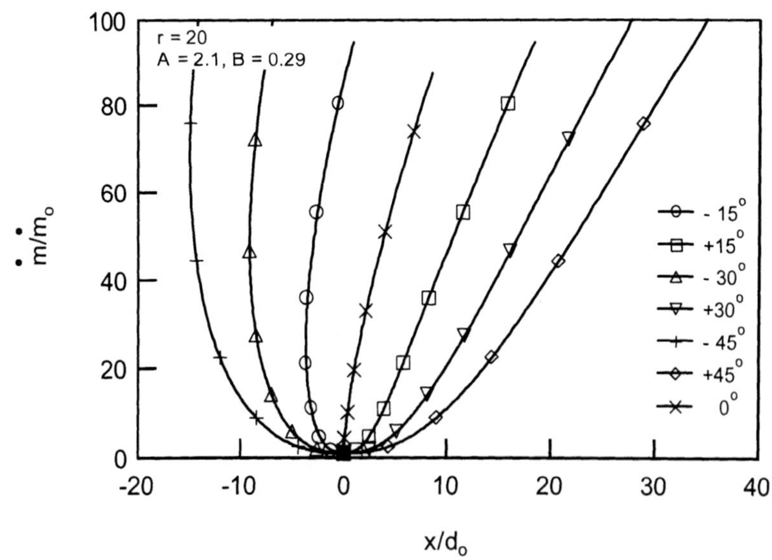

Fig. 2 Entrainment of jets injected obliquely in various angles to the crossflow following the x-direction for r = 20 jet with entrainment coefficient A = 2.1 and B = 0.29.

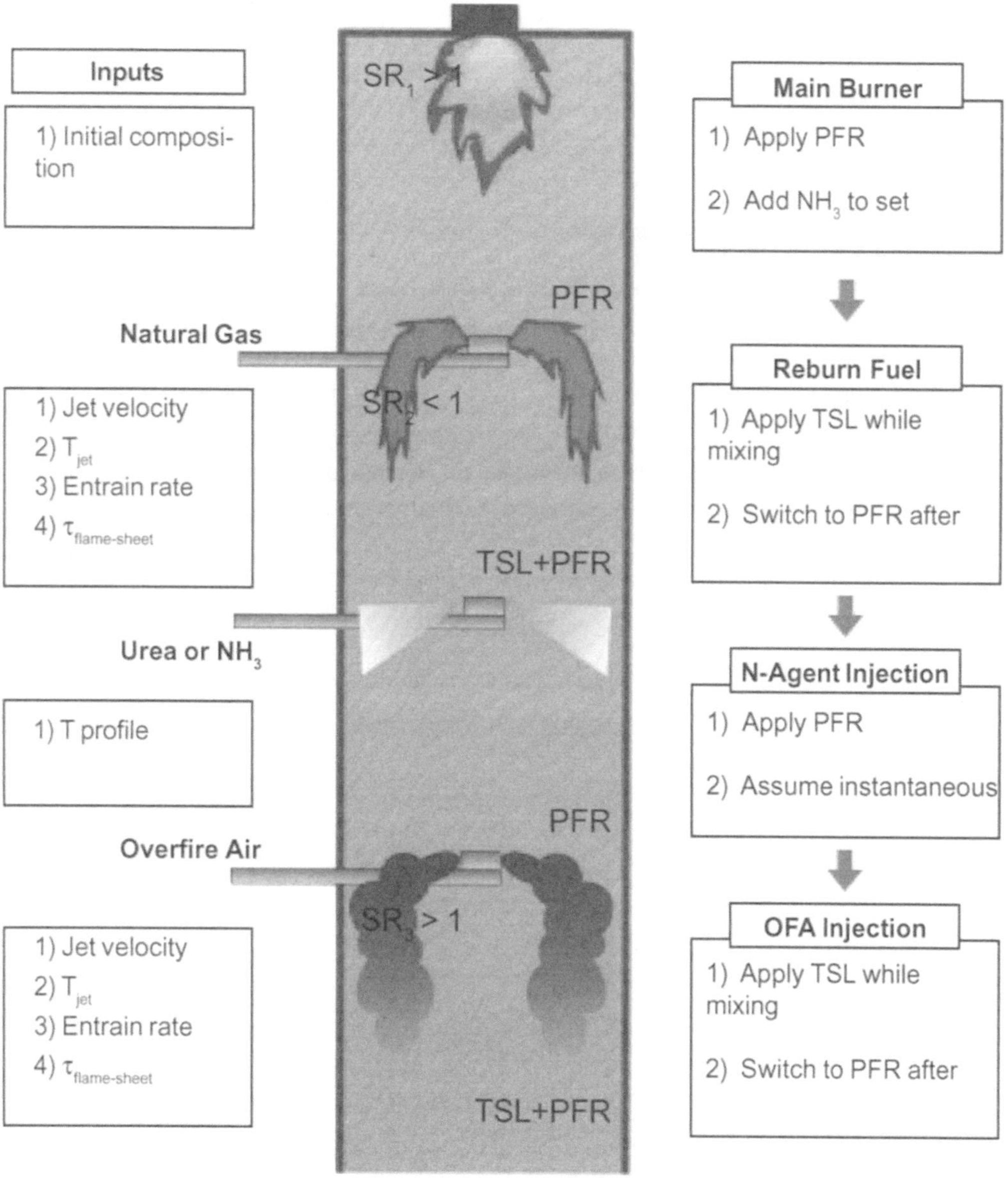

Fig. 3 The modeling scheme (right) applied to each stage with the model inputs (left) required

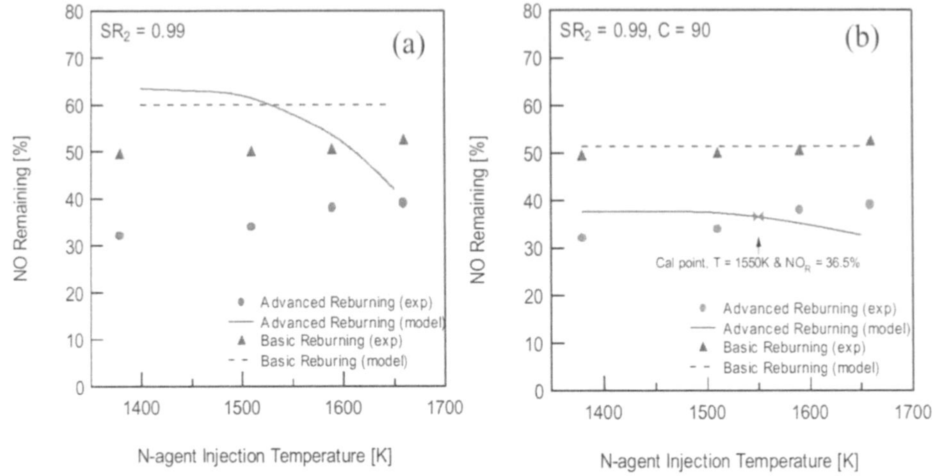

Fig. 4 The comparison between (a) time-distributed mixing model (ODF) and (b) TSL model result, for advanced and basic reburning prediction.

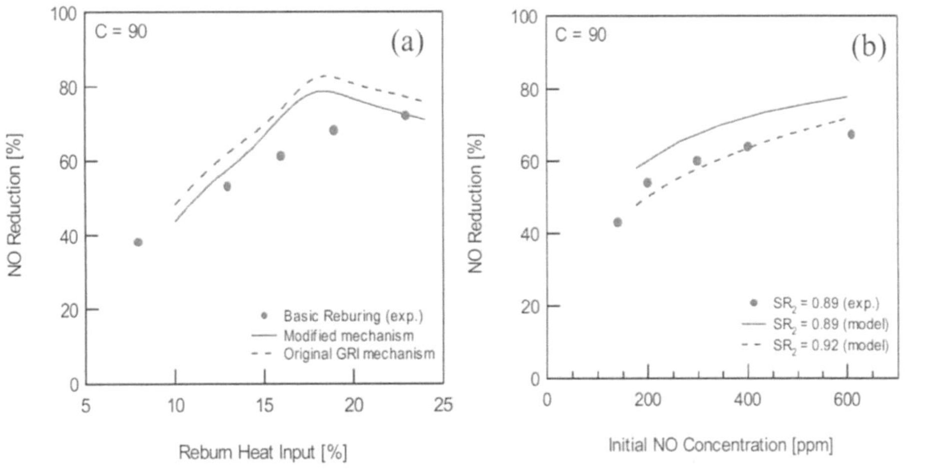

Fig. 5 The modeling results for the basic reburning process with the TSL mixing model. Each figure shows the prediction of NO level for (a) various heat input, (b) inital NO concentration

Modeling turbulent particle dispersion in transverse jets

Marina Campolo, Massimiliano Fornasier and Alfredo Soldati *

Centro di Fluidodinamica e Idraulica and Dipartimento di Scienze e Tecnologie Chimiche, University of Udine, 33100, Udine, Italy

Abstract Turbulent particle dispersion in transverse jets depends closely on the evolution dynamics of the large scale fluid structures developing from the interaction between the jet and the crossflow. In this work, we model turbulent particle dispersion in transverse jets using Lagrangian approach for particles and Large Eddy Simulation for the flow field. We show the effects of the large scales structures on particle segregation and dispersion, demonstrating the relationship between vorticity field and particle concentration. We also show the inadequacy of engineering-like approach such as steady state RANS, to model dispersion in this type of flow.

1 Introduction

Issuing a jet in a transverse flow produces a number of fluid dynamic structures which are known to mix effectively the jet stream together with the transverse flow. The extent of this mixing is crucial, for instance, to size combustion control devices, or pollution abatement devices.

In the present work, we focus on the dispersion of inertial particles entrained by a jet issued orthogonally into a transverse boundary layer. This is an introductory account of a research project with the twofold objective of *i)* characterizing the dispersion of inertial particles in a transverse jet, and of *ii)* evaluating the effect of the interaction of particle dynamics with the fluid dynamics.

The fluid dynamics of a transverse jet has been investigated in a number of studies, which focused on the identification of the large scale fluid dynamic structures characteristic of this type of flow (Fric and Rosko, 1994, Kelso et al., 1996, Papaspyros et al., 1997, Yuan et al., 1999, Rivero et al., 2001). In particular, these studies underlined the strongly time dependent nature of the vortex generation process (Kelso et al., 1996) and examined the fluid dynamic parameters controlling the generation of vortical structures (Andreopoulos and Rodi, 1984, Andreopoulos, 1985). It was observed that different structures are produced, depending on the jet-to-crossflow velocity ratio and on the Reynolds number of the crossflow (Kelso et al., 1996). Further studies focused on the role of these large scale structures in the mixing process (Chang and Vakili, 1995, Xia and Lam, 1997, Smith and Mungal, 1998, Gosh and Hunt, 1998, Hasselbrink and Mungal, 2001a and 2001b), and on the possibility to control the mixing process by fluid dynamic modifications (Lazaro and Lasheras, 1992, Davis and Glezer, 1999).

Particle dispersion and mixing was investigated with Lagrangian simulations in axi-symmetric jets (Chung and Troutt 1988) or plane mixing layers (Chein and Chung, 1988, Crowe and Troutt,

* Corresponding author. Email: soldati@uniud.it; Ph.: +390432558864; Facs: +390432558803; http://158.110.50.22

1989, Wen et al., 1992, Yang et al., 2000). These studies examined the role of the large fluid dynamic structures in the dispersion process. In particular, the Stokes number (St), which is the ratio between the particle aerodynamic response-time and a time scale characteristic of the flow field, was identified as the key factor determining enhanced dispersion. Most of the experimental studies (Chein and Chung, 1988, Crowe and Troutt, 1999, Wen et al., 1992) show that particles with $St \sim 1$ are dispersed more effectively.

In this work, we use a Lagrangian approach to model turbulent dispersion of particles in a transverse jet. In the Lagrangian approach, mixing is intrinsically related to the accurate characterization of the path history of tracers, which, in turn, is bound to rely on precise fluid dynamic data, sufficiently resolved in space and time to capture all the relevant scales and dynamics. To reproduce the time dependent flow field generated by the issuing jet, we use a Large Eddy numerical simulation. The calculated flow field is then used to simulate turbulent dispersion by Lagrangian tracking of entrained particles. We discuss the results of Lagrangian tracking based on LES calculated flow field versus the Lagrangian tracking based on steady state RANS solution of the flow field. The last approach is customary used for common engineering applications to evaluate mixing and dispersion characteristics of inertial particles. We show here the inadequacy of such approach even to reproduce the macroscopic features of the phenomenon.

2 Computational Methodology

2.1 Computational domain

We consider the reference configuration examined experimentally by Shandorov (Abramovich, 1963) in which a jet of air ($v_{jet} = 77.6\ m/s$) is injected through a pipe of diameter $D = 0.02\ m$ into a crossflow ($v_{cf} = 35.6\ m/s$).

In Figure 1, we show the computational domain investigated. The test section of the wind tunnel is represented by a box, and a pipe represents the nozzle and plenum. The reference coordinate system is centered at the jet-exit with x, y and z representing streamwise, spanwise and vertical directions, respectively. We limit the numerical analysis to a reduced portion of the wind tunnel, i.e. the region nearest to the jet exit. This allows to contain the computational cost and resource requirements necessary to perform a fine-resolved, time-dependent numerical simulation (Rudman, 1996, Yuan et al., 1999, Hsu and Guo, 2000), which can reproduce the transient flow structures characteristic of the jet in crossflow such as oscillating wakes and the roll-up of horseshoe vortices (Fric and Rhosko, 1994).

Using the pipe diameter D as the reference length, the box is 13 D high, 8 D wide and 10 D long. The pipe is 3 D long, and enters the box 3 D downstream the crossflow inlet. Results from our previous works (Degano et al., 2002, Campolo et al., 2002) based on steady state simulations show that modeling a 2 D long pipe allows to reproduce correctly the interactions between crossflow and pipe-flow. In this work, we fixed the pipe length to 3 D to account for possible flow oscillations not considered by the previous steady state simulations. Similar dimensions of the computational domain are considered in time dependent simulations of the jet in crossflow reported by Yuan et al. (1999) and Rudman (1996). These calculations show that transient simulations made for this region allow to capture the wake vortex system, the jet shear layer vortices and the counter-rotating vortex pair.

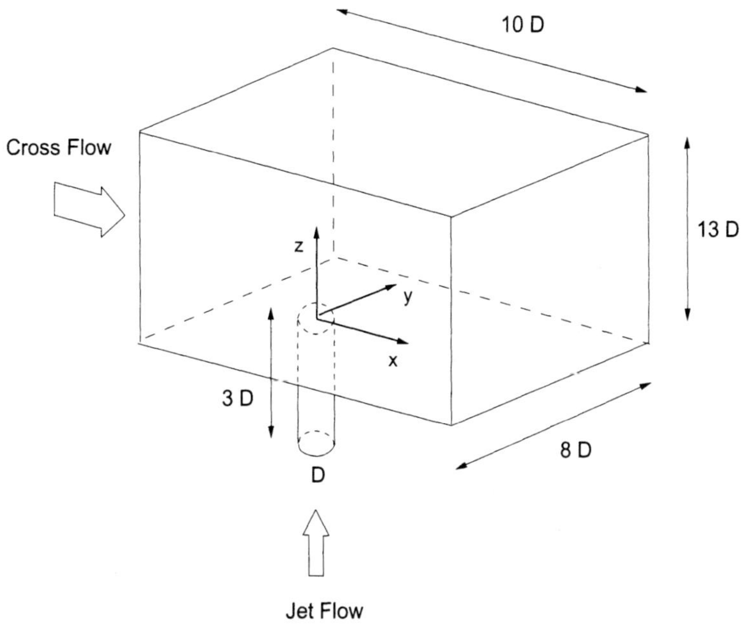

Figure 1. Geometry of considered experimental apparatus and reference coordinate system, centered at jet exit. Box represents test section of wind tunnel and pipe models jet outlet. Crossflow inlet section is at $-3\ D$.

We discretize the computational domain into finite volumes, generating nearly hexahedric cells and we refine the grid in the pipe and in the near-field of the jet adopting a hierarchical cell refinement. We determine the final geometry and size of the computational domain through a grid sensitivity analysis. The final computational mesh is made up of $256,752$ cells, of which $20,736$ for the pipe and $188,592$ in the region nearest to the jet exit. These regions correspond to about $1\ \%$ and $28\ \%$ of the overall volume, respectively.

2.2 Flow field calculation

We solve for the fluid balance equations using a commercial code (Star-CD). To simulate the flow, we impose *i)* a boundary layer velocity profile of the crossflow at the box inlet and *ii)* a fully developed velocity profile at the pipe inlet. Furthermore, we impose *iii)* a pressure condition at the lateral wall and at the outlet of the box, *iv)* a free stream condition at the upper wall and *v)* no-slip conditions at the lower wall. We use the pressure condition to avoid a dominant coupling between the outer flow and the jet field evolution (Sani and Gresho, 1994), and we use the upper free stream to avoid the formation of an upper boundary layer.

The reference conditions investigated experimentally by Shandorov (Abramovich, 1963) correspond to a jet-to-crossflow velocity ratio, $\alpha = v_{jet}/v_{cf}$, equal to 2.2 and to a jet Reynolds number, $Re_{jet} = \rho v_{jet} D/\mu$, of more than 10^5, well into the fully turbulent regime. For these conditions, we perform the calculation of the time dependent evolution of the jet using a large eddy simulation (LES) approach.

The rationale of LES is based on the full solution of the Navier-Stokes equations with a spatial resolution larger than the smallest turbulent vortices, which are modeled by sub-grid terms. The idea is that whereas larger scales must be resolved explicitly, since they are characteristic of the specific phenomenon under investigation, smaller turbulent scales are somehow more universal and easy to model (Germano et al., 1991, Salvetti and Banerjee, 1995). In this work, we use a Smagorinsky model to simulate turbulence scales not resolved by the grid resolution.

We choose the time step for our LES calculation considering the basic time dependent phenomena that characterize the jet in crossflow. Similarly to the flow past a solid body, we define the Strouhal number for the jet in crossflow as fD/v_{cf}, where f is the frequency for vortex shedding. In the case of the jet, f is related to the shedding of large scale structures associated with puffing or roll-up of vortex-rings. Andreopoulos (1985) reports experimental values equal to 0.41 for the Strouhal number when conditions are similar to the one investigated here ($\alpha = 2$, $Re_{cf} = 1 \div 4 \cdot 10^4$). Kelso et al. (1996) evaluate a Strouhal number equal to 0.65 close to the frequency of the unforced shear layer roll-up. Hsu et al. (2000) calculate a Strouhal number equal to 0.13 for jet-to-crossflow velocity ratios from 3 to 5 and crossflow Reynolds number from 3,800 to 11,400. We calculated the time-period for vortex shedding using the suggested values of Strouhal number and our geometrical and velocity data. Then, we fixed the time step equal to $\Delta t = 1.232 \cdot 10^{-5} s$ for the flow field calculation. This time step is small compared to the flow field characteristic period for vortex shedding and is appropriate for the subsequent Lagrangian tracking of particles.

The numerical simulation started from still fluid, and we let jet and crossflow evolve until an oscillating pseudo-periodic flow field was obtained. This was achieved after 2,450 time steps, corresponding to a time period of $0.687 \cdot 10^{-1} s$. At this point, we assumed that the calculated flow field described the fully developed, time dependent evolution of the jet.

We simulated the jet evolution for 1,000 time steps further, saving data every 10 time steps for subsequent particle tracking. The overall simulation required 300 hours of CPU time on a Sun Ultra 10, 400 Hz processor, 512 Mb RAM, and storage of flow field data required 2GB space. Computational costs and storage disk requirements for this simple configuration are large enough to make alternative approaches, less computationally intensive, more attractive.

2.3 Evolution of flow structures

To examine the time evolution of the flow structures generated by the interaction between jet and crossflow, we focus first on the velocity magnitude, considering the velocity variations in the $x - z$ plane ($y/D = 0$), and in a $y - z$ plane located at $x/D = 4$.

Consider first Figure 2 (a), which corresponds to time $0.718 \cdot 10^{-1} s$ from the start of the simulation. The domain shown is $-1 \div 7\ D$ long in the streamwise direction and $0 \div 8\ D$ high in the vertical direction. We show iso-contours of $x - z$ components of velocity, using a solid line for the crossflow velocity value ($35.6\ m/s$). Dashed lines represent larger velocity magnitudes (from 40 to 90, step 10 m/s), and dash-dotted lines represent lower velocity magnitudes (from 10 to 30, step 10 m/s). We observe that the crossflow velocity is reduced in the region upstream of the jet, because of the blocking effect, and is reduced also in the lower part of the jet (the wake region), where complex three-dimensional structures develop. The dashed contours identifying the jet issuing from the pipe show that the shape of the jet penetrating into the crossflow is fluctuating, with high velocity regions (A) detaching periodically from the tip of jet (B). For each high velocity region (A and B), a corresponding low velocity region is found (A' and B') in the lower part of the jet. High velocity regions are also present in the part beneath the jet, indicated as C. Likely, they are the outcome of the crossflow fluid which is entrained back under the jet.

In Figure 2 (a), we show also the trace of the vertical plane located at $x/D = 4$. The iso-contours of velocity magnitude for this plane are shown in Figure 2 (b). The domain shown is $-4 \div 4$ D wide in the spanwise direction and $0 \div 8$ D high in the vertical direction. Solid line is again the crossflow velocity, dashed lines represent larger velocity magnitudes (40, 45 and 50 m/s) and dash-dotted lines represent lower velocity magnitudes (from 5 to 30, step 5 m/s). The vertical plane corresponding to the domain centerplane ($y/D = 0$) is also shown for reference. We can observe that, at the time considered, the main high velocity region of the jet is shifted on the left half of the domain. Small high velocity regions can be observed in the right half of the domain, where the low velocity region located between the high velocity core and thc wall is the prevalent structure. Near to the lower wall, the high density of isocontours of lower velocity magnitudes indicates that in this zone the boundary layer profile of the crossflow is not modified by the jet.

To examine the time evolution of the flow field, we considered subsequent time-steps of the flow field simulation. Comparing subsequent snapshots of the velocity magnitude in the domain centerplane, we found that after a time equal to $1.5 \cdot 10^{-3}s$ the velocity field is most similar to the snapshot shown in Figure 2 (a). Therefore, we assumed this time period representative of jet tip oscillation. This is in good agreement with the time period for jet oscillation estimated by Andreopoulos (1985), which is $1.3 \cdot 10^{-3}s$. We chose to represent the flow field evolution of the jet with snapshots taken at time $0.723 \cdot 10^{-1}s$ (Figure 2 (c) and (d)) and $0.728 \cdot 10^{-1}s$ (Figure 2 (e) and (f)), which correspond to $1/3$ and $2/3$ of the jet oscillation time period, respectively.

Consider first the flow field evolution in the domain centerplane (Figure 2 (a), (c) and (e)). In the upper part of the jet, the region A which starts separating in sub-picture (a) moves downstream, stretching the high velocity contour of the jet (c) and finally separating completely from the jet tip (e). In the same time, region B, which corresponds to the tip of the jet at time (a), follows a very similar evolution. At time (e), the jet tip is identified by a newly born region D. If we consider the trajectory of high and low velocity structures in the upper part of the jet and their corresponding time of travel, we can estimate a convective velocity for these paired structures about 26.4 m/s, which is 0.74 v_{cf}, i.e. lower than the crossflow velocity. Turning now the attention to the lower part of the jet, we observe that also the high velocity region C travels downstream. In Figure 2 (e) we can observe that a newly born, high velocity region (E) appears in the lower part of the jet. The position at which this structure is born, indicates that its origin from the crossflow is most probable. If we evaluate now the trajectory of the high velocity structures in the lower part of the jet (C and E), we can calculate a lower convective velocity compared to the upper part of the jet, which is about 22.3 m/s and corresponds to 0.63 v_{cf}. As already observed by Kelso et al., (1996), convective velocities of the eddies vary widely over the velocity field.

If we consider now the flow field evolution in the plane orthogonal to the bent jet (Figure 2 (b), (d) and (f)), we can observe that the high velocity and low velocity puffs travelling downstream determine variation of the flow in the transverse section. We observe that the upper high velocity region of the jet, that is mainly located in the left half plane in Figure (b), moves toward the center (d) and then toward the right half plane (f). This indicates a spanwise alternate oscillation of the jet already observed in experiments (Smith and Mungal, 1998), which is likely due to the alternate shedding of structures from opposite jet sides. This alternate motion determines intense exchange of fluid between the right and the left plane which contributes to transverse mixing. A

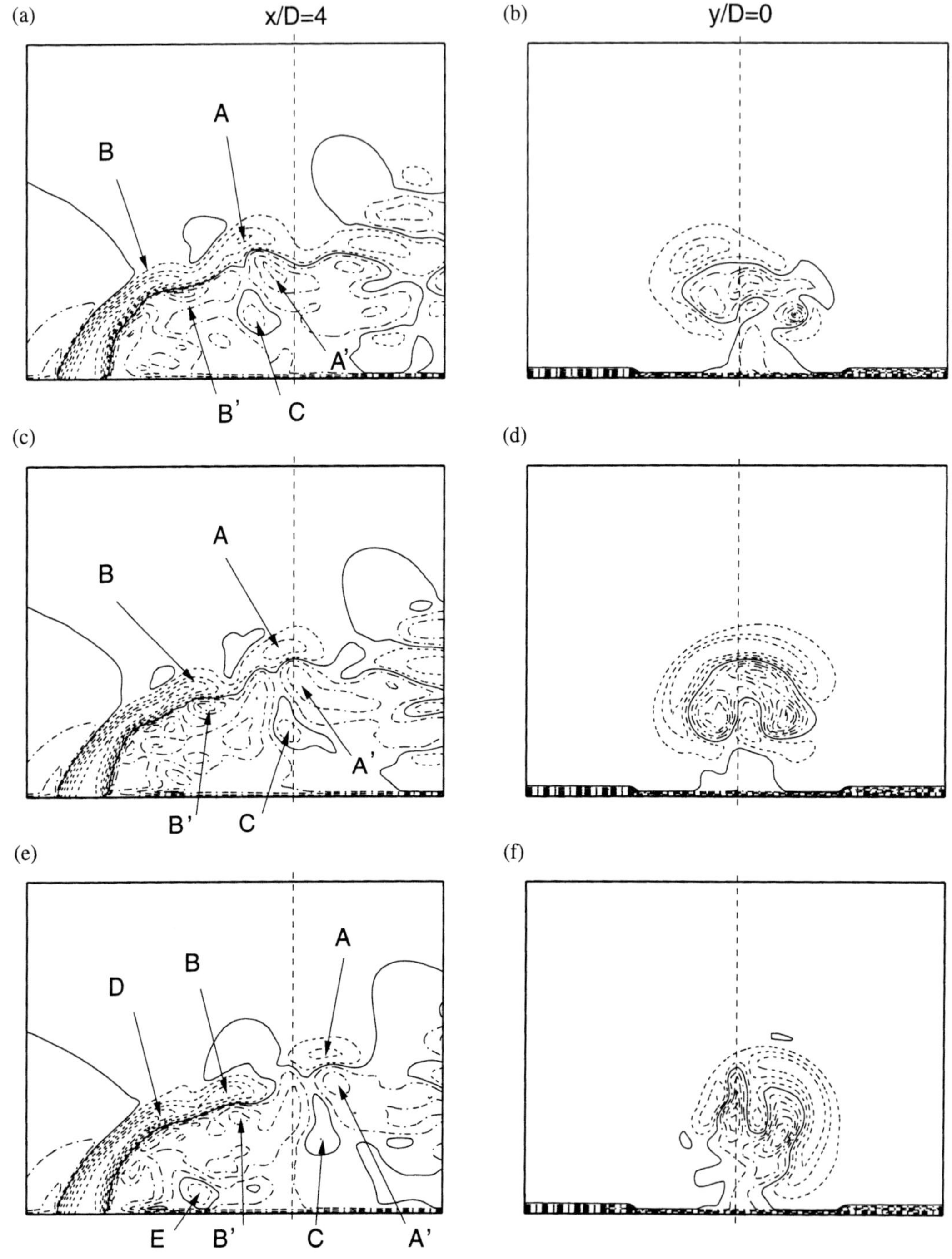

Figure 2. Velocity magnitude in plane $y/D = 0$, (a), (c) and (e) and in plane $x/D = 4$, (b), (d) and (f). Domain is $-1 \div 7\ D$ long in streamwise direction, $0 \div 8\ D$ high in vertical direction and $-4 \div 4\ D$ wide in spanwise direction. Snapshot times are $0.718 \cdot 10^{-1}\,s$, $0.723 \cdot 10^{-1}\,s$ and $0.728 \cdot 10^{-1}\,s$, from top to bottom. Velocity isocontours are from 10 to 30, step 10 m/s (dash-dotted line), 35.6 m/s (solid line) and from 40 to 90, step 10 m/s (dashed line) in (a), (c) and (e); from 5 to 30, step 5 m/s (dash-dotted line), 35.6 m/s (solid line) and from 40 to 50, step 5 m/s (dash-dotted line) in (b), (d) and (f).

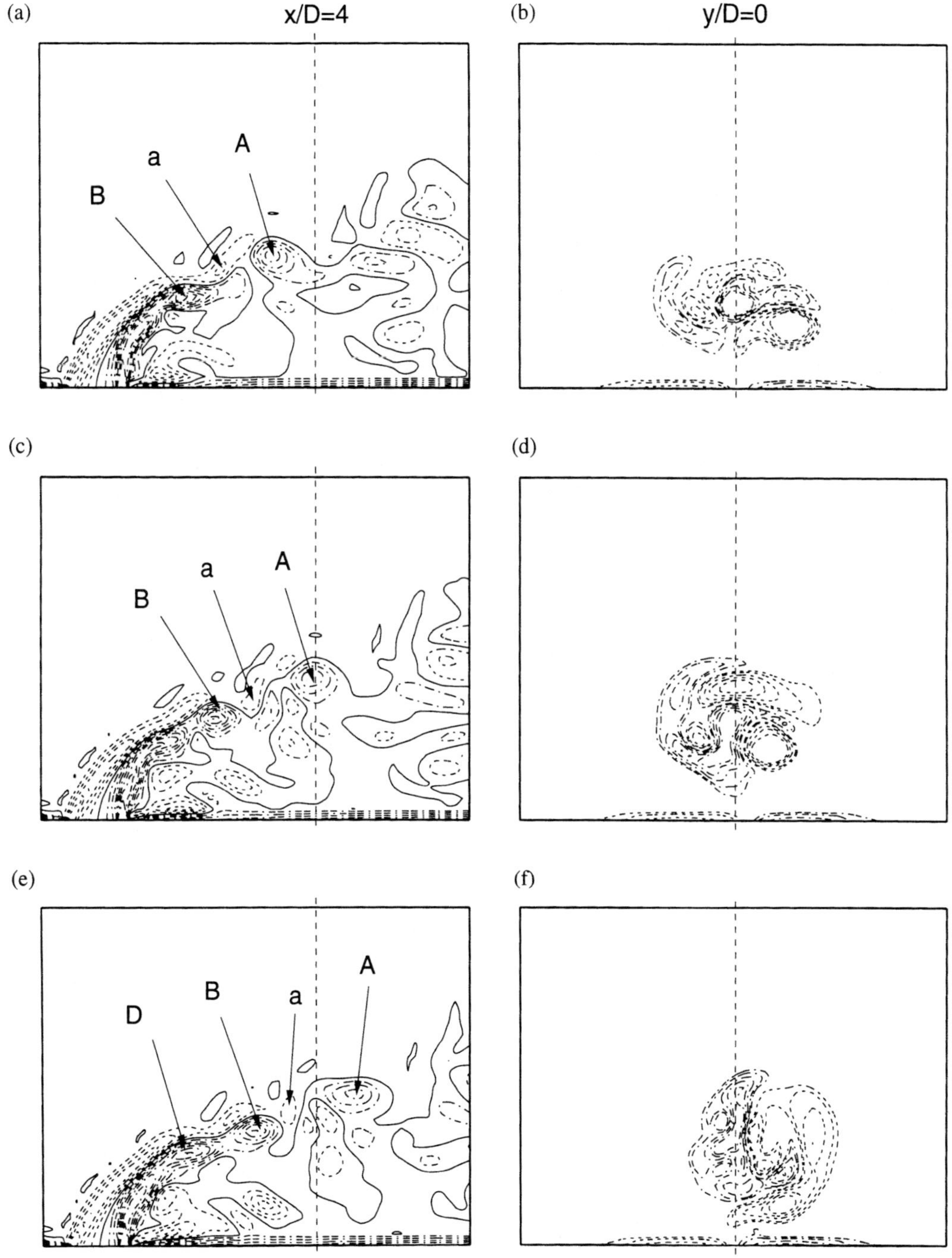

Figure 3. Spanwise vorticity in domain centerplane, $y/D = 0$, (a), (c) and (e) and streamwise vorticity in plane $x/D = 4$, (b), (d) and (f). Domain is the same as in Figure 2. Snapshot times are $0.718 \cdot 10^{-1} s$, $0.723 \cdot 10^{-1} s$ and $0.728 \cdot 10^{-1} s$, from top to bottom. Dash-dotted lines, solid line and dashed lines identify positive vorticity, zero vorticity and negative vorticity, respectively. Vorticity isocontours are from -350 to 350, step 50 s^{-1} in (a), (c) and (e) and ± 15, ± 25, ± 45, ± 65, and ± 85 s^{-1} in (b), (d) and (f).

limited portion of the domain, located under the jet and near the wall, is characterized by low velocity values during all the different phases of jet oscillation. However, this region is not left out from the enhanced fluid exchange region, since when the high velocity region moves toward the right or the left plane, low velocity fluid rises up to the high velocity core of the jet.

The periodical modifications in the velocity field are related to the development of vortical structure generated by the interaction between jet and crossflow. We evaluated the relationship between velocity variations and structures that form considering the evolution of the vorticity field. In Figure 3, we show the contours of spanwise vorticity in the same domain and at the same instants of Figure 2. In Figure 3 (a), dashed lines identify isocontours of positive vorticity, i.e. counter-clockwise rotation, dash-dotted lines identify isocontours of negative vorticity, i.e. clockwise rotation, and the solid line is the zero vorticity isocontour. Values for isocontours are from -350 to 350, step 50 s^{-1}. Pictures similar to Figure 3 (a) are reported by Kelso et al. (1996), who observed the jet development using a natural forcing to identify the vortical structures. Comparison between Figure 2 (a) and Figure 3 (a) shows that, for each couple of high velocity-low velocity regions, a closed contour of spanwise vorticity exists. Corresponding structures are indicated with same labels in Figures 2 and 3. We observe that, for each high velocity-low velocity region detaching from the jet tip, a spanwise oriented, clockwise rotating (negative vorticity) structure forms. These structures correspond to vortex roll-up generated by shear at the interface between the high speed jet and the crossflow and expand in the lower side of the jet. We observe that weaker, closed vorticity structures of opposite sign, indicated as a, exist in the upper side of the jet. These structures are weaker because the velocity difference between the jet and the upper crossflow is smaller than in the lower side of the jet.

We examined also streamwise vorticity distribution in the plane located at $x/D = 4$, which is indicated in Figure 3 (a) as a dashed, vertical line. The calculated isocontours of streamwise vorticity are shown in Figure 3 (b). Dashed lines identify negative vorticity, i.e. clockwise rotation (discrete values are -15, -25, -45, -65 and -85 s^{-1}), dash-dotted lines identify positive vorticity, i.e. counter-clockwise rotation (discrete values are 15, 25, 45, 65 and 85 s^{-1}). The trace of the domain centerplane is also shown for reference. Near the lower wall, we observe that the vorticity distribution is symmetric. As we move away from the wall, we observe that the vorticity in the jet region becomes rather complex, and asymmetric. As discussed by Smith and Mungal (1998), the instantaneous fingerprint of the counter-rotating vortex pair is not symmetric, and symmetry often observed for this structure is produced by the time-averaging process used in experimental observation. We observe that in the body of the bent jet the fluid is arranged in layers which rotate in opposite directions, contributing to enhance the transverse mixing within the stream.

To evaluate the evolution of the spanwise oriented and streamwise oriented vortical structures, we considered subsequent time steps representations of the vorticity field. Consider first the evolution of spanwise oriented structures in the domain centerplane (Figure 3 (a), (c) and (e)). Again, the snapshots describe the cycle of natural oscillation for the jet. We observe that negative vorticity regions A and B move downstream as soon as they detach from the jet tip. The positive vorticity regions indicated by a which generates from the upper part of the jet moves downstream, but is rapidly entrained into the jet, moving in between the two negative vorticity regions. This process reduces the vorticity content of positive vorticity structures which are damped by the flow. The role of negative, spanwise vorticity regions in the mixing process can be very important, since these structures contribute to flow field stretching in the streamwise direction. Moreover, since the convective velocity is larger for structures generated from the jet tip than from the crossflow, this may result in enhanced longitudinal dispersion within the lower part of the jet in crossflow.

Finally, we turn the attention to Figure 3 (b), (d) and (f), in which we show the evolution of the counter-rotating vortex pair. It is clear from the sequence of images that the counter-rotating vortex pair oscillates strongly between the two half planes. This oscillation is most probably related to the alternate vortex shedding from the opposite sides of the jet. The oscillation produces intense fluid exchange between the two half planes and justifies the significant asymmetry observed experimentally in scalar field concentrations (Smith and Mungal, 1998).

2.4 Calculation of particle dispersion

To describe particle motion, we used a Lagrangian approach, tracking swarms of particles in the flow field.

The trajectory of each particle is calculated by integrating explicitly over time the equation of motion. The basic assumptions in the particle motion analysis are: *i)* all particles are non-deformable solid spheres; *ii)* the density of the particle is large compared to the density of the fluid; *iii)* particle-particle interactions are neglected; *iv)* the effect of the particles on the flow is neglected; *v)* virtual mass force, pressure gradient force and Basset force are neglected. The analysis reported by Maxey and Riley (1983) supports the simplifications adopted in this work, since the particle-to-fluid density ratio (829.9 in the present simulation) is large.

The equation of particle motion includes inertial, drag and gravity force, and can be written as:

$$\rho_p \frac{\pi D_p^3}{6} \frac{d\mathbf{v}_p}{dt} = \frac{1}{2} C_D \frac{\pi D_p^2}{4} \rho (\mathbf{v} - \mathbf{v}_p)|\mathbf{v} - \mathbf{v}_p| + \frac{\pi D_p^3}{6} (\rho_p - \rho)\mathbf{g} \tag{1}$$

where, ρ_p, D_p and $\mathbf{v}_p$ are particle density, diameter and velocity, ρ and $\mathbf{v}$ are fluid density and velocity, t is time, $\mathbf{g}$ is gravity and $C_D = f(Re_p)$ is the Stokes coefficient for drag, where $Re_p = \rho D_p|\mathbf{v} - \mathbf{v}_p|/\mu$ is the particle Reynolds number.

To calculate the drag force, we need the fluid velocity, $\mathbf{v}$, at each particle position. Since the LES approach calculates the flow field for computational grid points only, first, we need an efficient algorithm for particle location that allows to identify the grid points nearer to the particle position; second, we need to perform grid-value interpolation to calculate fluid velocity at particle position.

We used the algorithm described in Zhou and Leschziner (1999) to locate particles in the unstructured mesh, and a linear interpolation among the four nodes nearest to the particle position for velocity interpolation. We used the Stokes coefficient for drag, with non linear correction for particle Reynolds number, Re_p, larger than unity (Chein and Chung, 1988), which can be written as:

$$C_D(Re_p) = \frac{24}{Re_p} \left(1 + 0.15 \cdot Re_p^{2/3}\right) . \tag{2}$$

We simulated the behavior of water particles (density $\rho_p = 1000\ kg/m^3$) with diameter equal to 1 and 5 μm, for which the characteristic times are $3.086 \cdot 10^{-6} s$ and $7.716 \cdot 10^{-5} s$. These values are small enough to allow a full interaction between particles and fluid dynamic structures, which makes possible to evaluate the effect of these structures on particle dispersion. The Stokes number, given by $St = \tau_p/\tau_f$, where $\tau_f = D/U_{jet}$, is equal to $1.2 \cdot 10^{-2}$ and $2.9 \cdot 10^{-1}$, indicating that particle behavior is similar to that of fluid tracers.

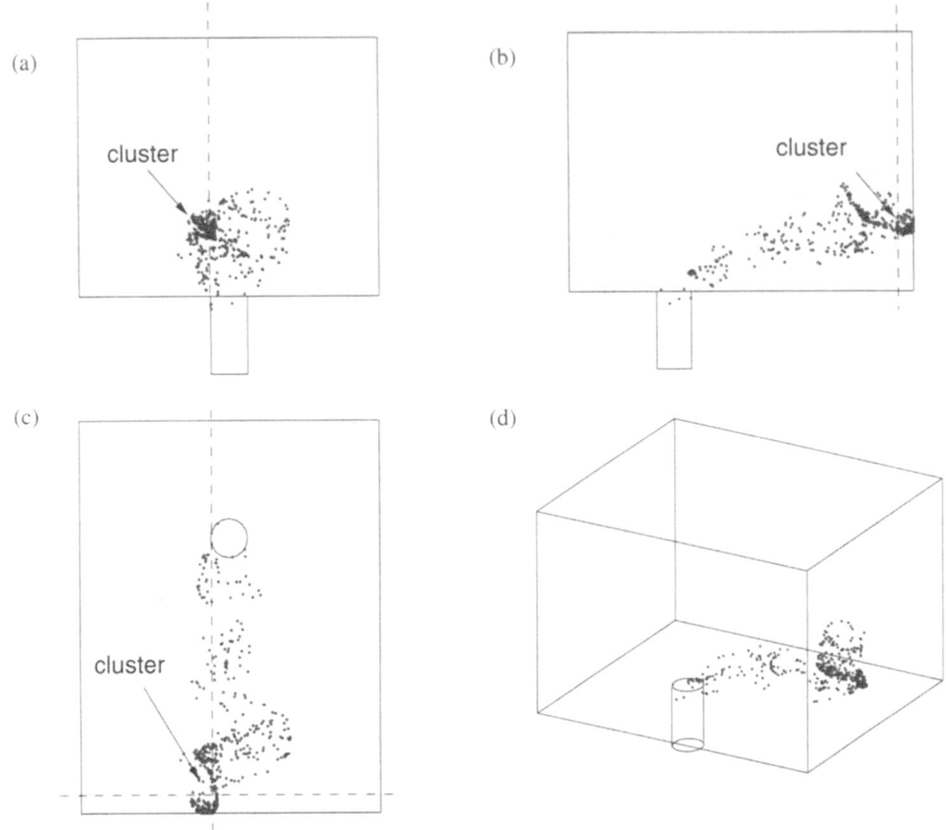

Figure 4. Particle tracking in LES flow field. Swarm of $1,000$ particles ($D_p = 1\ \mu m$) injected from $z/D = -3$. Box shown is $-3 \div 7D$ long in streamwise direction, $-4 \div 4D$ wide in spanwise direction and $0 \div 10D$ high in vertical direction. Snapshot corresponds to time $t = 3.696 \cdot 10^{-3}\ s$ after particle injection.

Particles were released at the pipe inlet ($z/D = -3$), uniformly distributed in the section of the pipe and with a starting velocity equal to the local fluid velocity. We fixed the time step for integration as one tenth of the particle characteristic time, $\tau_p = \rho_p D_p^2/18\mu$, to simulate in detail the variation in particle velocity. The flow field was updated every $1.232 \cdot 10^{-4} s$, i.e. the time chosen for flow field recording from LES. This stepwise variation of the flow is a satisfactory compromise which allows to save disk space and yet to evaluate the effect of the time dependent convective structures on particle dispersion directly.

2.5 Particle behavior

In Figure 4, we show a snapshot of the particle dispersion pattern in the computational domain. The domain shown is a box, which is $-3 \div 7\ D$ long in the streamwise direction, $-4 \div 4\ D$ wide in the spanwise direction and $0 \div 10\ D$ high in the vertical direction, and the pipe, which is $3\ D$ long. The snapshot refers to time $t = 3.696 \cdot 10^{-3} s$ after particle injection, when the first particle reaches the outlet section, located at $x/D = 7$.

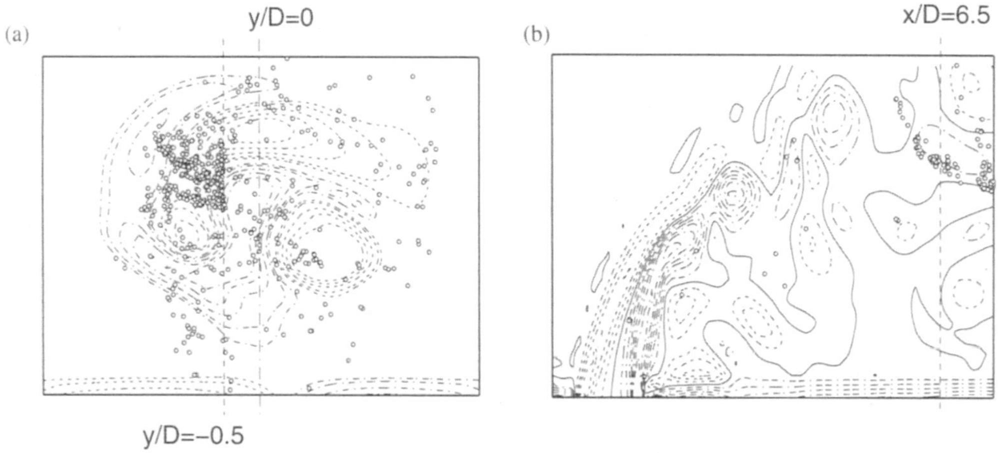

Figure 5. Particle distribution in time dependent flow field at time $t = 3.696 \cdot 10^{-3}$ s after injection. Domain is $-1 \div 7$ D long in streamwise direction, $0 \div 4$ D high in vertical direction and $-2 \div 2$ D wide in spanwise direction. Particles accumulate in region between vortical structures: (a) streamwise vorticity and particle distribution in plane $x/D = 6.5$; (b) spanwise vorticity and particle distribution in plane $y/D = -0.5$. Isocontours of vorticity are as in Figure 3.

Consider first Figure 4 (a), which is an end-side view of the tunnel. It is apparent that distribution of particles is not symmetric. The initial distribution of particles is uniform in the pipe cross section and the downstream asymmetric distribution is caused by the presence of time dependent structures. In Figure 4 (c) we show the top view of the computational domain at the same time step. We observe that particles distribute along a meandering structure which is similar to the wake past a solid cylinder. This again suggests that particles organize in preferential regions of the flow, which are most likely related to the flow structures that form. Moreover, we observe that even if all particles are injected in the pipe at the same time, they spread over a large distance in the streamwise direction. The side view (Figure 4 (b)) and the 3-D view (Figure 4 (d)) give an idea of the complex distribution of particles. We only observe here that velocity variations in both space and time are required for this complex distribution to form. Animations of particle dispersion available at our web site (http://158.110.50.22) show that the non uniform velocity profile inside the pipe causes particles to emerge from the pipe at different times. However, particles emerging symmetrically from the pipe at the same time step behave differently in the flow field, depending on the concurrently evolving fluid dynamic structures. A very similar behavior is observed also for 5 μm particles (not shown here).

To evaluate the role of time dependent structures on the preferential accumulation of particles, we examine in detail the cluster of particles indicated by arrows in Figure 4. We examine particles position together with the flow field structures, limiting our analysis to the planes containing the position of the center of mass of the cluster. These planes are shown in Figure 4 (c) as dashed lines and are located at $y/D = -0.5$ and $x/D = 6.5$, respectively. In Figure 5, we show streamwise and spanwise vorticity components and we superimpose the position of particles which are up to $\pm D/10$ away from the planes shown. Consider first Figure 5 (a). The domain is $-2 \div 2$ D wide in the spanwise direction and $0 \div 4$ D high in the vertical direction. Isocontours identify streamwise vorticity using the same scale of Figure 3 (b), (d) and (e). The domain centerplane

($y/D = 0$) is indicated by a dash-dotted line whereas the plane containing the center of mass of the cluster of particle is indicated by the dashed line. We can observe that the particles of the cluster in the plane $x/D = 6.5$ concentrate between the positive and negative vorticity contours and in the region between two negative vorticity islands. Rotation of the fluid associated to the vortical structures causes particles to be propelled away from the vortex core and to accumulate in the region between vortices. We turn now the attention to Figure 5 (b), in which the domain shown is $-1 \div 7$ D long in the streamwise direction and $0 \div 4$ D high in the vertical direction. We show spanwise vorticity isocontours, which are the same as indicated in Figure 3 (a), (c) and (e). The plane corresponding to Figure 5 (a) is indicated as a dashed vertical line for reference. The number of particles in this plane is small, and particles are mainly concentrated near the $x/D = 6.5$ plane. Particles accumulate along the boundary of a spanwise oriented negative vorticity region. The distribution is stretched in the streamwise direction for the effect of the spanwise vortex (the dash-dot line corresponds to clockwise rotation). This and similar spanwise structures are most likely responsible for the large dispersion of particles in the streamwise direction.

2.6 Particle tracking in time-averaged flow field

Our numerical results show that if the time evolution of the flow field is considered, particle behavior can be described accurately. Such a detailed flow field characterization is necessary even if the flow field is steady in a statistical sense. However, in most engineering applications, the time-dependence of the flow field is too costly to model, and particle dispersion is evaluated using steady state solutions of the flow field. In this work, we evaluated the adequacy and the limitations of this simplified approach for the specific case of particle dispersion in a jet in crossflow. We calculated the steady state RANS solution for the jet and performed again the Lagrangian tracking. The calculation of the time-averaged flow, performed on a Sun Ultra 10, 400 Hz processor, 512 Mb RAM, required a total time of six hours, which is a small fraction of the time required to perform the LES simulation.

The flow field calculated by steady state RANS is very different from the instantaneous LES results. Differences disappear when the LES solution is averaged over a time period longer than at least two mean through flow times. In other papers (Campolo et al., 2002, Degano et al., 2002), we described the steady state structures for a jet in crossflow and we observed that the steady state RANS solution is symmetric and filters out the couples of high velocity-low velocity regions periodically shed from the jet, which are reproduced by the LES calculation. The comparison between vorticity fields calculated from LES and RANS shows that the spanwise vorticity associated to the periodic roll-up of ring vortices is completely suppressed in the RANS solution. The main vortical structure which is left in the flow is the counter-rotating vortex pair, which is perfectly symmetric in each transverse plane.

We performed one simulation of the Lagrangian dispersion of particles into the RANS flow field integrating Equation 1. We performed a second simulation including an extra term in Equation 1. This term is a stochastic forcing function which we use to reproduce the effect on particle motion of the turbulent scales filtered out by the RANS approach. At each time step, the turbulent fluctuation forcing particle motion is randomly generated from a normal distribution with a mean square displacement proportional to the local turbulent kinetic energy of the fluid (Gosman and Ioannides, 1981). The turbulent fluctuation is assumed to be isotropic, and fluctuations in the three directions at each time step are independent variables. Particle behavior is similar

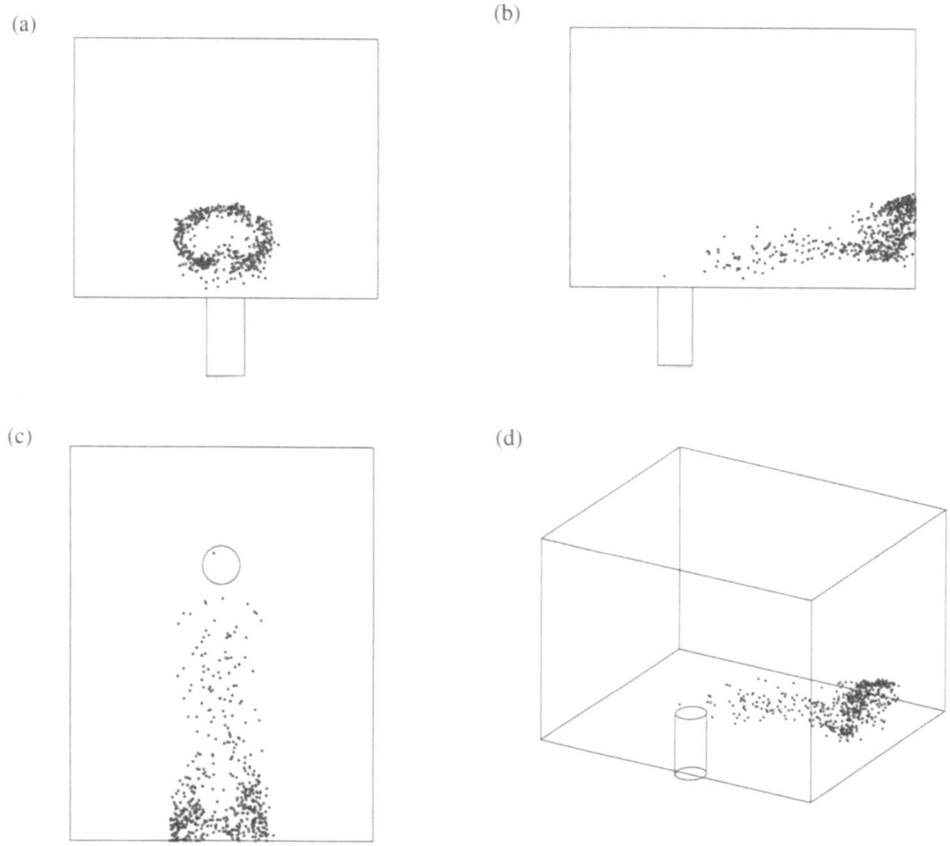

Figure 6. Particle tracking in RANS flow field with stochastic forcing proportional to local turbulent kinetic energy. Domain and conditions for injection are as in Figure 4. Snapshot corresponds to time $t = 4.22 \cdot 10^{-3}$ s after injection.

for the two simulations. However, large differences can be observed by comparing the RANS Lagrangian to the LES Lagrangian simulation. Here, we will present only the results relative to the RANS Lagrangian including the stochastic forcing for 1 μm particles, which are shown in Figure 6.

The domain shown is the same as in Figure 4 and the snapshot is taken at time $t = 4.22 \cdot 10^{-3} s$ after particle injection, i.e. the time at which the first particle exits from the computational domain. This time is slightly larger than the one found from the LES simulation. In the LES simulation, the wake structures and the roll-up vortices modulate the velocity of the jet as subsequent high and low velocity puffs. Particles in the high velocity puffs decelerate less whereas particles entrained in the low velocity puff decelerate more. This determines a reduced time of travel for the particles moving faster, and has also effect on the overall distribution of particles in the domain. If we compare particle dispersion in Figure 6 and in Figure 4, we observe that particles motion in the RANS flow field is more ordered than in the LES flow field. From the end-side

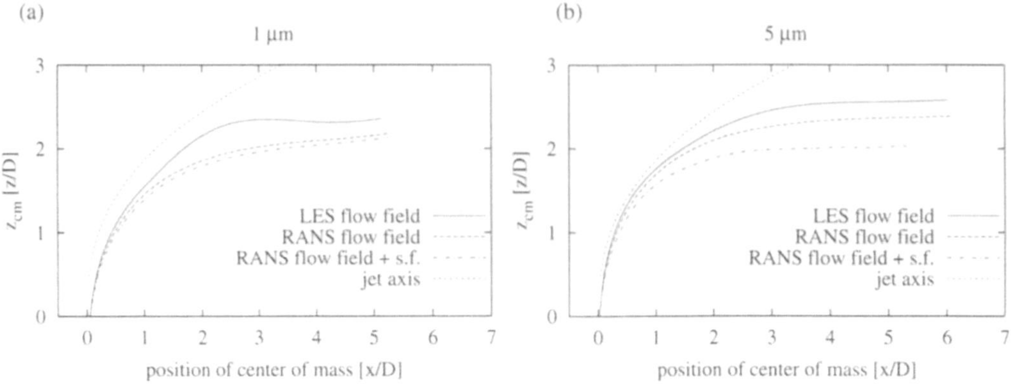

Figure 7. Trajectory of center of mass for 1 μm and 5 μm particles in RANS and in LES flow field. Dotted line is jet trajectory.

view of the computational domain (Figure 6 (a)), we observe that particles distribute along a curved shape which corresponds to the fingerprint of the time averaged, counter rotating vortex pair. Particles disperse only slightly in the radial direction, due to the turbulent fluctuations. The accumulation of particles along the counter rotating vortex pair surface is caused by the same mechanism already discussed in Figure 5. Particle with 1 μm diameter accumulate in the low vorticity regions and remain entrained at the periphery of the vortical structures embedded in the steady state flow field. Turning now the attention to the top view in Figure 6 (c), we observe that particle motion in the streamwise direction is characterized by a large number of particles moving downstream as a compact front and two streamwise almost symmetric dispersion tails. As can be observed from Figure 6 (b), the dispersion tails correspond to particles entrained in the lower part of the counter rotating vortex pair, which is characterized by low velocity values. The 3-D view of the particle swarm confirms that the concentration field is completely different than in the LES flow field. In particular, we observe that the particle dispersion in the streamwise direction is uniform, whereas is modulated by the spanwise structures in the LES flow field. A very similar behavior is observed also for 5 μm particles (not shown here).

For practical engineering applications, the relevant information about particle dispersion is relative to the statistics of particle dispersion. Here, we calculated the position of the center of mass for each swarm (comprising $1,000$ particles) as:

$$\mathbf{x}_{cm} = \frac{\sum_{i=1,N} \mathbf{x}_{p,i}}{N} \tag{3}$$

where $\mathbf{x}_{cm}$ is the position of the center of mass, $\mathbf{x}_{p,i}$ are the coordinates of the i-th particle, and N is the number of particles. In Figure 7, we show the trajectory of the center of mass for (a) 1 μm and (b) 5 μm particles calculated using *i)* the LES flow field, *ii)* the RANS flow field and *iii)* the RANS flow field with stochastic forcing. Horizontal and vertical coordinates are streamwise and vertical position of the center of mass in dimensionless units, respectively. We report also the trajectory of the jet – the jet axis – as a dotted line for reference.

Consider now the results for the LES Lagrangian tracking (solid line). We observe, first, that particles penetration is smaller than jet penetration (Figure 7 (a) and (b)), and, second, that

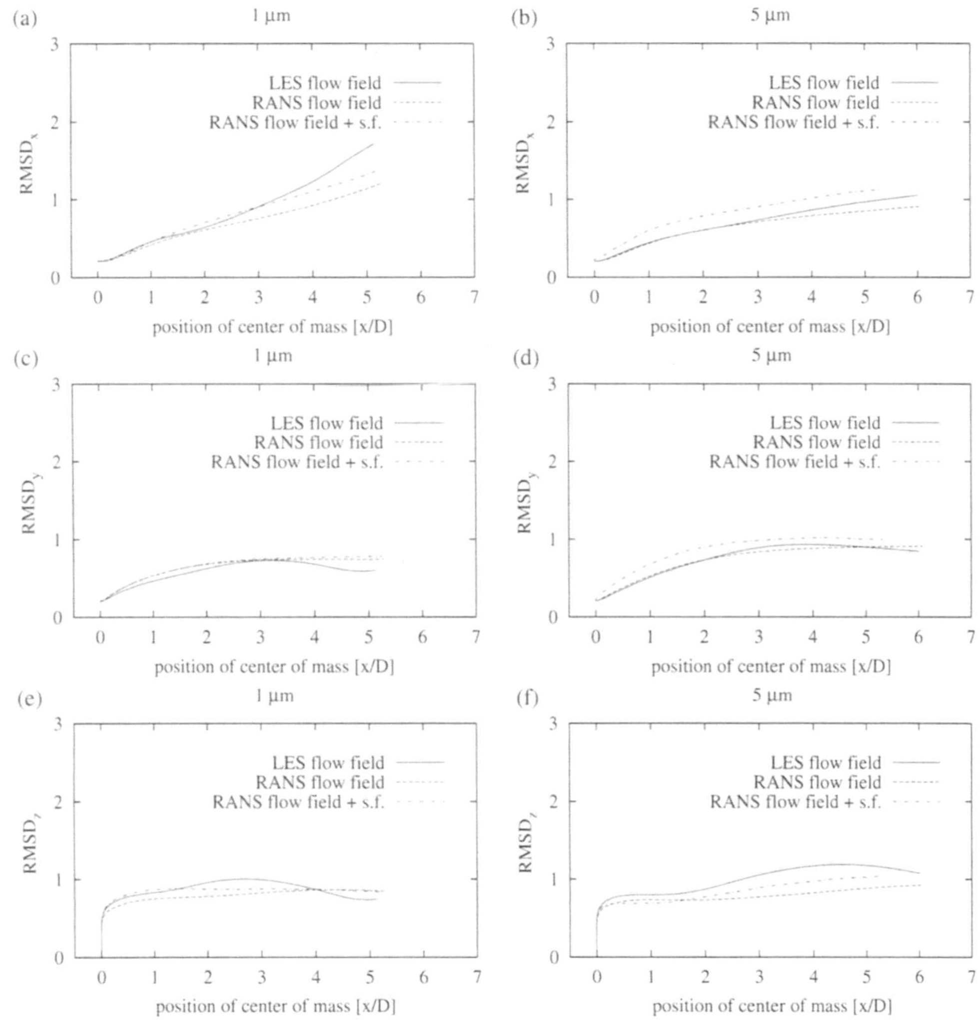

Figure 8. Components of root mean square displacement for 1 μm a nd 5 μm particles tracked in RANS and LES flow fields.

$5\mu m$ particles – characterized by larger inertia – penetrate deeper than 1 μm particles into the crossflow. If we compare Lagrangian tracking in LES and RANS, we observe that the trajectory of the center of mass calculated by RANS lies always under that calculated by LES. This is due to the different flow field sampled by the particles in the LES and in the RANS simulation as they are issued from the pipe. When the stochastic term is used, the trajectory of the center of mass is further lowered, especially for 5 μm particles. This effect is likely due to particle distribution in the pipe. The stochastic forcing induces particles to accumulate toward the pipe wall (turbophoresis, see Cerbelli et al., 2001), where axial velocity is lower.

To evaluate the differences in particle dispersion in the LES Lagrangian and RANS Lagrangian simulations, we calculated the components of the mean square displacement for the swarms of particles as

$$MSD_p = \frac{\sum_{i=1,N}(\mathbf{x}_{p,i} - \mathbf{x}_{cm})^2}{N} \tag{4}$$

In Figure 8, we show the components of the root mean square displacement ($RMSD$) for 1 μm and 5 μm particles, using the streamwise position of the center of mass of particles as independent variable. Streamwise $RMSD$ calculated by LES is larger for 1 μm than for 5 μm particles, because smaller particles are better entrained by the vortical flow structures. Consider now the RMSD in the streamwise direction for 1 μm particles, shown in Figure 8 (a). We observe that particle dispersion calculated by LES Lagrangian tracking is larger than that calculated by RANS Lagrangian tracking. When we use the stochastic forcing, the streamwise dispersion increases and the RANS profile approaches LES results. For 5 μm particles, the $RMSD$ for the RANS without stochastic forcing is similar to that calculated for the LES, and the use of the stochastic forcing determines streamwise dispersion overestimation.

Consider now the $RMSD$ in the spanwise (Figure 8 (c) and (d)) and vertical directions (Figure 8 (e) and (f)). The values of the $RMSD$ in the spanwise direction are similar for the LES and the RANS without stochastic forcing, whereas the $RMSD$ is increased by the stochastic forcing. The increase is negligible for 1 μm particles whereas determines overestimation for the 5 μm particles. The values of the $RMSD$ in the vertical direction are different for the LES and the RANS without stochastic forcing, and the stochastic term contributes to reduce deviation.

3 Summary and Conclusions

In this work we simulated numerically the turbulent dispersion of inertial particles in a jet in crossflow. We calculated the flow field of the transverse jet using a LES approach coupled to a Lagrangian tracking. A fine-mesh, time-resolved flow field is necessary to simulate particle dispersion numerically, since particle behavior depends closely on the dynamics of the large scale vortical structures and on the interaction between the particle dynamics and the fluid dynamics.

We examined in detail the vortical structures in the time-dependent flow field to understand their effect on particle dispersion. Qualitative comparison between our simulation and previous results show that the numerical simulation reproduces closely the time-dependent large scale vortical structures characteristic of the jet. We identified pairs of high and low velocity regions which detach for the periodic oscillation of the jet tip and the complex structure for the instantaneous fingerprint of the counter-rotating vortex pair. These features of the flow field determine the complex behavior of particles that is reproduced by the Lagrangian tracking. Detailed analysis of particle dispersion in the LES flow field shows that particles move following the vortical structures. The high correlation between the vorticity field and the particle distribution demonstrate that vortical structures are responsible for the preferential distribution of particles, which segregate in the low vorticity regions. In particular, particles are propelled away from the vortex core and accumulate in the region between subsequent vortex cores. Spanwise oriented vortical structure are found to be responsible for enhanced streamwise dispersion.

Since the Large Eddy simulation of the jet in crossflow is computationally costly, we evaluated the adequacy of a simplified approach, based on Lagrangian tracking in the time-averaged flow field. We calculated the time-averaged flow field using a steady state RANS approach and we quantified the different results that can be obtained using Lagrangian tracking in a LES or in a RANS flow field. We compared the trajectory of the center of mass of particles and the components of RMSD calculated by the LES and the RANS. In the RANS flow field, spanwise vorticity is completely suppressed and vortical structures are primarily streamwise oriented. This determines a reduced particle dispersion. We used a stochastic forcing based on the local turbulent kinetic energy to model the effect on particle dispersion of the turbulent scales filtered out in the RANS approach. We find that the stochastic forcing improves only marginally the results of the RANS approach.

References

Abramovich, G., (1963). *The theory of turbulent jets* , Physical Sciences Engineering.

Andreopoulos, J., and Rodi, W., (1984). Experimental investigation of jets in a crossflow *J. Fluid Mech.*, **138**, 93-127.

Andreopoulos, J., (1985). On the structure of jets in a crossflow *J. Fluid Mech.*, **157**, 163-197.

Campolo, M., Degano, G. M., Soldati, A., and Cortelezzi. L., (2002). Numerical appraisal of jet-to-crossflow coupling in a transverse jet (this book)

Cerbelli, S., Giusti, A., and Soldati, A., (2001). ADE approach to predicting dispersion of heavy particles in wall-bounded turbulence *Int. J. Multiphase Flow*, **27**, 1861-1879.

Chang, Y. K., and Vakili, A. D., (1995). Dynamics of vortex rings in crossflow *Phys. Fluids*, **7**, 1583-1597.

Chein, R., and Chung, J. N., (1988). Simulation of particle dispersion in a two-dimensional mixing layer *AIChE J.*, **34**, 946-954.

Chung, J. N., and Troutt, T. R., (1988). Simulation of particle dispersion in an axisymmetric jet *J. Fluid Mech.*, **186**, 199-222.

Crowe, C. T., Chung, J. N., and Troutt, T. R., (1989). Particle mixing in free shear flows *Progress of Energy Combustion Science J.*, **14**, 171-194.

Davis, S. A., and Glezer, A., (1999). Mixing control of fuel jets using synthetic jet technology: velocity field measurements *AIAA Paper 99-0447*, 37th Aerospace Sciences Meeting, Reno, NV, 1-15.

Degano, G. M., Campolo, M., Soldati, A., and Cortelezzi, L., (2002). Influence of inlet boundary conditions on the numerical calculation of a jet in crossflow *Int. J. Heat and Fluid Flow*, (submitted).

Fric, T. F., and Rhosko, A., (1994). Vortical structure in the wake of a transverse jet *J. Fluid Mech.*, **279**, 1-47.

Germano, M., Piomelli, U., Moin, P., and Cabot, W. H., (1991). A dynamic subgrid-scale eddy-viscosity model *Phys. Fluids*, **3**, 1760-....

Gosh, S., and Hunt, J. C. R., (1998). Spray jets in a cross-flow *J. Fluid Mech.*, **365**, 109-136.

Gosman, A. D., and Ioannides, E., (1981). Aspects of computer simulation of liquid-fueled combustors *AIAA Paper 81-0323.*

Hasselbrink, E. F., and Mungal, M. G., (2001a). Transverse jets and jet flames. Part 1. Scaling laws for strong transverse jets *J. Fluid Mech.*, **443**, 1-25.

Hasselbrink, E. F., and Mungal, M. G., (2001b). Transverse jets and jet flames. Part 2. velocity and OHfield imaging *J. Fluid Mech.*, **443**, 27-68.

Hsu, A. T., and Guo, Y. (2000). Unsteady simulation of a jet-in-crossflow *Int. J. Comp. Fluid Dyn.*, **14**, 41-53.

Kelso, R. M., Lim, T. T., and Perry, A. E. (1996). An experimental study of round jets in a cross-flow *J. Fluid Mech.*, **306**, 111-144.

Lazaro, B. J., and Lasheras, J. C. (1992). Particle dispersion in the developing free shear layer. Part 1. The natural flow *J. Fluid Mech.*, **235**, 143-178.

Maxey, M. R., and Riley, J. J., (1983). equation of motion for a small rigid sphere in a nonuniform flow *Phys. Fluids*, **26**, 883-889.

Papaspyros, J. N. E., Kastrinakis, E. G., and Nychas, S. G. (1997). Coherent contribution to turbulent mixing of a jet in crossflow *Appl. Sci. Res.*, **57**, 291-307.

Rivero, A., Ferrè , J. A., and Giralt, F. (2001). Organized motions in a jet in crossflow *J. Fluid Mech.*, **444**, 117-149.

Rudman , M. (1996). Simulation of the near field of a jet in a cross flow *Exp. Thermal & Fluid Sci.*, **12**, 134-141.

Salvetti, M. V., and Banerjee, S., (1995). A priori tests of a new dynamic subgrid-scale model for finite-difference large-eddy simulations *Phys. Fluids*, **7**, 2831-2847.

Sani, R. L., and Gresho, P. M. (1994). Resume and remarks on the open boundary condition minisymposium *Int. J. Num. Meth. Fluids*, **18**, 983-1008.

Smith, S. H., and Mungal, M. G. (1998). Mixing, structure and scaling of the jet in crossflow *J. Fluid Mech.*, **357**, 83-122.

Wen, F., Kamalu, N., Chung, N. J., Crowe, C. T., and Troutt, T. R., (1992). Particle dispersion by vortex structures in plane mixing layers *J. Fluid Eng.*, **114**, 657-666.

Xia, L. P., and Lam, K. M. (1997). A vertical round jet issuing into an unsteady crossflow consisting of a mean current and a sinusoidal fluctuating component *J. Wind Eng. Ind. Aerodyn.*, **67&68**, 843-857.

Yang, Y., Crowe, C. T., Chung, N. J., and Troutt, T. R., (2000). Experiments on particle dispersion in a plane wake *Int. J. Multiphase Flow*, **26**, 1583-1607.

Yuan, L. L., Street, R. L., and Ferziger, J. H. (1999). Large-eddy simulations of a round jet in crossflow *J. Fluid Mech.*, **379**, 71-104.

Zhou, Q., and Leschziner, M. A., (1999). An improved particle-locating algorithm for Eulerian-Lagrangian computations of two-phase flow in general coordinates *Int. J. Multiphase Flow*, **25**, 813-825.

Particle/Vortex Interactions in Free and Impinging Air Jets

Ellen K. Longmire[1] and Steven L. Anderson[2*]

[1]Department of Aerospace Engineering & Mechanics, University of Minnesota, USA
[2]TSI Inc., Shoreview, Minnesota, USA

Abstract. Particle-laden gas jets with Reynolds number of order 20000 were forced at fixed frequencies in order to excite periodic vortex ring structures. Particle-to-air mass loading ratio was varied from 4% to 50%, and the flow behavior was documented by flow visualization, number density measurements, and particle and gas velocity measurements. The experiments showed that a Stokes number based on particle inertial and vortex time scales was important in determining the degree of particle clustering and the rate of particle dispersion in free jets. The same Stokes number determined the degree of lag occurring between particles and vortices in the stagnation zone of impinging jets. Vortex rings were found to persist in the free jet when the mass loading ratio was increased to 50%. However, PIV measurements demonstrated that downstream of the stagnation zone in impinging jets, vortices were weakened by mass loading ratios of only 4%.

1 Introduction

Particle-laden gas jets occur frequently in industrial applications where fine solids or liquid droplets must be injected into combustion chambers, process streams, or reaction vessels. Particle-laden jets are inherently more complex than single-phase flows because the particles represent a discrete phase that must interact with the continuous fluid phase. Since liquid or solid particles are much denser than the gas phase, they have much more inertia than fluid elements moving at similar velocities and are generally unable to follow every scale of fluctuation in the flow. When particle and fluid velocities do not match, particles exchange energy and momentum with the fluid through viscous drag. Faster moving particles can increase local fluid velocities by dragging the fluid along with them and vice versa. Particles may damp or amplify local fluctuation levels and interfere with eddy coherence. The degree of particle response to fluid eddies is commonly characterized in terms of a Stokes number which relates the particle inertial time scale to the appropriate eddy time scale. Experiments and computations documenting particle response to specific eddy time scales include the work of Crowe et al. (1985), Chung and Troutt (1988), and Lazaro and Lasheras (1989, 1992), for example. In some flow locations, the particle Reynolds number can also become significant, and this then becomes an additional parameter of importance.

* The ideas and support of Professor John Eaton in initiating and carrying out the free jet experiments at Stanford University are gratefully acknowledged. In addition, the authors thank Sanjay Pahuja who developed the phase separation software employed in the flow modification studies. The work was supported by the Electric Power Research Institute (RP-2154-10 and RP-8034-01) and the National Science Foundation (CTS-9457014).

Eddy response to particles, which has been difficult to quantify experimentally or numerically, is as yet poorly understood.

In this paper, experiments on free and impinging round jets are described where periodic forcing was used to make vortex ring structures (and their associated time scales) repeatable and periodic. Particle distribution and dispersion are described in relation to particle Stokes and Reynolds numbers. In addition, modifications to vortices caused by the presence of particles are discussed.

2 Facilities and Methods

2.1 Facilities

For the experiments described, several jet facilities were employed (see Longmire and Eaton, 1992, and Anderson and Longmire, 1995, 1997 for details). Each facility was designed to produce an axisymmetric air jet with an initially thin shear layer and low core turbulence level. The impinging jet facility is shown in Figure 1. The air flow enters the jet plenum through two opposing ports. The plenum contains a honeycomb flow straightener to eliminate swirl and a screen to disperse particles. Flow is accelerated through a round contraction with area ratio of 12.1:1 yielding an exit diameter D of 3 cm. An audio speaker located at the upper end of the plenum was used to force the flow. The impingement surface was a circular glass plate located 5D downstream of the nozzle exit. Particles were added to the flow upstream of the plenum from a slightly pressurized variable rate screw feeder. The facilities used in the free jet studies had similar designs, but the nozzle outlet diameters measured 2 cm.

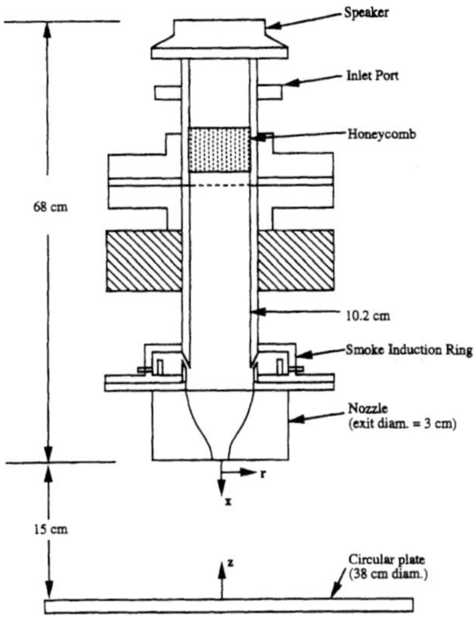

Figure 1. Impinging jet facility.

In the free jet experiments, one set of spherical glass beads, sized between 50 and 60 μm, was used. In the impinging jet experiments, three sets of solid and hollow glass beads with smaller diameters and time constants were employed. The particle characteristics are summarized in Table 1.

A commercial fog generator provided tracer particles for both visualization and seeding of the gas flow. The fog machine was placed in a slightly pressurized steel drum where glycerin droplets were allowed to mix with the surroundings. An outlet tube located at the top of the drum carried the resulting 'smoke' downstream. To mark the entire jet core, smoke was fed into the plenum through opposing ports. To mark only the jet shear layer, smoke was fed through an induction ring just upstream of the nozzle. PDA measurements of smoke particles in the impinging jet indicated diameters ranging from 1-2 μm. The smoke particles were therefore assumed small enough to track the air flow faithfully.

In all experiments, the jet Reynolds number was on the order of 20000, where mean nozzle outlet velocities U were 14.4 m/s and 17.0 m/s in the free jet and 10.6 m/s in the impinging jet. Hot wire measurements performed on unforced jet flows yielded top-hat shaped mean velocity profiles with initial turbulence levels of 1% or less. The jets were forced at Strouhal numbers (S_D = fD/U) in the range 0.3-1.1 to encourage the early formation of vortex rings at the forcing frequency. Forcing amplitudes were originally large (as in Figures 2 and 3), but were later reduced to the minimum levels required to induce vortex formation at the forcing frequency. In figure captions, amplitudes are characterized by the normalized rms fluid velocity at the nozzle exit (u'/U).

Table 1. Particle characteristics.

Particle Group	*T21*	*T7*	*T3*	*T2*
ρ_p (kg/m^3)	2400	2500	2500	340
d_p (μm)	50-60	25-35	15-25	35-45
τ_p (ms)	21	6.9	3.1	1.7

2.2 Measurement Techniques

Flow Visualization. In the free jet experiments, a Plasma Kinetics 151 copper vapor laser was used to illuminate both smoke and particles. The laser generated green light pulses of approximately 30 ns duration. Pulse energy varied from 2-4 mJ. The beam passed through cylindrical and spherical lenses of 400 mm and 1 m focal length to form a thin sheet of light that intersected the jet axis. The normal operating pulsing frequency of the laser of 5600 Hz could be reduced or altered as needed for several minutes at a time.

For the smoke and particle visualization studies discussed in Section 3.1, the laser pulsing and jet speaker forcing frequencies were coupled to create a strobing effect. Strobed flow was captured by a JVC video camera framing at a rate of 30 Hz. Therefore, each frame of videotape contained a superposition of a number of images depending on the laser pulsing frequency.

In the impinging jet, instantaneous images of gas and solid phases were obtained by illuminating the flow with a pulsed frequency-doubled Nd:YAG laser. The beam passed through spherical and cylindrical lenses of 1 m and 12.7 mm focal length respectively to form a diverging sheet with thickness less than 1 mm. Light scattered perpendicular to the laser sheet was photographed with a Nikon 35 mm camera and a 105 mm micro lens. The laser pulsing was phase-locked to the forcing signal so that specific phases of the forcing cycle could be examined.

PIV measurements. Particle image velocimetry (PIV) was employed to obtain two-dimensional velocity fields of the gas and solid phases. Photographs of a single double-pulse event were taken with the Nikon camera mentioned above. Extender rings were inserted between the camera and lens to increase magnification. To eliminate directional ambiguity and to improve measurement quality, an artificial image shift was imposed on the flow field by placing a rotating mirror between the flow and the camera. To obtain full gas velocity fields, both the jet and the surrounding enclosure were seeded with smoke. Each film negative was digitized with a high-resolution scanner, and the resulting files were processed on a SGI Indy workstation using a spatial auto-correlation method.

Simultaneous two-phase PIV measurements were performed by capturing images of the flow seeded with both smoke and glass particles. This was done in order to measure instantaneous slip velocities between the particles and the fluid and to determine modification of the fluid field caused by particles. T-MAX 400 film processed with T-MAX developer was found to yield clear images of both smoke and glass particles. Typically, the glass particles appeared brighter and slightly larger than the smoke particles in the digitized images.

An image separation scheme was developed to separate the scanned images into smoke and glass particle files (see Pahuja and Longmire, 1995). All glass particle images were identified by a "region-growing" algorithm with a threshold intensity set at 250 on a gray-scale of 0 - 255. A subroutine determined the size of each region of connected pixels and accepted only those regions that contained 4 or more pixels. These two criteria assured that unusually bright smoke particles were not included and that unusually small glass particles were not omitted. All regions that were identified as glass particles were written to a file in which the background intensity was set to 1 to ensure good contrast.

To identify smoke particles, a "local maximum" algorithm, which searched the original image file line-by-line, was chosen. If the intensity value of the local maximum was within a specified range, then this pixel was considered the center of a smoke particle and was written to a 'smoke' file. Adjacent pixels with intensity values corresponding to smoke particles were also written to this file. The length of the local search domain was set at 3 pixels, and the acceptable intensity range was set at 1 - 250 on a gray-scale of 0 - 255.

The code performed well in separating the two phases, losing very little information in the process. In regions where clusters of glass particles were present, the maximum undercount of smoke particles was estimated as 5%. However, because these regions were heavily seeded with smoke in the first place, the local loss of smoke particles had no significant effect on the quality of the resulting auto-correlations that were computed.

After image separation, the particle-only and smoke-only image files were processed independently to determine their respective vector fields. The interrogation spot size was set at 80 x 80 pixels which corresponded to a spatial resolution of 1.3 mm x 1.3 mm in real space. The interrogation spot overlap was set at 25% so that a 34 x 22 vector field was produced for each image file. In this paper, only the results from gas vector fields are presented.

3 Results

3.1 Free Jets

Effects of vortices on particles. Experiments on natural and forced free jets demonstrate the importance of Stokes number in determining the particle distribution and dispersion in the flow. In a free jet, a relevant fluid time scale can be derived from length and velocity scales of the dominant vortex rings that form. In previous publications, we have used axial vortex spacing (L_v) and vortex propagation speed (U_v) to derive a fluid time scale τ_f. When the jets are forced, the fluid time scale can be represented as the reciprocal of the forcing frequency.

Examples of single-phase and particle-laden flow forced at two Strouhal numbers, $S_D = 0.43$ and 0.76 are shown in Figures 2 and 3. The corresponding Stokes numbers $St' = \tau_p f$ are 7.6 and 13.6 for the lower and higher forcing frequency respectively. All of the images, which result from superpositions of multiple laser pulses, demonstrate that the flow structure is locked to the forcing frequency. In single-phase flow, vortex rings form at the forcing frequency, and their propagation velocity is approximately constant over the axial range depicted. For the Stokes numbers shown, the images of particle-laden flow demonstrate that the particles tend to become concentrated in the zones of deceleration downstream of the vortex rings. As the Stokes number decreases, the particle concentration effect in these zones becomes stronger. Also, the clustering becomes stronger with increasing axial distance until most of the particles disperse. Additional cases in free and impinging jets document cluster formation for $0.2 < St' < 14$. The lower limit listed represents the minimum value that we have tested (see Longmire and Anderson, 1995). Above the upper limit, the relative inertia of the particles becomes too large for them to react to the variations in the axial velocity component caused by the vortices.

An additional observation is that, for a given fluid time scale, particles with larger time constant are focused slightly downstream of those with smaller time constant (see Figure 6 below and Longmire and Anderson, 1995). This occurs because larger particles respond more slowly to local decelerations, and particles with large enough St′ overshoot local saddle points (see also Martin and Meiburg, 1994).

The $St' = 7.6$ case depicted in Figure 2 also gives a good indication of the dominant means of particle dispersion. Particles traveling downstream of vortex rings are accelerated outward by the outwardly moving fluid in those locations. Therefore, the particle clusters form arrowhead shapes, and the radial breadth of the clusters increases with downstream distance. Particle concentration and velocity measurements show that low-frequency forcing, which accelerates the formation of larger vortex length scales, enhances particle dispersion. For the two values of St′ shown, dispersion in the jet near field was enhanced by forcing, and dispersion occurred more quickly for the lower St′ (Longmire and Eaton, 1994). The initial radial particle distribution is also an important parameter in determining particle dispersion rates. Particles located initially near the jet axis are dispersed more slowly than those located initially near $r/D = 0.5$. Particles near the jet edge encounter outwardly moving fluid downstream of the vortices sooner and therefore are accelerated outward sooner than particles near the jet axis (see also computations by Chung and Troutt, 1988).

In the free jet experiments, it should be noted that particle Reynolds numbers based on local slip velocity (Re_p) were generally small. The largest slip velocities occurred near the nozzle exit where particle acceleration lags fluid acceleration. In the case of the largest particles examined (d_p = 55 μm), the mean fluid velocity at the nozzle exit was 14.4 m/s, and the local slip velocity was on the order of 6.4 m/s, yielding Re_p of 24. The slip velocity and the associated Re_p decrease as particles in the jet core continue to accelerate with downstream distance. The initial particle lag has two ramifications. First, it is likely that the axial particle velocity affects a given particle's residence time in outwardly moving flow zones downstream of vortices which propagate at approximately 0.5U. Therefore, the particle dispersion rate is affected. Second, the jet must give up momentum in order to accelerate the particles. As the particle mass loading is increased, the ratio of particle to gas momentum will increase faster than the mass loading ratio.

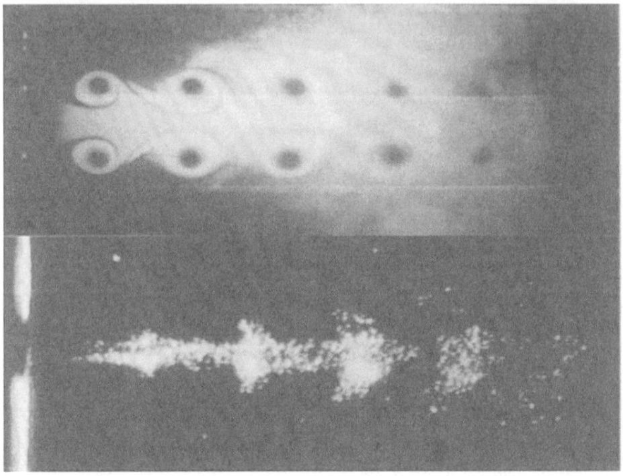

Figure 2. Single-phase and particle-laden flow, $Re_D = 23000$, $S_D = 0.43$, $St' = 7.6$, $u'/U = 0.26$. (taken from Longmire and Eaton, 1992).

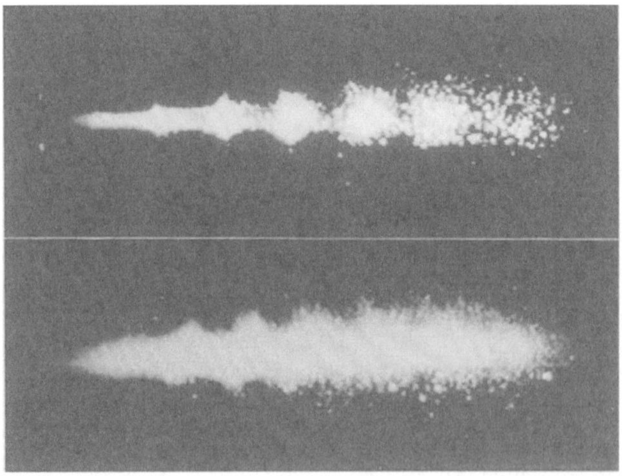

Figure 3. Single-phase and particle-laden flow, $Re_D = 23000$, $S_D = 0.76$, $St' = 13.6$, $u'/U = 0.04$. (taken from Longmire and Eaton, 1992).

Effects of particles on vortices. All of the results discussed thus far were obtained for low particle-to-air mass loading ratios on the order of 4%. At these levels, we assumed that the effect of the particles on the gas flow structure was very small. (Subsequent PIV experiments in free jets measured insignificant modification at these loadings, Pahuja and Longmire, 1995). Additional visualization experiments were performed on jets forced over a range of frequencies for larger mass loadings up to about 80%. Images of strobed flow at two frequencies are shown in Figures 4 and 5. In each case, images at two mass loadings are compared. To avoid saturating the images, the laser and camera settings were adjusted at the higher mass loadings. Therefore, absolute scattering intensities cannot be compared directly. The main point observable from the results depicted is that periodic axial variations in particle concentration persist at the larger mass loadings. This suggests that vortices continue to be formed at the forcing frequency and that the vortices persist as the flow moves downstream. Therefore, the particles have not damped out the flow structures completely. From the flow visualization, it is impossible to tell if the particles have damped the vortices at all.

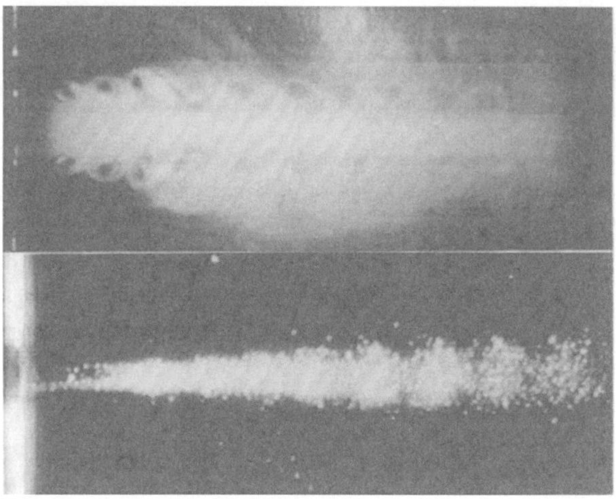

Figure 4. Variation of particle loading, $Re_D = 19000$, $S_D = 0.51$, $St' = 7.6$, $u'/U = 0.11$, M=0.04 (top), M=0.45 (bottom), (taken from Longmire and Eaton, 1992).

3.2 Impinging Jets

Effects of vortices on particles. In the impinging jet experiments, hot wire measurements were performed on single-phase flow to determine the effects of the wall (located 5D downstream of the nozzle exit) on the flow structure upstream. Time-averaged and spectral measurements showed that the flow resembled a free jet until 1D upstream of the impingement surface (Anderson and Longmire, 1995). In the free jet region, the effects of vortices on particles are the same as those described in the previous section. In the stagnation region, however, particles tend to accumulate (again because of fluid deceleration), and particle clusters can become decoupled from the associated vortices upstream.

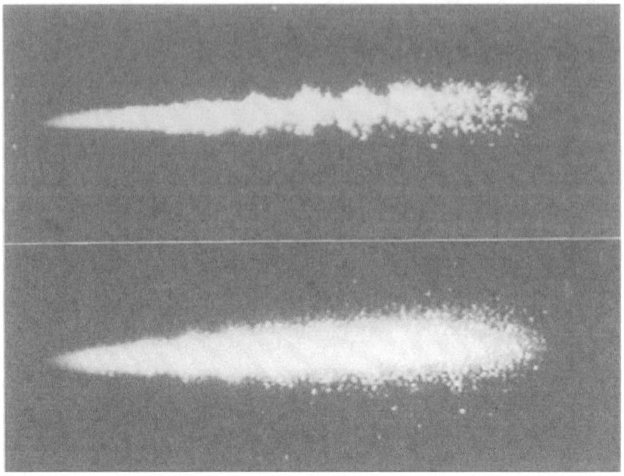

Figure 5. Variation of particle loading, $Re_D = 19000$, $S_D = 0.71$, $St' = 10.7$, $u'/U = 0.02$, $M=0.05$ (top), $M=0.39$ (bottom), (taken from Longmire and Eaton, 1992).

Instantaneous images of smoke-marked and particle-laden flow at one forcing condition ($S_D = 0.51$, $Re_D = 21000$) are shown in Figure 6. Three sets of particles were tested, and all of the images were taken at the same phase in the forcing cycle ($\Phi = 180°$). Note that the initial particle distribution varies with particle size. Upon exiting the contraction, the larger particles (T7) are focused closer to the jet centerline while the smaller particles (T2) are distributed more evenly throughout the jet core. This variation in distribution and the smaller response time of these particles to radial acceleration combine to cause earlier and stronger dispersion of the smaller particles.

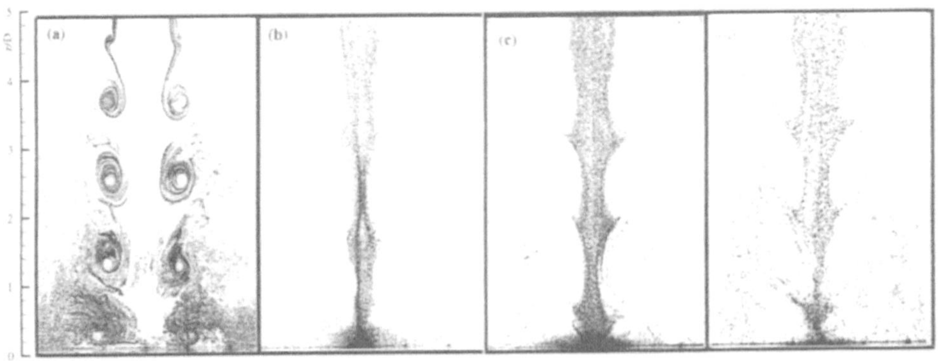

Figure 6. Instantaneous images of impinging flow, $Re_D = 21000$, $S_D = 0.51$, $\Phi = 180°$, $u'/U = 0.10$. (a) fluid; (b) $St' = 1.25$; (c) $St' = 0.56$; (d) $St' = 0.3$.

At the forcing phase depicted ($\Phi = 180°$), a vortex ring has impacted the wall and is beginning to expand radially while another approaches from upstream. For the two smaller particle sets, corresponding to $St' = 0.56$ and 0.3, it is clear that a cluster riding ahead of the upstream vortex ring is approaching the wall, while significant concentrations of particles exist downstream. For

St'= 1.2, an approaching cluster is no longer apparent, but a dense mound of particles exists at the wall.

In general, the particle number density increases in the stagnation zone where the locations of increased number density are taller and larger for increasing Stokes number. While the fluid approaching the wall must decelerate to zero, the particles can decelerate only by drag interactions. Therefore, they tend to rebound at the wall, moving upward into the oncoming flow until they are eventually turned downward again and accelerated outward. Calculations on a simplified flow show that particle rebound heights and residence times in the stagnation zone increase with increasing Stokes number.

The 'decoupling' of a particle cluster is shown in Figures 6c and 6d when St' = 0.56 and 0.3. Note that although the cluster approaching the wall lies downstream of the approaching vortex core, the volume of particles at the wall lies upstream of the core that is expanding radially. In subsequent phases, most of the particles remain upstream of the core and therefore, they must be accelerated outward by the following ring. These cases as well as additional ones at a lower Strouhal number demonstrate that for St' > 0.3, a collection of particles can skip one or more cycles before exiting the stagnation zone. For St' = 1.2, it is also possible that vortices approaching the wall break up concentrations of particles in the stagnation zone.

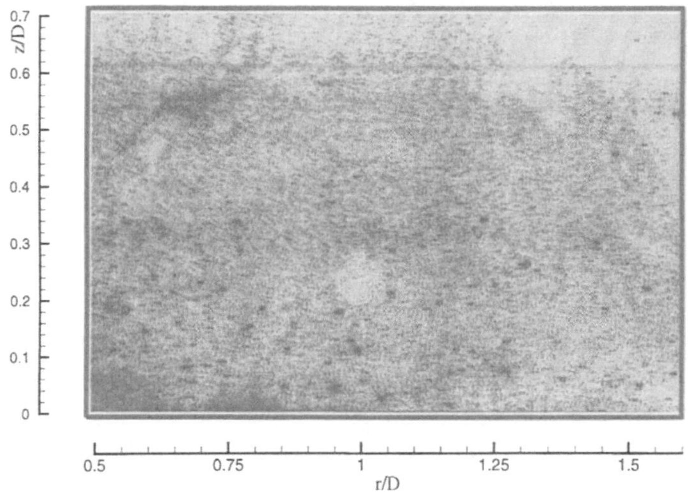

Figure 7. Typical two-phase PIV image above the impingement surface, $Re_D = 21000$, $S_D = 0.51$, $\Phi = 180°$, St' = 1.25.

Effects of particles on vortices. Figure 7 shows a typical two-phase PIV photograph of the region $0.5 \leq r/D \leq 1.6$ for the same flow as in Figure 6 ($S_D = 0.51$, $\Phi = 180°$). The flow is laden with T7 particles (St' = 1.25) with an initial mass loading ratio of 3.8%. The resulting image files were separated into particle and smoke files before velocity fields were determined. A set of 111 phase-locked images was processed and averaged to produce an ensemble-average vector field of the fluid phase in particle-laden flow. A set of 51 phase-locked images of the same field and forcing phase of gas-only flow was processed and averaged for comparison.

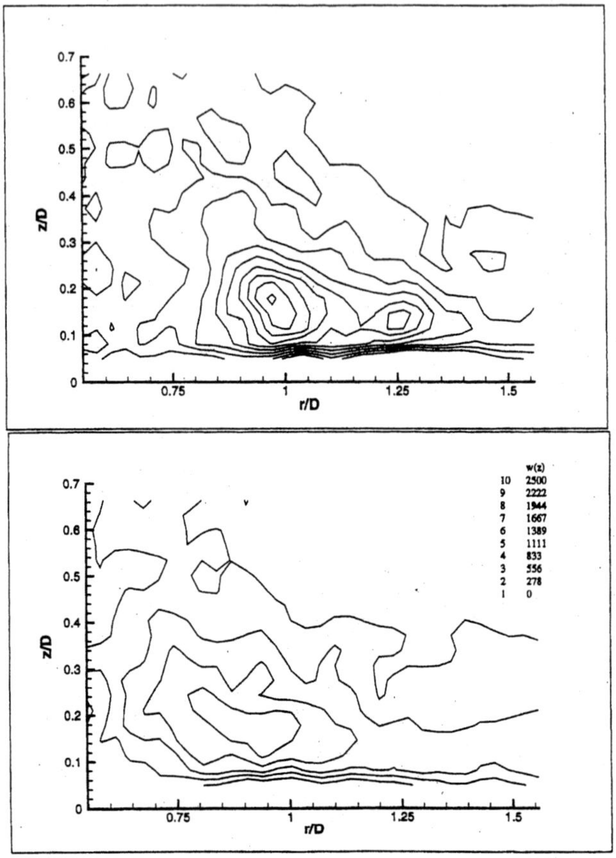

Figure 8. Vorticity contours above the impingement surface, $Re_D = 21000$, $S_D = 0.51$, $\Phi = 180°$, M=0.04. (a) Single-phase flow; (b) Particle-laden flow with St' = 1.25.

Vorticity fields were derived from the resulting vector fields using a local circulation method. The results are shown in Figure 8a and 8b for gas-only and particle-laden flow respectively. It is clear from a comparison that the vorticity is reduced significantly and shifted toward the jet centerline when particles are added to the flow (i.e. the vortex in particle-laden flow lags behind the vortex in single-phase flow). Also, the maximum vorticity for particle-laden flow is spread radially compared to that in single-phase flow indicating that the vortex core location varies more in particle-laden flow.

The circulation of each averaged vortex was computed using a line integral around the box defined by z/D = 0.05, r/D = 1.3, z/D = 0.4, and r/D = 0.65. The results for single-phase and particle-laden flow were $\Gamma = 0.44\ m^2/s$ and $0.31\ m^2/s$ respectively. The addition of particles thus reduced the circulation by 30%. Further downstream, it is expected that the vortex core in particle-laden flow would lag further behind the vortex core in single-phase flow due to this reduction in circulation and the associated propagation velocity.

The difference velocity field resulting from the point-by-point subtraction of the fluid velocity field of the particle-laden flow from the velocity field of the single-phase flow gives an idea of how the particles decrease the vortex circulation (see Figure 9). In the region adjacent to the stagnation zone and everywhere around the vortex, the particles damp the vortex-induced fluid velocities. Fluid radial velocities adjacent to the stagnation zone (i.e. near r/D = 0.5) are reduced by an average of 25% compared to single-phase flow, due presumably to considerable momentum exchange between the fluid and rebounding particles in the stagnation zone. The effects of particles are even greater at larger radii, near the vortex core. On the upstream side of the vortex, the difference vectors are significant in both the radial and axial directions. The magnitudes are 25 - 30% of the single-phase velocity vectors. Underneath the core and close to the wall, where fluid velocities and slip velocities are highest, the difference vectors, which are primarily in the radial direction, measure as high as 35% of single-phase velocities.

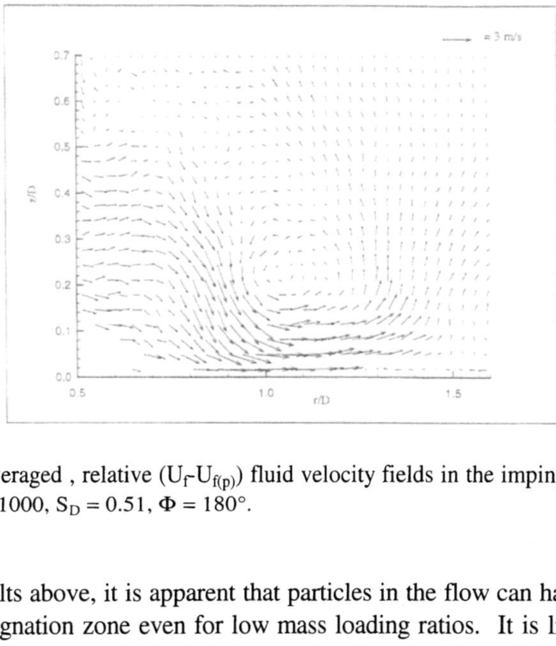

Figure 9. Ensemble averaged , relative (U_f-$U_{f(p)}$) fluid velocity fields in the impinging shear layer region of a forced jet, $Re_D = 21000$, $S_D = 0.51$, $\Phi = 180°$.

Based on the results above, it is apparent that particles in the flow can have significant effects in and around the stagnation zone even for low mass loading ratios. It is likely that rebounding particles in the stagnation zone (which attain large axial slip velocities and Reynolds numbers ~ 25 for T7 particles) remove significant momentum from the fluid. In addition, the fluid must transfer momentum to the particles in order to accelerate them radially and drag them out of the stagnation zone. Both of these factors would act to decrease radial fluid velocities near the wall and therefore to decrease the circulation about vortices occurring there.

4 Conclusions

This paper attempts to summarize key results that have appeared in several previous papers and unpublished reports. Because of space limitations only a limited number of figures could be presented. Experiments were performed on a number of cases for forced jets laden with particles of a range of time constants. Fluid time constants were varied by varying forcing frequencies.

It was found that the influence of vortex rings on particle concentration and dispersion in free jets could be characterized by a Stokes number relating the particle inertial time scale to the vortex time scale. In impinging jets, this Stokes number also determines the degree of reorganization occurring in the stagnation zone in terms of local concentrations of particles and the vortex cores located immediately upstream. The reorganization occurs in particular when the mean flow changes direction, and the vortex core lies between the particles and the center of curvature of the mean flow. For $St' > 0.3$, particle clusters 'skip' at least one structure and become associated with a new vortex.

Our studies of vortex modification by particles have been more limited. They showed that moderate particle loadings do not tend to eliminate the axisymmetric rings that develop. In impinging jets, however, the fluid must decelerate particles in the axial direction and accelerate them in the radial direction. Slip velocities and particle Reynolds numbers can become significant in the stagnation zone such that the fluid loses momentum. In the forced flow, this reduction in momentum resulted in reduced circulation in vortex rings downstream of the stagnation zone.

References

Anderson, S.L., and Longmire, E.K. (1995) Particle motion in the stagnation zone of an impinging air jet. *Journal of Fluid Mechanics* 299:333-366.

Anderson, S.L. and Longmire, E.K. (1997) Particle Motion in the Stagnation Region of Natural and Forced Impinging Round Air Jets. *Report 1997-4, Dept. of Aerospace Engineering & Mechanics, University of Minnesota.*

Chung, J.N. and Troutt, T.R. (1988) Simulation of particle dispersion in an axisymmetric jet. *Journal of Fluid Mechanics* 186:199-222.

Crowe, C.T., Gore, R. and Troutt, T.R. (1985) Particle dispersion by coherent structures in free shear flows. *Particulate Science and Technology* 3:149-158.

Lazaro, B.J. and Lasheras, J.C. (1989) Particle dispersion in a turbulent plane shear layer. *Physics of Fluids A.* 1:1035-1044.

Lazaro, B.J. and Lasheras, J.C. (1992) Particle dispersion in the developing free shear layer. Part 2. Forced flow. *Journal of Fluid Mechanics* 235:179-221.

Longmire, E.K., and Anderson, S.L. (1995) Effects of vortices on particle motion in a stagnation zone. *Gas-Particle Flows, ASME FED* 228:95-101

Longmire, E.K., and Eaton, J.K. (1992) Structure of a particle-laden round jet. *Journal of Fluid Mechanics.* 236:217-257

Longmire, E.K., and Eaton, J.K. (1994) Active Open-Loop Control of Particle Dispersion in Round Jets. *AIAA Journal.* 32:555-563.

Martin, J.E., and Meiburg, E. (1994) The accumulation and dispersion of heavy particles in forced two-dimensional mixing layers. Part I: The fundamental and subharmonic cases. *Physics of Fluids.* 6:1116-1132.

Pahuja, S., and Longmire, E.K. (1995) Simultaneous two-phase PIV measurements in a particle-laden air jet. *Report 1995-3, Dept. of Aerospace Engineering & Mechanics, University of Minnesota.*

Characteristics of Droplet Pinch-off in Liquid/Liquid Jet Systems

Ellen K. Longmire*

Department of Aerospace Engineering & Mechanics, University of Minnesota, USA

Abstract. Natural and periodically forced jets of water/glycerin solution were examined as they flowed into an ambient of immiscible Dow Corning 200 fluid. The dynamics of jet pinch off was investigated for jet Reynolds numbers of order 50. Refractive index matching and laser induced fluorescence were employed to obtain clear images of the interface and the jet interior. Planar velocity fields were obtained by particle image velocimetry. The results showed that the pinch off mode, including whether or not satellites form, depends on the viscosity ratio, jet Reynolds number, and jet Strouhal number. Pinch off angles appear relatively insensitive to Reynolds and Strouhal number, but sensitive to the fluid viscosity ratio.

1 Introduction

The flow of a liquid jet into a second immiscible liquid has been studied extensively due to its relevance in many practical applications (see, e.g. Bright, 1985, Meister and Scheele, 1969, and Skelland and Johnson, 1982). For example, it is important for the designers of chemical reactors to be able to predict the location of jet pinch-off as well as the size and frequency of the droplets that result. Such quantities have a significant effect on the resulting degree of heat and mass transfer between the fluids.

The work described here is motivated by the more general problem of the need to quantify and understand the dynamics of pinch-off and reconnection in liquid/liquid flows where interfacial tension is significant. Such an understanding is essential in order that generalized numerical models for such transitions can be developed. The models can then be applied to many practical flows involving emulsification and separation of liquid mixtures. In the current study, the jet flow is chosen for its relatively simple geometry as well as its practical significance.

A major objective of the work is to quantify the local interface shape and curvature as well as the flow dynamics surrounding pinch-off transitions. As will be described below, we apply an index-matching method in order to make such measurements possible. An additional objective (not discussed in this paper) is to compare the experimental results with the results of numerical simulations incorporating different 'pinch-off' models to assess their accuracy in approximating the flow physics.

* This work was supported by the National Science Foundation under grant CTS-9457014 and by the Engineering Research Program of the Office of Basic Energy Sciences at the Department of Energy (Grant DE-FG02-98ER14869). The author thanks Professor Donald Webster, Derek Gefroh, Timothy Norman, and Ilija Milosevic who performed the experiments.

2 Experimental Facility and Methods

2.1 Facility

The experiments were performed in the closed-loop liquid tank facility depicted in Figure 1. The test section is 56 cm in height and 20.3 cm x 20.3 cm in cross section. The flow (a water/glycerin mixture) is driven by a magnetic-drive pump with constant frequency, and the mean velocity is set by a needle valve. Upstream of the test section, the flow passes through a honeycomb straightener before it is accelerated through a nozzle with 16.5:1 area ratio and 1 cm exit diameter (*D*). The flow exits the nozzle directly into a liquid ambient. A piston-driven forcer in the flow circuit imposes a sinusoidal variation on the nozzle exit velocity. The forcer, which is driven by an amplified signal from a function generator, was employed to generate repeatable, phase-locked pinch-off conditions. The initial jet velocity (*U*) is determined by a flow meter calibrated for the circulating fluid.

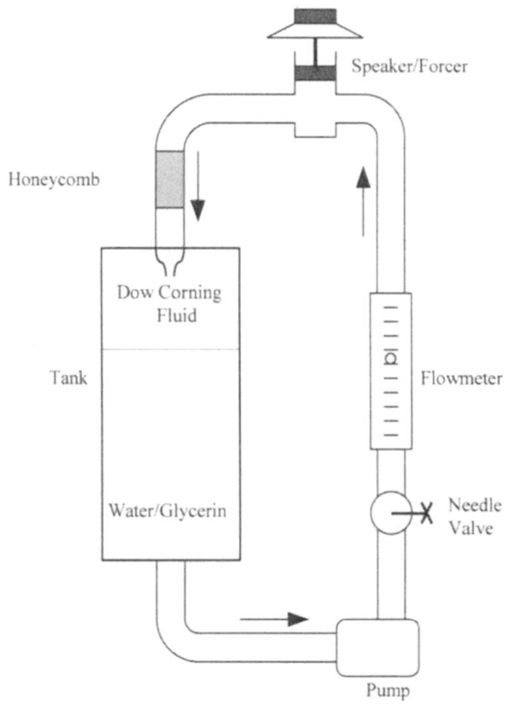

Figure 1. Jet facility (taken from Longmire et al., 2001)

The test section contained a layer of Dow Corning 200 fluid of either 19 cm depth (50 cS or 1000 cS) or 33 cm depth (5 cS) lying on top of the water/glycerin solution. The glycerin volume fraction was chosen so that the indices of refraction of the jet and ambient fluids were nearly matched. (See Table 1 for fluid properties). For visualization, Rhodamine 6G, which fluoresces under laser illumination, was added to the jet fluid, in order to distinguish it from the ambient. For PIV measurements, titanium dioxide seed particles (d ~ 1-3 μm) were added to both the jet and ambient fluids. Both fluids were seeded at similar number densities, and a small quantity of Rhodamine 6G dye was added to the jet fluid to mark the interface location. The interfacial tension between water/glycerin solution and 50 cS Dow Corning fluid was measured with a spinning drop tensiometer to be 29.5 mN/m for both seeded and unseeded fluids. Measurements of water/glycerin solution and 5 cS Dow Corning fluid yielded an interfacial tension of 29 mN/m.

Fluid Property	Density ρ (kg/m^3)	Dynamic Viscosity μ (cP)
Dow Corning 200, 50 cS	960	48.0
Dow Corning 200, 5 cS	920	4.6
Dow Corning 200, 1000 cS	970	970
Water/glycerin mixture (52/48 percent by volume)	1136	7.37

Table 1. Fluid properties.

2.2 Methods

In visualization studies, the flow was illuminated by sheets of light from a pulsed frequency-doubled Nd:YAG laser. Phase-locked images were captured by either a Kodak DCS 420M single frame camera (1012 x 1054) or a Kodak Megaplus dual-frame camera (1008 x 1018 pixels). For PIV, pairs of images (illuminated by sequential pulses from two Nd:YAG lasers) were captured with the Megaplus dual-frame camera. Timing and control signals were generated by a National Instruments NB-MIO-16X board installed in a Macintosh computer. These signals were employed to drive the forcer as well as an external timing circuit that controlled the laser pulsing and image acquisition.

Acquired images were transferred to a Pentium-style PC and subsequently analyzed with a dual frame cross-correlation routine. The images were preprocessed with custom software to remove the ‘dyed’ background as well as any coagulated particles. Interrogation areas were 64 by 64 pixels (equivalent to 0.73 mm x 0.73 mm in real space), and the areas were overlapped by 50%. The percentage of valid vectors in each image ranged from 98 to 100 percent. Vorticity fields were determined by calculating line integrals using the 8 vectors surrounding each position, and dividing by the area within the closed contour. All vorticity contours were normalized by the ratio of jet exit velocity to jet exit diameter (U/D).

3 Results

The results will be characterized in terms of five dimensionless parameters, including the density and viscosity ratios, a jet Reynolds number, an Ohnesorge number, and a Bond number:

$$r = \frac{\rho_i}{\rho_o}, \quad \lambda = \frac{\mu_i}{\mu_o}, \quad Re = \frac{\rho_i U D}{\mu_i}, \quad Oh = \frac{\mu_i}{\sqrt{\rho_i D \sigma}}, \quad Bo = \frac{\Delta \rho g\, D^2}{\sigma}.$$

In the expressions above, the subscripts i and o denote the jet and ambient fluids, ρ is the density, μ is the viscosity, $\Delta\rho$ is the density difference between the fluids, g is the gravitational constant, and σ is the interfacial tension.

Effect of periodic forcing. Images of natural and forced jets are shown in Figure 2. The jet fluid fluoresces under excitation by the laser sheet. The length of each image is given as L, and the nozzle exit is slightly upstream of the top of each image. Immediately after the jet fluid exits the nozzle, it accelerates due to gravity. As a consequence, the jet diameter decreases. Thus, the jet width at the top of each image is typically less than D.

In the natural jet, a series of drops with relatively constant diameter and spacing may form as in Figure 2a, or the drop size and spacing may vary significantly as in Figure 2b. The largest drops formed have diameter on the order of $0.8D$. Note the instability developing upstream of the jet tip in Figure 2b and the variations in wavelength that must be responsible for the formation of drops of differing diameter. In this Reynolds number regime, an increase in the jet velocity moves the pinch-off location downstream. Therefore, in Figure 2c where Re has been increased to 50, the jet column remains stable over a distance of $20D$ at which point it intersects the reservoir interface downstream.

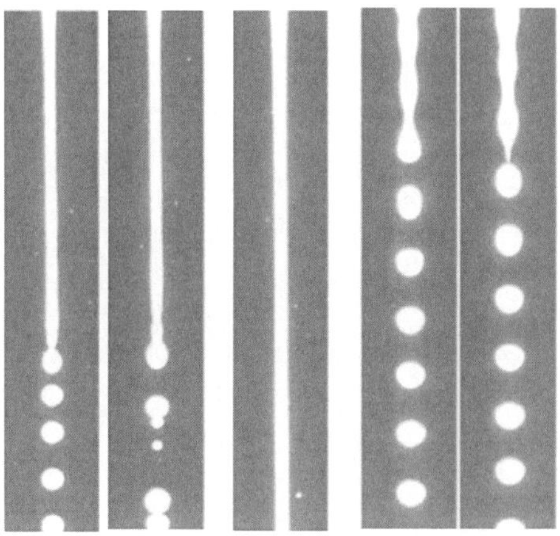

(a) (b) (c) (d) (e)

Figure 2. LIF visualization for $\lambda = 0.15$, $r = 1.18$, Oh = 0.013, Bo = 6.0: (a,b) Natural jet at Re = 16, $L = 17.5D$. (c) Natural jet at Re = 50, $L = 18.1D$. (d,e) Forced jet at Re = 50, St = 3.1, $L = 18.1D$. (parts taken from Webster and Longmire, 2001).

Forcing the jet by superposing periodic velocity oscillations onto the mean amplifies the wave growth along the jet/ambient interface and moves the pinch off location upstream. This is demonstrated in Figures 2d and 2e which show the Re = 50 jet forced at a Strouhal number St = fD/U of 3.1. In this example, the pinch off location occurs 5.5*D* downstream of the nozzle exit. Increasing the forcing amplitude moves the pinch-off location further upstream until eventually a 'dripping' mode results at the nozzle exit. The effect of forcing amplitude on pinch-off location can be explained by considering a reference frame traveling with the instability waves on the jet interface. The velocity oscillations cause fluid to move into thicker sections and out of thinner sections thus accelerating the jet's tendency to pinch off.

For a given fluid combination, the choice of forcing frequency determines the pinch off mode. In Figures 2d and 2e, the jet is forced at a dimensionless frequency such that one drop forms for every forcing cycle. In this forcing mode, then, the drops that form have a uniform size and spacing. By reducing the forcing frequency sufficiently, the associated wavelength becomes much longer than the most unstable Rayleigh wavelength, and it is possible to obtain 'subharmonic' modes wherein multiple drops form during each cycle. Examples of this behavior are given in Figure 3 for Re = 50. The upper sequence (St = 1.6) yields a mode with two drops per cycle where larger and smaller drops form approximately 180° apart in the forcing cycle. In this case, the volume of fluid associated with one forcing cycle is approximately twice that of the St = 3.1 case in Figure 2. Evidently, this volume is too large to sustain the formation of a single drop. This result can be explained partially by considering the forces on a forming drop and their dependence on the drop diameter *d*. The downward gravitational force on a drop scales with d^3, while the upward drag force scales approximately with *d* at this Reynolds number. Therefore, when the drop diameter becomes sufficiently large (see the last and first images in the sequence), the drop is encouraged to accelerate downward and detach from the jet above. In this flow, the larger drop in each cycle has a diameter of 0.95*D* while the jet neck above it has a diameter of only 0.25*D*.

In the bottom sequence shown, the Strouhal number has been reduced to 0.8, and five drops are formed per forcing cycle. In a given set of drops, the sequence shows that the diameter decreases in order of formation with the largest drop having a diameter of 1*D*. In addition, it is interesting to note that two of the drops form from a single volume that breaks off of the jet column and then splits (see bottom right and bottom left images).

Both of the sequences in Figure 3 demonstrate that larger drops tend to propagate faster than smaller ones, which is also explained by the scaling arguments given above. Note the differences in propagation distance observed between neighboring images of smaller and larger drops. In both sequences, larger drops eventually catch up to smaller ones and push them forward as they continue to move downward. Coalescence between drops was not observed in any of the cases examined. Instead, individual drops persisted until they eventually coalesced at the interface beneath the layer of Dow Corning fluid.

Effect of viscosity ratio. The ratio of jet to ambient fluid viscosity (λ) has a significant effect on the flow dynamics. Examples of pinch-off in the 'single drop' mode are given in Figure 4 for three viscosity ratios. A range of values was obtained by varying the viscosity of the ambient Dow Corning fluid. The values of the Bond and Ohnesorge numbers and the density ratio were held relatively constant. The Strouhal number was varied to achieve pinch-off of one droplet per forcing cycle.

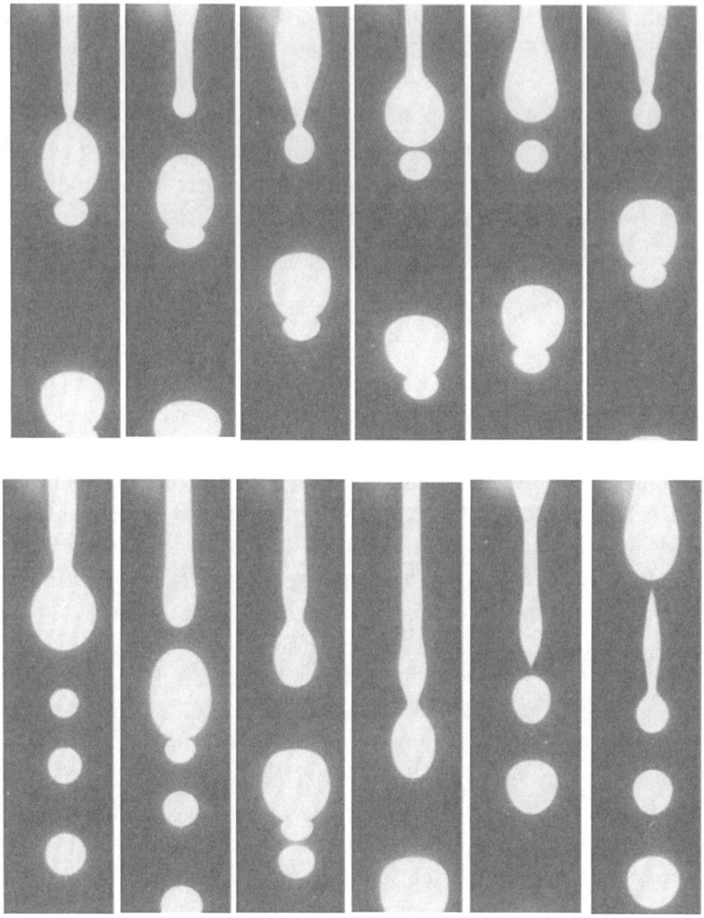

Figure 3. Forced jet at $\lambda = 0.15$, Re = 50, $r = 1.18$, Oh = 0.013, Bo = 6.0: St = 1.6 (top), St = 0.8 (bottom). $L = 6.7D$ in all images. (Taken from Webster and Longmire, 2001).

In general, the case with $\lambda = 0.15$ yields a stable jet over a range of Reynolds numbers. The drops that break off from the jet are initially prolate, but they oscillate at a low frequency such that their shape becomes oblate over the field of view observed (see also Figure 2). When the ambient viscosity is decreased (e.g. $\lambda = 1.6$ in Figure 4), the drops that form have much stronger oscillations. The drop interface can exhibit locations of inverted curvature, as in the drop forming in Figure 4. Also, the upstream surface of the drops can invert strongly to produce a bowl shape. When the ambient viscosity is increased to a very large value ($\lambda = 0.008$ in Figure 4), it becomes impossible to form an axisymmetric jet. At low Reynolds numbers (Re < 200), drops form immediately downstream of the nozzle as in the image shown. When the Reynolds number is increased, a jet forms, but its direction is unsteady such that it deviates from the geometric centerline in a random fashion (see Webster and Longmire, 2001).

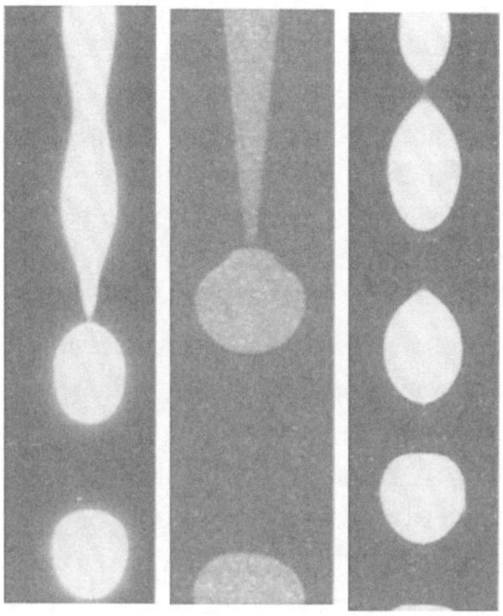

Figure 4. Forced jets with varying ambient viscosity. Left: $\lambda = 1.6$, Re = 39, $r = 1.23$, Oh = 0.013, Bo = 7.3, St = 3.5, $L = 4.5D$. Center: $\lambda = 0.15$, Re = 50, $r = 1.18$, Oh = 0.013, Bo = 6.0, St = 3.1, $L = 6.5D$. Right: $\lambda = 0.008$, Re = 180, $r = 1.17$, Oh = 0.013, Bo = 5.7, St = 0.03, $L = 13.5D$. (Parts taken from Webster and Longmire, 2001).

It is interesting to compare the interface angles just before pinch off in the three cases shown. For $\lambda = 1.6$, the interface upstream develops into a pointed cone, while the upstream end of the forming drop flattens as pinch off is approached. The angle between the fluid interface upstream of the pinch off and the horizontal plane measures 78-84° given experimental uncertainty, and the angle between the horizontal and the downstream interface is 4-7°. When the viscosity ratio is decreased to 0.15, the cone becomes slightly broader near its tip, and the upstream end of the drop is slightly pointed. The upstream interface angle decreases to 68-74°, while the downstream angle increases to about 11° (see Longmire et al, 2001 and Webster and Longmire, 2001). Both of these results match within uncertainty with slope values found by Zhang and Lister (1999) who performed boundary integral calculations of a straining flow and Cohen et al. (1999) who performed experiments on a dripping flow. These results suggest that, at least for these values of λ, the pinch off angles are relatively independent of gravitational, inertial, and surface tension force values, and that the interface shape near pinch off is self similar.

The behavior for $\lambda = 0.008$ is more difficult to characterize. At the top of the image shown, a pinch-off appears imminent. At this location, the upstream and downstream interface angles appear symmetric, and their value is 45-50°. Further downstream, however, a very thin filament of water/glycerin solution connects two drops. At the upstream end of the filament, the upstream and downstream interface angles are nearly 0° and 90° respectively. At the downstream end of the filament, the upstream and downstream interfaces have angles of about 90° and 41° respectively. Thus, it is unclear from these results how and where exactly the pinch off occurs. Additional photographs yielded similar angles near the nozzle exit as well as similar thin filaments. By comparison, Zhang and Lister's simulations did not extend to viscosity ratios that were this small.

Cohen et al. performed experiments at similar viscosity ratios and found 'steep' and 'shallow' angles on the order of 52° and 2° which would appear like a wide cone attached to a flat drop end. Thus, it is unclear from the various results what conclusions can be drawn.

For the two larger viscosity ratios, jet Reynolds and Strouhal numbers appear to have little effect on the upstream and downstream interface angles at pinch off for 'single drop' modes. In cases where multiple drop modes exist, our results are inconclusive thus far. In Figure 3, the two images on the bottom right show that the upstream and downstream angles can reverse in nature. The exact values of these angles (as well as those corresponding to the pinch off of the small drop at St = 1.6) appear to vary. These transitions have not been measured in detail, however.

Effects of parameter variation on satellite formation. Several Reynolds/Strouhal combinations were observed in order to determine conditions under which multiple drop modes and satellites occurred. The results of these tests are plotted in Figure 5 for viscosity ratios of 1.6 and 0.15. In general, the lower ambient viscosity was associated with stronger and more complicated perturbations to the interface. When the jet was forced at Strouhal numbers smaller than those associated with the Rayleigh instability, higher frequency (or shorter wavelength) perturbations would develop along the jet interface. Under certain conditions, these perturbations allowed the formation of small 'satellite' droplets immediately upstream of larger primary droplets. By contrast, the higher ambient viscosity tended to stabilize the interface. Although the points corresponding with $\lambda = 0.15$ match the topology trends for $\lambda = 1.6$ in terms of the number of primary drops formed per cycle, no satellites formed. In fact, no 'harmonic' instabilities or zones of concave curvature were observed on developing primary drops for this fluid combination.

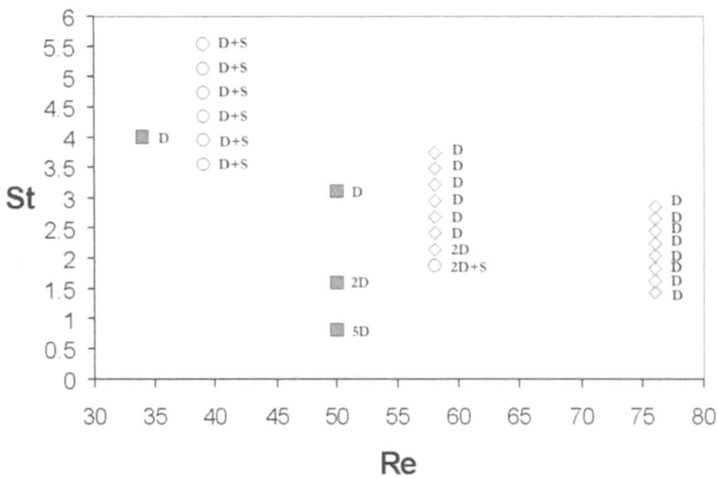

Figure 5. Pinch off topology for various Re-St combinations. All points were obtained using $\lambda = 1.6$, except for those marked by "■" where $\lambda = 0.15$. The symbol D denotes a drop of significant volume, and S denotes the presence of a satellite (taken from Milosevic and Longmire, 2002).

The plot demonstrates that lower Reynolds numbers as well as lower Strouhal numbers encouraged satellite formation for λ = 1.6. The Strouhal trend is consistent with the 'preferred' wavelengths for jet instability estimated by Meister and Scheele (1967) using Tomotika's theory for cylindrical jets. Smaller Strouhal numbers clearly support longer wavelengths. If the wavelength becomes significantly longer than the preferred wavelength, the formation of harmonics resulting in multiple drops or satellites is possible. The relation between satellite formation and Reynolds number is more subtle. From Meister and Scheele's calculations, the most unstable dimensionless wave number for the λ = 1.6 is (ka) = 0.664 where k and a represent the wave number and jet radius respectively. Measurements from our experiments yielded (ka) = 0.652 for St = 3.5, Re = 39, and (ka) = 0.679 for St = 3.5, Re = 58. The lower Re, which supported a wavelength longer than the most unstable one, also supported satellites. The higher Re, which supported a wavelength shorter than the most unstable one, did not (see Figure 5).

The mechanism by which satellites form can be understood by examining the two flow cases given above as examples. In Figures 6 and 7, vorticity contours and the corresponding interface locations are plotted for phase-locked sequences of these two cases. When Re = 58 (Figure 6), the interface shape approaching pinch off is characterized by a very sharp jet tip and nearly flat upstream end of the drop as was seen in Figure 4. The flat section eventually detaches from the pointed tip. When Re = 39, the jet tip is blunter than in the previous case, and the upstream end of the drop is relatively flat except for a cone-shaped structure centered on the axis. The interface slope changes sharply where the cone intersects the drop (see $\Phi = 334°$ and 347°).

In fact, the jet does not pinch off above the cone. Instead, the cone evolves into a small bulge as the interface of the primary drop downstream begins to invert. At this phase, the interface upstream of the eventual break point is rotating in the opposite direction. The primary drop then breaks off of the small bulge. Finally, a thin filament develops between the eventual satellite and jet tip before the satellite pinches off. (Details of this sequence can be seen in Milosevic and Longmire, 2002).

Now we compare the vorticity patterns. In Figure 6, the plot at $\Phi = 313°$ contains vorticity maxima at the jet neck, at the upstream end of the drop, and at the location of inverted curvature halfway down the drop. The vorticity near the upstream end of the drop acts to rotate and flatten the interface and hence to encourage pinch off. As the flow progresses to $\Phi = 350.6°$, the circulation at the jet neck weakens, and the downstream maximum moves toward the neck. The largest vorticity values are found outside of the drop interface at the location of the inverted curvature. This area of circulation persists through the pinch off, remaining attached to the drop that forms. After the drop pinches off, a small ring of inverted vorticity develops at the jet tip due to the recoiling interface there.

When Re = 39 (Figure 7, $\Phi = 327.5°$), the location of inverted curvature lies closer to the jet neck, and the three vorticity maxima are spaced more closely. These differences ultimately result in a different pinch off topology. The strong maximum at the jet neck quickly dominates the center maximum which disappears upon formation of the cone ($\Phi = 334°$). The downstream maximum is focused at the junction of the cone and drop. This core and the core at the jet neck have larger maximum values and larger circulation than their counterparts at Re = 58. While the center at the neck dissipates, the center at the cone/drop interface intensifies ($\Phi = 347°$). The strong rotation helps to drive fluid downward out of the cone, decreasing its volume, and to rotate the drop interface inward. As the vortex core radius decreases, its propagation speed increases so that the center of vorticity moves inside of the developing drop ($\Phi = 353.5°$). The streaming flow inside of the core acts to stretch and drag the drop interface downward, encouraging the eventual pinch off.

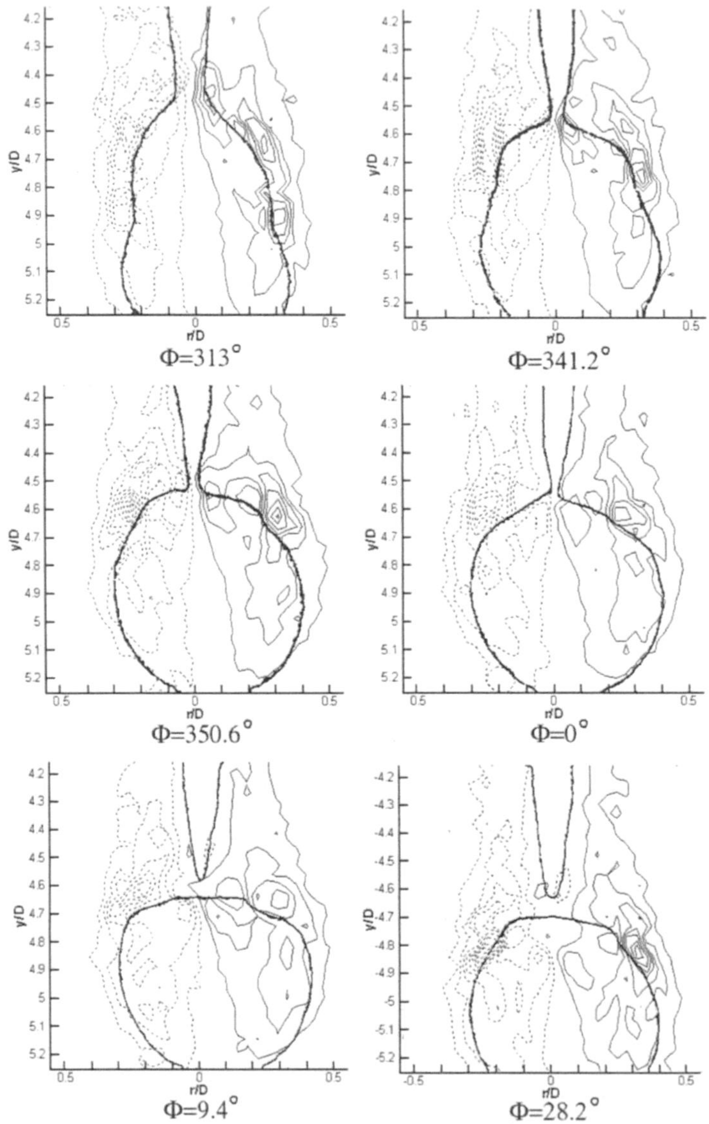

Figure 6. Normalized vorticity fields of forced flow at St = 3.5, Re = 58, r = 1.23, Oh = 0.013, Bo = 7.3. Lowest contour level is 4, and succeeding levels are incremented by 4. (taken from Milosevic and Longmire, 2002).

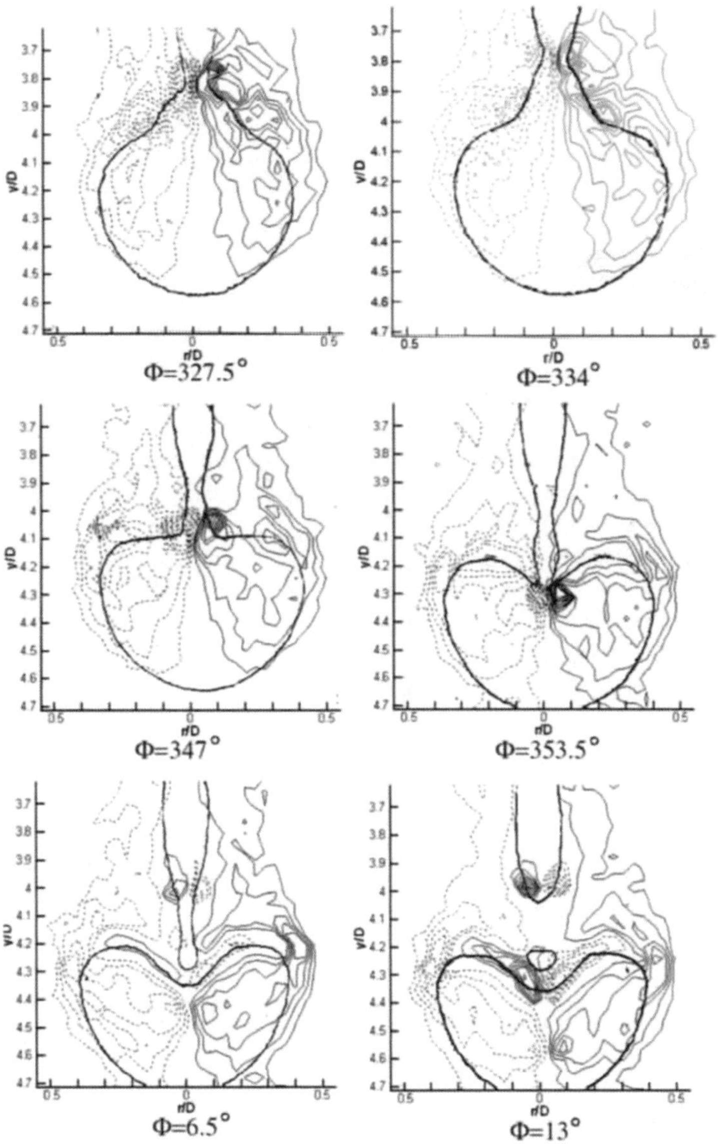

Figure 7. Normalized vorticity fields of forced flow at St = 3.5, Re = 39, r = 1.23, Oh = 0.013, Bo = 7.3. Lowest contour level is 4, and succeeding levels are incremented by 4. (taken from Milosevic and Longmire, 2002).

After the drop pinch off, a ring of inverted vorticity develops upstream of the satellite, encouraging the break between satellite and jet tip. This vorticity persists as the jet tip recoils. Inverted vorticity is also present along the upstream drop interface indicating that the interface has begun reverting toward a convex shape.

4 Conclusions

In this paper, liquid/liquid jet systems are discussed where interfacial tension causes instabilities in and eventual break up of the jet column. Flow visualization demonstrated that forcing amplifies the instabilities so that the jet pinches off at earlier axial locations. The ratio of the wavelength associated with the forcing frequency to the wavelength of the Rayleigh instability ultimately determines the pinch off mode. When this ratio is low, multidrop modes result. When λ is greater than 1, small satellite drops can occur (as well as multiple primary drops). Vorticity fields derived from PIV measurements revealed the kinematics behind satellite formation. Satellite occurrence depends on the interface shape approaching pinch off as well as the local vorticity patterns. Satellite formation requires a section of concave curvature on the interface downstream of the eventual jet tip. The concave section must lie close to the jet neck, so that the resulting velocity and vorticity pattern deform the interface into an upstream-facing cone attached to the main drop. This cone ultimately breaks off as a satellite.

Measurements of interface angles shortly before pinch off in single drop modes agreed with a similarity solution when $\lambda = 0.15$ and 1.6. When $\lambda = 0.008$, however, the results differed from measurements reported in Cohen et al. (1999) and appeared to diverge from extrapolations of the similarity solution in Zhang and Lister (1999). Additional measurements of these angles are underway for pinch off of primary and satellite drops.

References

Bright, A. (1985) Minimum drop volume in liquid jet breakup. *Chemical Engineering Research & Design.* 63:59-66.

Cohen, I., Brenner, M.P., Eggers, J., & Nagel, S.R. (1999) Two fluid drop snap-off problem: experiments and theory. *Physical Review Letters.* 83:1147-1150.

Longmire, E. K., Lowengrub, J. S. and Gefroh, D. L. (1999) A comparison of experiments and simulations on pinch-off in round jets. ASME FEDSM99-711, 3rd ASME/JSME Joint Fluids Engineering Conference, San Francisco.

Longmire, E.K., Norman, T.L., and Gefroh, D.L. (2001) Dynamics of pinch-off in liquid/liquid jets with surface tension. *International Journal of Multiphase Flow.* 27:1735-1752.

Meister, B. J. & Scheele, G. F. (1967) Generalized solution of the Tomotika stability analysis for a cylindrical jet. *AICHE Journal.* 13:682-688.

Meister, B. J. & Scheele, G. F. (1969) Predictions of jet length in immiscible liquid systems. *AICHE Journal.* 15:689-699.

Milosevic, I.N. and Longmire, E.K. (2002) Pinch-off modes and satellite formation in liquid-liquid jet systems. *International Journal of Multiphase Flow*, to appear.

Skelland, A. H. P. & Johnson, K. R. (1982) Jet break-up in liquid-liquid systems. *Canadian Journal of Chemical Engineering.*52:732-738.

Webster, D. R. and Longmire, E. K. (2001) Jet pinch-off and drop formation in immiscible liquid/liquid systems. *Experiments in Fluids.* 30:47-56.

Zhang, W.W. & Lister, J.R. (1999) Similarity solutions for capillary pinch-off in fluids of differing viscosity. *Physical Review Letters.* 83:1151-1154.

Part IV

Controlled Jets in Crossflow and Control via Jet Systems

A Framework to Design Controllers for Engineering Applications of Transverse Jets

Luca Cortelezzi[1], Robert T. M'Closkey[2] and Ann R. Karagozian[2] ⋆

[1]Department of Mechanical Engineering, McGill University, Montreal, H3A 2K6, Canada

[2]Department of Mechanical and Aerospace Engineering,
University of California, Los Angeles, California 90095-1597, U.S.A.

Abstract. We propose a hypothetical framework to develop control strategies to optimize the mixing characteristics associated with the actively driven jet in crossflow. This framework is a three-step design procedure. As a first step, we form a preliminary controller with a feed-forward loop in order to develop controllers able to compensate for the dynamics of actuators and sensors. As a second step, we augment the feed-forward loop with a feedback loop in order to develop a plant controller based on a hierarchy of reduced-order models. As a final step, we remove the feed-forward loop and concentrate on the robustness of the plant controller with respect to unmodeled and unpredictable uncertainties.

It is often desirable to control mixing in shear flows because of the potential benefits to engineering applications. Numerous experiments have shown that enhanced mixing in shear flows is achievable by periodically pulsing the flow near the shear-layer separation point. In particular, it has been shown that jet penetration and mixing can be improved (e.g., Hermanson *et al.*, 1998; Schuller *et al.*, 1999; Johari *et al.*, 1999; M'Closkey *et. al*, 2002(a); M'Closkey *et. al*, 2002(b)), combustion efficiency can be increased and emissions therefore reduced, (e.g., Vermeulen *et al.*, 1992; Lee &Morris, 1997; Sato *et al.*, 1999; Cohen *et al.*, 2001), and separation can be controlled with the effect of improving airfoil lift or enhancing pressure recovery in diffusers (e.g., McCormik & Gursul, 1996; Ng & Lang, 1999; Kiya *et al.*, 2001).

Most of these experiments are a proof-of-concept. Yet few studies have used control theory to design successful controllers. This is due to the fact that enhancing mixing by means of control poses a non-standard control problem: how to create a non-equilibrium dynamics with control. This goal is very different from the classical (and well-studied) problem of how to stabilize equilibrium with control. In recent years, efforts have been made to apply control theory and dynamical systems analysis for designing controllers to enhance mixing in various engineering applications (e.g., D'Alessandro *et al.*, 1999; Fountain *et al.*, 2000; Stroock *et al.*, 2002; M'Closkey *et. al*, 2002(a); M'Closkey *et. al*, 2002(b); D'Alessandro *et al.*, 2002).

In general the problem of feedback control for an unsteady fluid flow is nonlinear. Especially challenging is the control of mixing processes in a hot, potentially reactive environment. Because of the high temperatures present, realistic sensors can only operate mainly in cool regions, e.g., away from flames and hot exhaust regions. Consequently, there is often a large time lag between the time at which an actuator would modify, for example, a fuel or dilution air jet's characteristics in a combustion chamber, and the time at which a sensor would measure the effect of this action on the jet's mixing and/or reaction processes. During this time lag, flow dynamics, mixing, and combustion chemistry, if present, are dominated by nonlinear effects (Karagozian, 1999). One of the challenging aspects of controlling such mixing processes is

⋆ This work has been supported by NASA Dryden Flight Research Center under Grant NCC 2-374, by the California Energy Commission/EISG under Grant 51240A/99-01-56, by the University of California Energy Institute, and by NSERC Grant RGPIN217169.

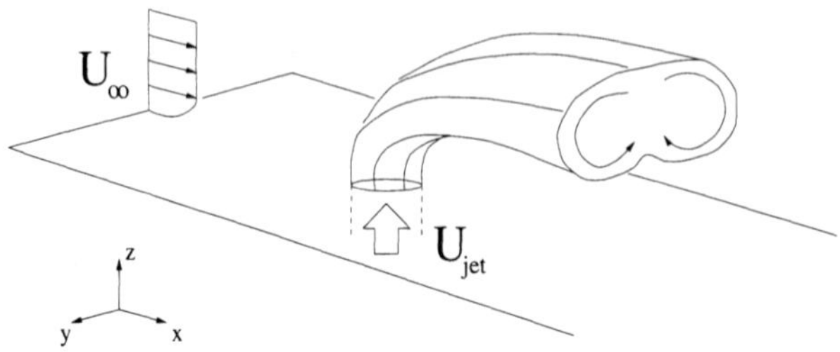

Figure 1. Flow geometry for a jet of velocity U_{jet} in a crossflow of freestream velocity U_∞. Figure taken from Cortelezzi & Karagozian (2001).

that there are usually both fast and slow characteristic time-scales present. In the presence of a combustion reaction, there will be additional reaction time scales that couple to the fluid mechanical time scales.

Consider, for example, the canonical flowfield of the non-reactive jet in crossflow, or transverse jet, as shown in Figure 1. In this flowfield there is a "fast" time-scale T_f over which the Kelvin-Helmholtz instability induces the formation of (secondary) vortex ring structures, but there is also a "slow" time-scale T_s over which the evolution of these vortical structures affects the mean properties of the flow. A striking feature of the mean flow of the transverse jet is the counter-rotating vortex pair (CVP), shown in Figure 1, which is associated with the jet cross-section. This vortical structure exists only in an average sense, since instantaneous measurements of the jet's vorticity field do not always clearly reveal the existence of the CVP (Cortelezzi & Karagozian, 2001; Cortelezzi & Karagozian, 2002). Mean measurements of the flow field suggest that the CVP is responsible for the entrainment of the crossflow into the injectant and, consequently, for the excellent mixing characteristics of the transverse jet as compared with the jet injected into quiescent surroundings.

A goal of our recent research has been to develop control strategies to optimize the mixing characteristics associated with the actively driven jet in crossflow. Active control of the mixing processes associated with the transverse jet (open as well as closed loop) may be accomplished through excitation of the jet flow issuing from the orifice shown in Figure 1. Experiments on pulsed or acoustically driven transverse jets (Kelso *et al.*, 1996; Schuller *et al.*, 1999; Johari *et al.*, 1999; Eroglu & Breidenthal, 2001; M'Closkey *et. al*, 2002(a)) demonstrate that temporally varying the jet velocity allows jet penetration and spread to be enhanced, at specific conditions of excitation, due to control of the generation of jet vorticity and hence control of the fast time scale T_f. In practice this type of actuation can be achieved for gaseous injectants using a loudspeaker placed in the plenum upstream of the jet orifice, while for liquid injectants the excitation can be achieved using a rotating valve placed within the line which delivers injectant to the orifice. In either case a signal generator may be used to produce the desired temporally varying jet velocity. It has been observed through experimental (Johari *et al.*, 1999; Schuller *et al.*, 1999; M'Closkey *et. al*, 2002(a); M'Closkey *et. al*, 2002(b)) and numerical evidence (Cortelezzi & Karagozian, 1999; Campolo *et al.*, 2002) that the temporal variation of the velocity $\mathrm{U_{jet}}$ at the orifice can be quite different from that imposed by the signal generator. Although the dominant frequency of excitation is retained by the jet, differences in actual duty cycle, phase, and waveform shape can arise due to the unsteady flow evolution within the jet plenum and nozzle, as well as due to the jet's nonlinear coupling to the crossflow. Most significantly, the difference between the input function and the resulting temporal jet velocity variation is apparatus-dependent; this is represented in Figure 2 by the block "Actuator Dynamics". As a consequence of the differences between signal generator input function and jet exit velocity temporal variation (which is the actuation for the flowfield), it becomes necessary to design a feedback controller for the jet actuator, shown in Figure 2, which is

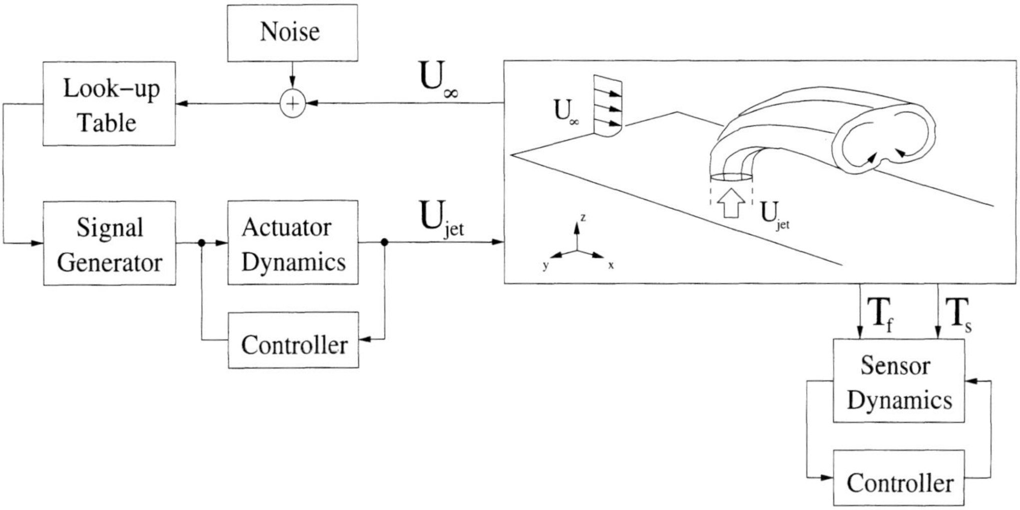

Figure 2. Block diagram of the feed-forward control strategy based on a "Look-up Table". This configuration allows for the development of controllers able to compensate for actuator and sensor dynamics.

distinct from the "Plant Controller" (see Figures 3 and 6) used for the overall experiment (Cortelezzi *et al.*, 2000).

In developing a controller for the transverse jet problem there is an interesting trade-off between the complexity of actuating and sensing. On the one hand, while the temporal excitation of the jet velocity is clearly a good candidate for manipulating the flow, there are a number of quantities which could be sensed or measured in order to provide timely feedback information to the controller, quantities which are potentially relevant to mixing and/or combustion efficiency. Further, while actuation is necessarily associated with manipulating the vortical structures generated by the jet on fast time scales T_f, there is the possibility of distributing different sensors which can measure quantities evolving on both fast and slow time scales (e.g., pressure and temperature, respectively). Hence in designing a control algorithm for this problem, as shown in Figure 2, the block "Sensor Dynamics" could represent a sophisticated mixture of sensors (pressure, temperature, species, flame luminosity, etc.) which may be used to extract information about flow properties evolving on both time scales. Sensing thus also becomes a separate control problem in itself, requiring a separate controller because sensors operate under extreme conditions and must be capable of self-testing and self-calibration.

We propose a three-step designing procedure as a framework for developing a robust feedback controller for this problem. As a first step, we form a preliminary controller with a feed-forward loop, based on a "Look-up Table", as described in the block-diagram shown in Figure 2. This architecture has been selected to allow for the initial effort of developing feasible controllers based on actuator and sensor dynamics. The "Look-up Table" is constructed by searching the parameter space for the combination of quantities that produce optimal mixing for given crossflow conditions. The input to the "Look-up Table" is the crossflow velocity U_∞, and the output parameters for the jet are the settings for the signal generator which modulate the jet velocity U_{jet}. The measurement of the crossflow velocity could also provide information about the background level of turbulence or the type of boundary layer formed upstream of the jet, yet this information could also be classified as contributing to uncertainties. The uncertainties are represented by the block "Noise" in Figure 2. The "Look-up Table" should be viewed from the perspective of a block that could be

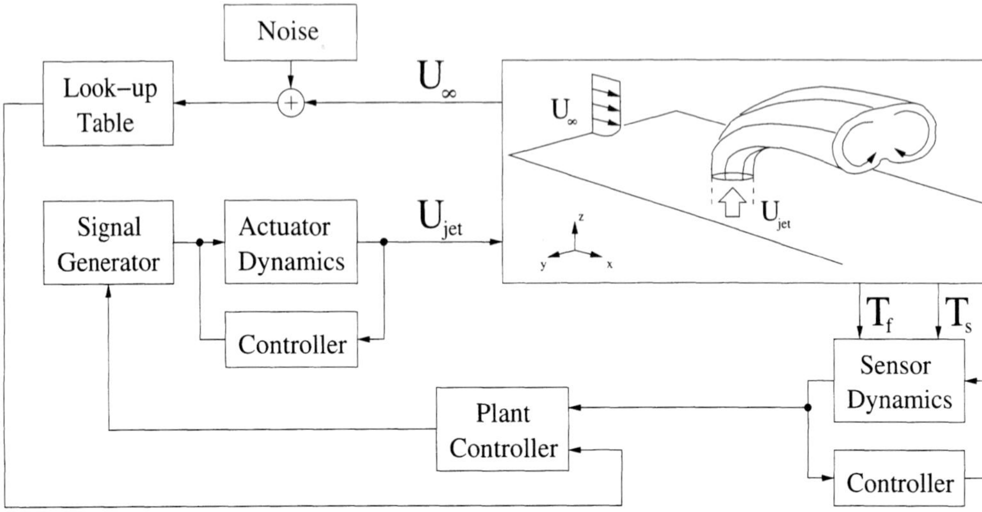

Figure 3. Block diagram of the feed-forward control strategy augmented by a feedback loop. This configuration allows for the development of a model-based controller.

replaced by a neural network or by a genetic algorithm which could provide more accurate input settings for the wave generator.

A feed-forward loop based on a "Look-up Table" does not use any information about the flow processes affected by the action of the actuator. Consequently, this configuration allows us to develop the relationships among measured flow variables, actuator dynamics (jet responses to imposed excitation) and sensor dynamics. The knowledge of these relationships will be crucial for developing robust feedback controllers. In our research, the actuator and sensor dynamics are characterized through gas-phase experimental studies.

In order to design high-authority robust controllers, it is necessary to employ high-bandwidth linear-response actuators and sensors. Hence, it is important to compensate actuator and sensor dynamics before designing a controller. Since our goal is to control transverse jets in engineering applications, actuator and sensor dynamics are more effectively characterized in an experimental set up. Relevant gas phase experiments have been conducted at UCLA in a blow-down wind tunnel in which a round air jet, seeded with smoke, is introduced perpendicularly into a crossflow of air (M'Closkey *et. al*, 2002(a)). Acoustic excitation of the jet is accomplished by placement of a loudspeaker (subwoofer) at the bottom of the jet plenum, which is driven by a function generator and amplifier. Sinusoidal as well as square wave excitation can be imposed for a wide range of frequencies, amplitudes of excitation, and duty cycles. The functional form of the variable speaker voltage thus becomes the input to the "Actuator Dynamics" block in Figure 2. The temporal response of the jet to such actuation, through its exit plane velocity U_{jet}, is quantified via hot wire anemometry. Hence the input-output relationships which characterize the actuator dynamics in the experiments can be quantified and used to design a feedback controller or shaping filter with the objective of gaining more authority over the temporal shape of the jet velocity (M'Closkey *et. al*, 2002(a); M'Closkey *et. al*, 2002(b)).

The second step of our design procedure will be to augment the block diagram shown in Figure 2 by a feedback loop as shown in Figure 3. This configuration allows for the development of a model based "Plant Controller". Since the derivation of a reliable reduced-order model could be difficult, in the early stages of the control development the controller will improve the output of the "Look-up Table" using the sensed

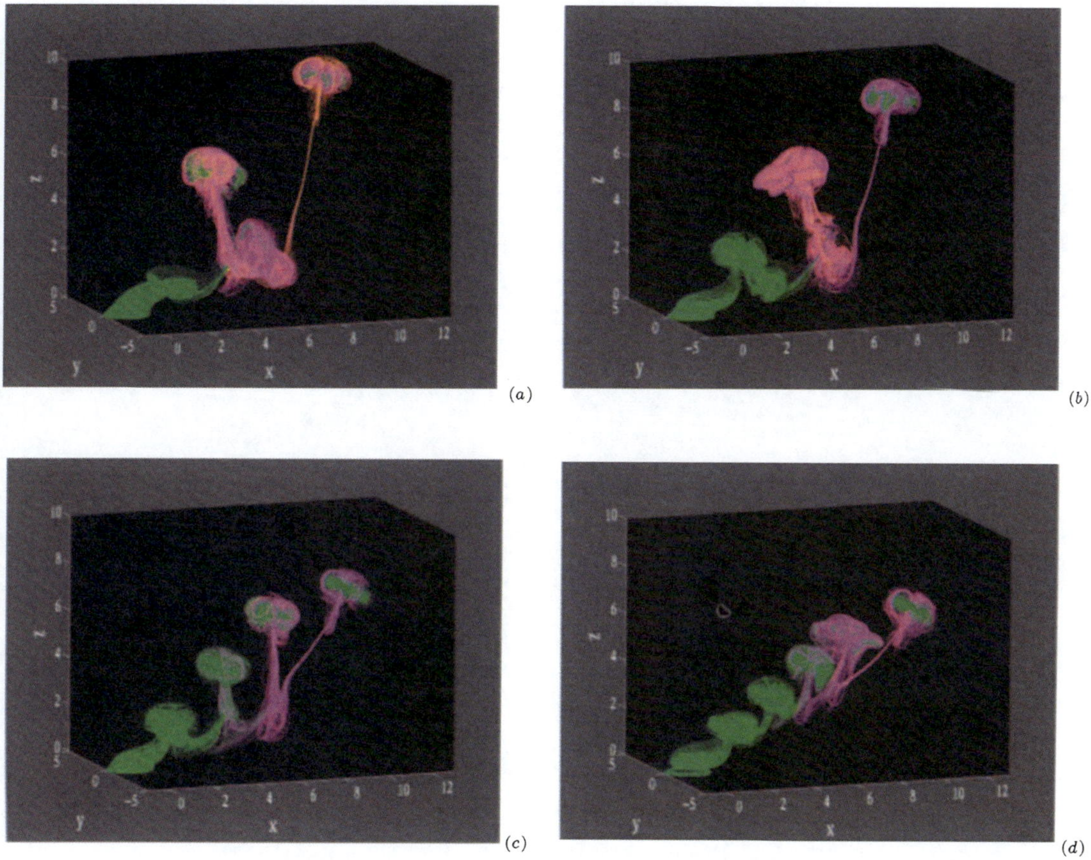

Figure 4. Simulations of three-dimensional vortex structure for the forced jet in crossflow with square wave excitation, approximate velocity ratio $U_j/U_\infty = 2.5$, amplitude of forcing 80%, duty cycle 10%, and non-dimensional frequencies 0.25 (a), 0.35(b), 0,50(c) and 0.70(d) , at non-dimensional flow time 8. The level of stretching to which a filament underwent during the simulation is represented by different colors: green is minimum stretching and orange is maximum stretching.

information. As the plant model will be developed and the controller will gain more authority, the "Look-up Table" will be used as a diagnostic.

The successful design of a controller for this nonlinear flow is possible only if a model can be identified which is able to capture the dynamics of the flow processes. The size of the controller, however, is a crucial parameter in any engineering application because of the amount of hardware and computer power necessary to compute a real-time control law. Thus, it is essential to derive a reliable hierarchy of reduced-order, or low-order, models of the nonlinear flow processes to be able to develop a practically implementable controller. In this book, the techniques to derive reduced-order models and design reduced-order controllers for fluid mechanics applications are reviewed by Cortelezzi (2002). Two primary reduced-order models, one "low-order" and one "high-order" may be used as a means of transferring an evolved control strategy to the flowfield of the pulsed jet injected normally into crossflow. Details on the models are outlined below.

Experimental observations of the circular jet in a crossflow indicate the presence of a pair of bound vortices initially associated with the lee side of the jet boundary, which gradually dominates the actual jet

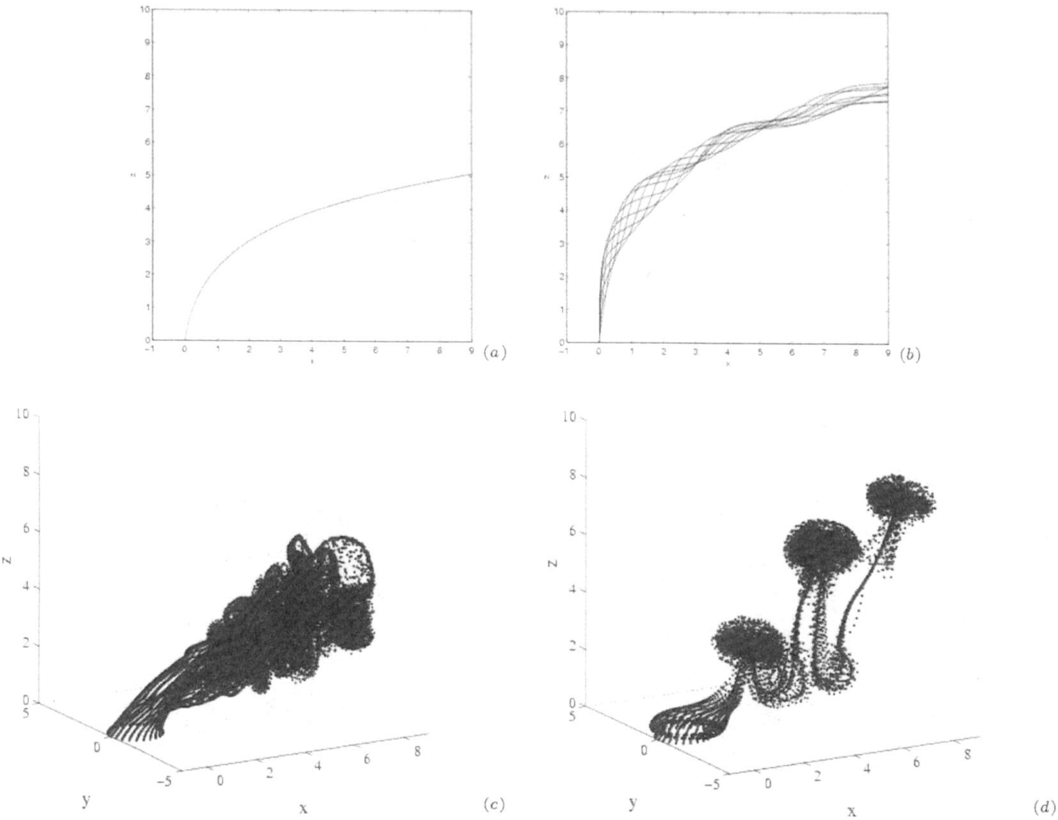

Figure 5. Comparison between unforced transverse jet cases (a), (c) and sinusoidally forced cases (b), (d), for an approximate velocity ratio of $U_j/U_\infty = 2.5$. Results are shown for the two-dimensional model in (a) and (b), and the three-dimensional model in (c) and (d). For the forced cases, the non-dimensional frequency is 0.4.

cross section, persisting well into the farfield (see Figure 1). A precise mathematical description of these deflected counter-rotating vortices would consider two semi-infinite, curved vortex lines emanating from points outside of the jet orifice and with image vortex lines represented in the lower wall. However, a reduced-order model can be derived assuming the vortex pair to be locally two-dimensional (Karagozian, 1986(a); Karagozian, 1986(b); Karagozian *et al.*, 1996). This assumption is clearly appropriate in the far field of the jet and, based on recent experimental results (Smith & Mungal, 1998), is not unreasonable in much of the near field. The concept of a flow time associated with the local (axial) jet velocity is introduced into the model as a means of describing the location of the vortex pair along the vortex curve. Also inherent in this formulation is the assumption that each vortex structure consists of a Gaussian distribution of vorticity. The concept of the diffusion of vorticity is particularly important to our model in that the downstream growth of each viscous vortex core gives rise to forces that act to separate the vortices. This "low-order" model, evolved from Karagozian (1986(a)) and Karagozian (1986(b)), is modified to accommodate a periodically varying jet velocity in order to represent the pulsed jet in a uniform crossflow. Preliminary controller development may be based on this two-dimensional model.

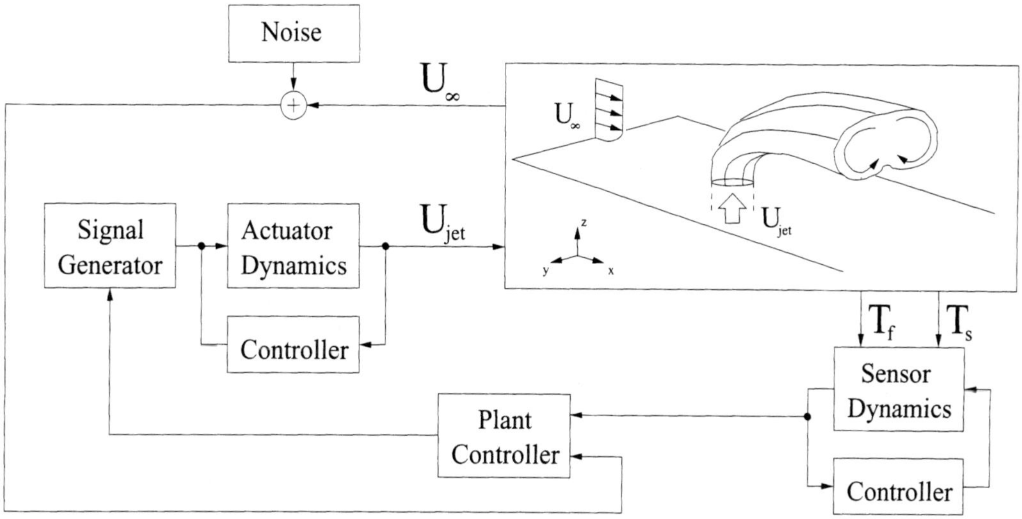

Figure 6. Block diagram of control strategies for the pulsed transverse jet.

A three-dimensional vortex filament simulation of the actively forced transverse jet (Cortelezzi & Karagozian, 1999; Cortelezzi & Karagozian, 2001; Cortelezzi & Karagozian, 2002) may be used to investigate the mechanisms for mixing enhancement that occur. This "high-order" model represents the mean jet flow using a semi-infinite, cylindrical vortex sheet, where the jet velocity is a function of the circulation per unit length of the vortex sheet. In general the jet velocity can be expressed as $U_{jet} = U_j + a_j f(2\pi\omega t)$, where U_j is a constant mean velocity and a_j is the amplitude of forcing at frequency ω. Different functional forms of f for jet forcing are currently being explored. The jet shear layer is modeled by introducing a three-dimensional vortex filament or ring near the jet exit at each time step of the computation. The velocity field induced by the evolution of the vortex filaments is then computed from the modified Biot-Savart integral. Sample results of the three-dimensional simulations, for the pulsed jet in crossflow, are shown in Figure 4. Parametric studies of the jet in crossflow, with and without jet pulsation (Cortelezzi & Karagozian, 1999; Cortelezzi *et al.*, 2001), show that the most critical parameters to be modified include the jet-to-crossflow velocity ratio, the upstream boundary layer thickness, the jet pulsation frequency, its forcing amplitude, and the "duty cycle" of the oscillations (for square wave excitation).

This three-dimensional "higher-order" model approach will be to characterize mixing processes by implementing numerically a technique similar to that of passive particle tracking, often used experimentally. The tracking of passive (massless) fluid particles can be computed efficiently by a Lagrangian scheme such as the present vortex code. The unmixedness associated with particle concentrations in different cross-sectional slices of the jet flowfield will be determined and compared with mixing characterizations as obtained by corresponding gas phase experiments at UCLA for the acoustically driven jet in crossflow. Global quantities such as the mixing length of the jet can also be obtained from these simulations and compared with experiments. Hence the optimal conditions for jet mixing may be predicted by the current "high-order" model.

These two-dimensional ("low-order") and three-dimensional ("higher-order") models constitute a two layer hierarchy of models of increasing complexity for the non-reactive mixing processes associated with the acoustically driven transverse jet. These models can be qualitatively validated at present by smoke visualizations performed in the low-speed wind-tunnel at UCLA (M'Closkey *et. al*, 2002(a)). The "lower-order" model will be used to develop "low-order" flow controllers associated with the pulsed jet in crossflow. The controllers, together with the "higher-order" model, will then be used to develop "higher-order" controllers for non-reactive and reactive processes associated with the driven transverse jet. These controllers will be tested computationally as well as experimentally in the laboratory.

Preliminary results from the hierarchies of models and from the corresponding experiments are shown in Figure 5. Figures 5(a) and (b) show the unforced and sinusoidally forced case predicted by the "low-order" two-dimensional model. The forced case shows that the jet responds periodically to pulsations in velocity at the jet orifice. Averaged over time, this periodic response creates a jet with greater penetration and much greater spreading than is observed for steady jet injection, a confirmation of experimental observations. The frequency of the acoustic forcing has an influence on how rapidly the vorticity is generated locally, which in turn affects mixing (and, if present, reaction) rates. In this "low-order" model, characterization of jet behavior in terms of the dynamics of the counter-rotating vortex pair actually produces governing ordinary differential equations which, similar to the approaches in Cortelezzi *et al.* (1994) and Cortelezzi (1996), lend themselves well to controller development. Figures 5(c) and (d) show corresponding results for the unforced and forced cases obtained from the "higher-order", three-dimensional model (Cortelezzi & Karagozian, 1999; Cortelezzi *et al.*, 2001; Cortelezzi & Karagozian, 2002), comparable to those in Figures 5(a) and (b).

The third and final step of our framework consists of designing a model-based robust feedback "Plant Controller", shown in Figure 6. This can be done only when the non-reactive mixing processes associated with the acoustically driven transverse jet are correctly predicted by the reduced-order models derived in the previous step. At this stage the "Look-up Table" shown in Figure 3 will be removed because it would be redundant. To perform properly in an experimental environment, the "Plant Controller" has to be made robust with respect to the unmodelled uncertainties. The iteration of the robustness process over different perturbations, (e.g., chemical reactions, background noise, etc.) could produce a "Plant Controller" able to optimize mixing and combustion in an experiment. The final ambitious step of feedback controlling engineering applications may be undertaken only if the final product of our framework is a controller which is robust enough to perform in presence of the unmodeled and unpredictable uncertainties contained in the engineering system, and which is fast enough to control the flow in real-time.

The present framework describes a proposed methodology for the complete closed-loop control of the jet in crossflow. The methodology is the subject of ongoing research, from the point of view of the actuation system as well as nearfield and farfield sensing. In subsequent papers in this section of the text, different aspects of the control of transverse jets, particularly through jet actuation systems, will be described.

References

CAMPOLO, M., DEGANO, G.M., CORTELEZZI, L. AND SOLDATI, A., 2002 Numerical appraisal of jet-to-crossflow coupling in a transverse jet. *Manipulation and Control of Transverse Jets*, Springer-Verlag.

COHEN, J. M., STUFFLEBEAM, J. H. AND PROSCIA, W., 2001 The effect of fuel/air mixing on actuation authority in an active combustion instability control system, *ASME Trans: J. Engng Gas Turb. Power* **123**, (3), 537.

CORTELEZZI, L., LEONARD, A., AND DOYLE, J. C. 1994 An example of active circulation control of the unsteady separated flow past a semi-infinite plate *Journal of Fluid Mechanics* **260**, 127.

CORTELEZZI, L. 1996 Nonlinear feedback control of the wake past a plate with a suction point on the downstream wall *Journal of Fluid Mechanics* **327**, 303.

CORTELEZZI, L. AND KARAGOZIAN, A. R., 1999 On the evolution of vorticity in pulsating jets in crossflow, *Bull. Amer. Phys. Soc.* **44**, (8).

CORTELEZZI, L., KING, J., LOBBIA, R., M'CLOSKEY, R., AND KARAGOZIAN, A. R., 2000 Actively controlled transverse gas injection, 39th *IEEE Conference on Decision and Control*, Sydney, Australia, Dec 12-15.

CORTELEZZI, L. AND KARAGOZIAN, A. R., 2001 On the formation of the counter-rotating vortex pair in transverse jets, *J. Fluid Mech.* **446**, 347.

CORTELEZZI, L., KING, J., KARAGOZIAN, A. R. AND M'CLOSKEY, R. T., 2001 Simulations and experiments on pulsed jets in crossflow, *Bull. Amer. Phys. Soc.* **45**, (9).

CORTELEZZI, L., 2002 Reduced-order controllers for fluid mechanics applications, In *Manipulation and Control of Transverse Jets,* Springer-Verlag.

CORTELEZZI, L. AND KARAGOZIAN, A. R., 2002 Three-dimensional vortex modeling of unforced transverse jets, In *Manipulation and Control of Transverse Jets,* Springer-Verlag.

D'ALESSANDRO, D, DAHLEH, M. AND MEZIC, I., 1999 Control of mixing in fluid flow: A maximum entropy approach, *IEEE Transaction on Automatic Control* **44**, (10), 1852.

D'ALESSANDRO, D, MEZIC, I. AND DAHLEH, M., 2002 Statistical properties of controlled fluid flows with applications to control of mixing, *Systems & Control Letters* **45**, (4), 249.

EROGLU, A. AND BREIDENTHAL, R. E., 2001 Structure, penetration, and mixing of pulsed jets in crossflow, *AIAA J.* **39**, 417.

FOUNTAIN, G. O., KHAKHAR, D. V., MEZIC, I. AND OTTINO, J. M., 2000 Chaotic mixing in a bounded three-dimensional flow, *J. Fluid Mech.* **417**, 265.

HERMANSON, J.C., WAHBA, A., AND JOHARI, H., 1998 Duty-cycle effects on penetration of fully modulated, turbulent jets in crossflow, *AIAA J.* **36**, (10), 1935.

JOHARI, H., PACHECO-TOUGAS, M., AND HERMANSON, J.C., 1999 Penetration and mixing of fully modulated turbulent jets in crossflow, *AIAA J.* **37**, (7), 842.

KARAGOZIAN, A. R., 1986(a) An analytical model for the vorticity associated with a transverse jet, *AIAA J.* **24**, 429.

KARAGOZIAN, A. R., 1986(b) The flame structure and vorticity generated by a chemically reacting transverse jet, *AIAA J.* **24**, 1502.

KARAGOZIAN, A. R., WANG, K. C., LE, A-T. AND SMITH, O. I., 1996 Transverse gas jet injection behind a rearward-facing step, *J. Propulsion and Power* **12**, (6), 1129.

KARAGOZIAN, A. R., 1999 Control of mixing and reactive flow processes (invited) *AIAA Paper 99-3572, 30th AIAA Fluid Dynamics Conference,* 28 June - 1 July 1999, Norfolk, Virginia.

KELSO, R. M., LIM, T. T., AND PERRY, A. E., 1996 An experimental study of round jets in cross-flow, *J. Fluid Mech.* **306**, 144.

KIYA, M., MOCHIZUKI, O., ISHIKAWA, H., 2001 Interaction between vortex rings and a separated shear layer: Towards active control of separation zones, *J. Fluids and Structures* **15**, (3-4), 399.

LEE, L.S. AND MORRIS, P.J., 1997 Absolute instability in a supersonic shear layer and mixing control, *J. Propulsion and Power* **13**, (6), 763.

M'CLOSKEY, R. T., KING, J., CORTELEZZI, L. AND KARAGOZIAN, A. R., 2002(a) The actively controlled jet in crossflow, *J. Fluid Mech.* **452**, 325.

M'CLOSKEY, R. T., KING, J., CORTELEZZI, L. AND KARAGOZIAN, A. R., 2002(b) Active control of jets in crossflow, In *Manipulation and Control of Transverse Jets,* Springer-Verlag.

MCCORMIK, S. AND GURSUL, I., 1996 Effect of shear-layer control on leading-edge vortices, *J. Aircraft* **33**, (6), 1087.

NG, T. T. AND LANG, Y., 1999 Shear layer instability induced separation control, *AIAA Journal* **37**, (3), 386.

SATO, N., IMAMURA, A., SHIBA, S., TAKAHASHI, S., *et al.*, 1999 Advanced mixing control in supersonic airstream with a wall-mounted cavity, *J. Propulsion and Power* **15**, (2), 358.

SCHULLER, T., KING, J., MAJAMAKI, A., AND KARAGOZIAN, A. R., 1999 An Experimental study of acoustically controlled gas jets in crossflow, *Bull. Amer. Phys. Soc.* **44**, (8), 111.

SMITH, S. H. AND MUNGAL, M. G. 1998 Mixing, structure and scaling of the jet in crossflow. *J. Fluid Mech.* **357**, 83–122.

STROOCK, A. D., DERTINGER, S. K. W., AJDARI, A., MEZIC, I., *et al.*, 2002 Chaotic mixer for microchannels, *Science* **295**, (5555), 647.

VERMEULEN, P. J., GRABINSKI, P. AND RAMESH, V., 1992 Mixing of an acoustically excited air jet with a confined hot crossflow, *ASME Trans: J. Engng Gas Turb. Power* **114**, 46.

Active Control of Jets in Crossflow

Robert T. M'Closkey[1], Jonathan King[1], Luca Cortelezzi[2], and Ann R. Karagozian[1*]

[1] Department of Mechanical and Aerospace Engineering, University of California, Los Angeles
[2] Department of Mechanical Engineering, McGill University

Abstract. The present study quantifies the dynamics of actuation for the temporally forced, round gas jet injected transversely into a crossflow, and incorporates these dynamics in developing a methodology for open loop jet control. A linear model for the dynamics of the forced jet actuation is used to develop a dynamic compensator for the actuator. When the compensator is applied, it allows the jet to be forced in a manner which results in a more precisely prescribed, temporally varying exit velocity whose RMS amplitude of perturbation can be made independent of the forcing frequency. Use of the compensator allows for straightforward comparisons among different conditions for jet excitation. Clear identification can be made of specific excitation frequencies and characteristic temporal pulse widths which optimize transverse jet penetration and spread through the formation of distinct, deeply-penetrating vortex structures. Further details on this work may be found in M'Closkey *et al.* (2002).

1 Introduction

While the vast majority of transverse jet studies have focused on the steadily injected jet (Kamotani & Greber (1972), Fric & Roshko (1994), Kelso *et al.* (1996), Broadwell & Breidenthal (1984), Smith & Mungal (1998)), active control of the mixing processes associated with the jet in crossflow often requires temporal excitation of the jet flowfield. Recent experiments on pulsed transverse jets (Vermeulen *et al.* (1992), Kelso *et al.* (1996), Johari *et al.* (1999), Schuller *et al.* (1999), Eroglu & Breidenthal (2001)) suggest that temporally varying the jet velocity U_j allows jet penetration and spread to be enhanced at specific conditions of excitation, likely due to control of the temporal generation of jet vorticity.

In gas and liquid pulsed transverse jet experiments a temporally varying jet velocity is achieved by driving the amplifier of a speaker or piston actuator with a signal generator at a fixed frequency f. Experimental evidence (Johari *et al.* (1999), Schuller *et al.* (1999)) suggests that the temporal variation of the velocity at the jet exit orifice (of diameter D) can be quite different from the input waveform from the signal generator. This difference is apparatus-dependent, although the dominant frequency of excitation is generally retained by the jet. Differences in the actual waveform shape, duty cycle, and phase can arise from the unsteady flow evolution within the jet plenum-nozzle and/or the jet's nonlinear coupling to the crossflow in the vicinity of the jet orifice. Hence it becomes difficult not only to identify scaled excitation conditions (e.g., a Strouhal frequency

* Research on controlled transverse jets has been supported by the Energy Innovations Small Grants Program of the California Energy Commission under grant 51240A/99-01-56, by NASA Dryden Flight Research Center under Grant NCC 2-374, by NSERC Grant RGPIN217169, by NSF grant ECS-9985046, and by the University of California Energy Institute.

$St_j \equiv fD/\bar{U}_j$) that lead to optimal transverse jet penetration and spread, but it is also difficult to compare results from one experimental configuration to another and from an experiment to a numerical simulation.

The aims of the present study are to understand and quantify the dynamics of the temporally forced jet in crossflow, in particular, the dynamics of jet *actuation*, and to use this information as a means of more precisely controlling jet behavior. These dynamics are characterized in experiments involving a gaseous round jet injected normally into a uniform gaseous crossflow, but the procedure outlined could apply to other experimental configurations or even to a numerical simulation. A simple linear model for the dynamics of the forced jet allows the development of a dynamic compensator which, when applied, permits the jet to be forced in a manner that results in a more precisely prescribed, temporally varying exit velocity. More importantly, the compensator allows the jet response at the nozzle exit to become essentially independent of the specific characteristics of the jet actuation apparatus, so that a more accurate identification of optimal forcing conditions may be made.

2 Experimental Setup and Baseline Jet Response

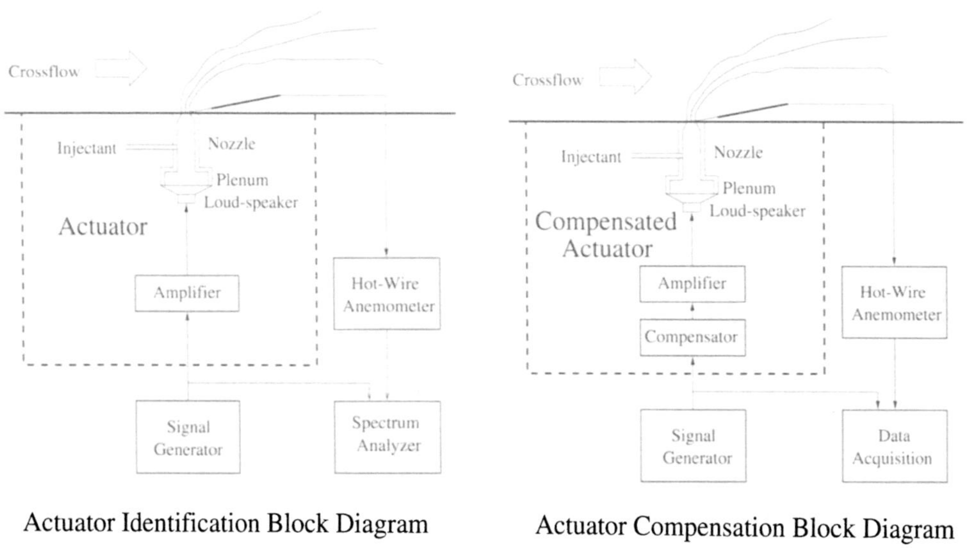

Figure 1. Schematic diagrams of the jet actuation system used for identification of jet dynamics (left) and for compensation of jet behavior (right). A hot wire anemometry system is used to measure temporal jet response at the exit plane of the jet nozzle. The "actuator" consists of the jet nozzle, plenum, loudspeaker, and amplifier; the "compensator" alters the amplifier input to control temporal jet behavior at the nozzle exit plane, as described in Section 3.

The acoustically pulsed, transverse gas jet is studied here in a small-scale, blow-down wind tunnel. A compressor is used to drive air through the tunnel at speeds that range from 0.5 to

20 m/s in the test section. The turbulence levels in the tunnel are quantified using a hot wire anemometry system (Dantec StreamLine with a 55P15 probe) to be less than 3%. The tunnel test section of 12cm $\times$ 12cm consists of four sets of ceramic walls (each of length 26 cm) which can be replaced by quartz or Plexiglas windows for optical access. In the present studies, the windows are placed so that only nearfield visualization of the jet is attained.

The jet nozzle, situated 9.51 cm downstream of the end of the tunnel contraction, has an inner diameter of 0.75 cm, with a 16:1 (area) contraction ratio and a length of 14 cm. The nozzle is connected to a small plenum with an approximate volume of 150 cm^3, sketched as part of the actuator shown in figure 1. The jet fluid (nitrogen) may be introduced in a steady mode or it may be temporally pulsed in an oscillatory fashion by a loudspeaker (Radio Shack 9.16 cm woofer) which is placed at the bottom of the plenum. A Tylan mass flowmeter is used to measure and verify the mean bulk velocity in the jet during both steady and unsteady operations. The hot wire anemometry system is used to quantify the spatial and temporal velocity profiles at the exit plane of the jet. In all results shown here, the mean jet velocity $\bar{U}_j$ is maintained at 3.1 m/s and the mean crossflow velocity U_∞ is 1.2 m/s, for a value of the jet-to-crossflow velocity ratio of R = 2.58 and a mean jet Reynolds numbers of the order 1500. A virtually flat spatial distribution in velocity is quantified at the jet exit plane during steady jet operation.

Jet visualization in the present experiments is achieved via smoke generation upstream of the jet nozzle/plenum. A heated seeder with a liquid parafin solution is used for this purpose. A 500W white light is fixed above the test section to illuminate the jet, while the ceramic test section walls are sooted to improve resolution. A Nikon N2000 SLR camera is used to obtain images of the evolution of smoke in the acoustically pulsed jet. When the jet is acoustically forced, the loudspeaker is driven by an amplifier that accepts input from a signal generator (Keithly Model 3940 Multifunction Synthesizer) indicated in figure 1. The operation of the compensated actuator shown in the right side of figure 1 will be described in Section 3.

A sample result is shown in figure 2 for a square wave input (dash-dotted line) delivered to the amplifier and loudspeaker from the signal generator, given in volts, with the mean subtracted. The response measured by the hot wire anemometer at the center of the jet exit plane is shown by the solid line, converted to velocity via King's Law (also with the mean subtracted). The square wave input is characterized by its frequency f and its duty cycle α, defined as the temporal pulse width τ during a given cycle divided by the period T, taken as a percentage. For square wave excitation the shape of the input signal is not accurately recovered at the jet exit plane, but rather resembles a somewhat peaked, nonuniform sine wave. As a consequence, the actual jet excitation has an effective duty cycle that differs significantly from that of the input signal; this phenomenon was also observed by Johari *et al.* (1999) in liquids. Moreover, as discussed in M'Closkey *et al.* (2002), the amplitude of the output waveform is strongly dependent on the frequency of excitation, for both sine waves and square waves. This difference may be seen by comparing the amplitudes of the solid lines in figures 2(a) and (b). The dashed lines in these figures represent the compensated case discussed below.

The frequency response of the jet/actuator may be characterized more extensively using a spectrum analyzer (HP Model 3856) as shown in the left part of figure 1. The signal generator is used to produce a broad-band excitation signal that drives the amplifier. The spectrum analyzer computes cross-spectra of the hot wire measurement and excitation signal for multiple data records and then averages these results to remove the effects of measurement noise and disturbances. This result is normalized by the excitation signal autospectrum to produce a frequency

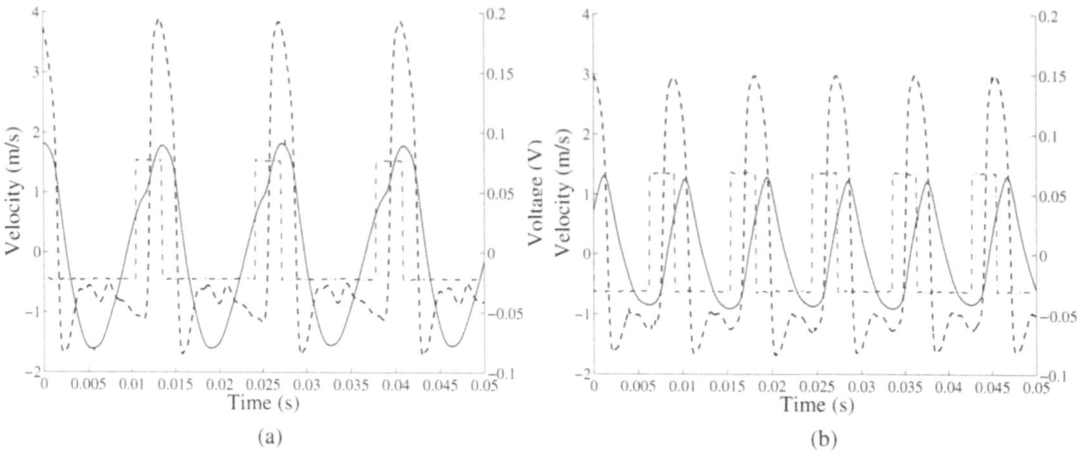

Figure 2. Comparison of the input waveform to the actuator (dash-dotted line), in volts, with the measured velocity perturbations (i.e., with the mean subtracted) at the jet exit, in m/s. Results shown are for velocity perturbations without compensation (solid line) and with compensation (dashed line). The compensated results correspond to the case in which $U'_{j,RMS}$, the RMS of the velocity perturbations, is fixed at a value of 1.7 m/s. All data correspond to cases for which the mean jet velocity $\bar{U}_j$ = 3.1 m/s and the crossflow velocity U_∞ = 1.2 m/s, with a square wave input at (a) frequency f = 73.5 Hz and duty cycle $\alpha = 22\%$ and (b) frequency f = 110 Hz and duty cycle $\alpha = 31\%$.

response estimate. The frequency response between 40 and 1640 Hz is shown by the thick dashed lines in figure 3. The two plots represent the amplitude (upper) and phase (lower) of the hot-wire voltage output relative to the input signal. In this frequency range, the hot wire anemometer dynamics have almost no phase lag or amplitude variation, according to the manufacturer, and thus the hot wire output voltage is a representation of the jet velocity via King's Law.

The Bode plots in figure 3 reveal that there is a maximum amplitude of the jet response at an excitation frequency of approximately 73.5 Hz, and that the response of the jet rolls off substantially at frequencies above 100 Hz. The peak amplitude at 73.5 Hz occurs at about one-third the value of the nearfield vortex shedding frequency (approximately 220 Hz) for the *unforced* jet in crossflow at the present values of $\bar{U}_j$ and U_∞. This nearfield vortex shedding frequency is quantified in the absence of any acoustic excitation from hot wire measurements (taken at both the center of the jet exit and in the upstream shear layer) as well as from smoke images. The nonuniform frequency response shown by the thick dashed lines in figure 3 is undesirable; a frequency response with flat magnitude and linear phase lag over a broad frequency band is required in order to be able to reproduce the desired square waveform excitation at the jet exit.

3 Linear System Model and Compensator

A linear model representing the dynamics of the nozzle-plenum-speaker-amplifier (herein referred to as the *actuator*) may be developed based on these experimental measurements of jet

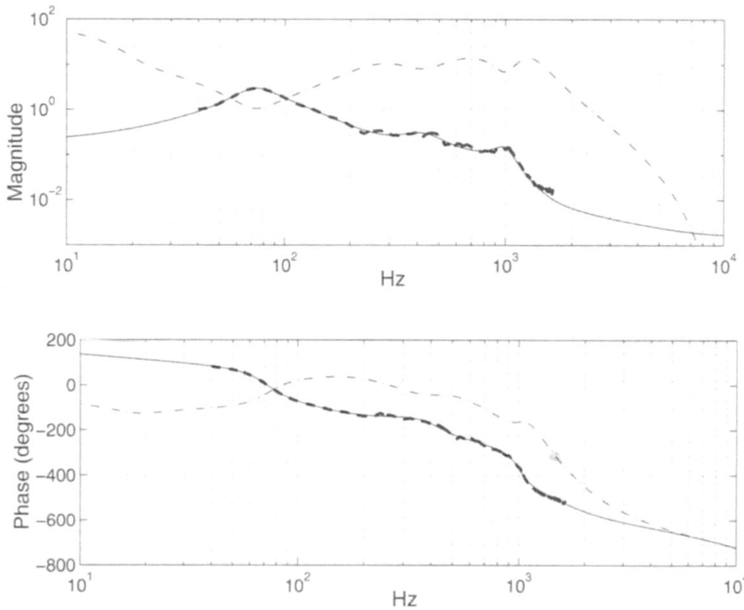

Figure 3. Bode diagrams describing the experimental frequency response of the jet/actuator at the nozzle exit (thick dashed line); these represent the magnitude and phase of the ratio of the output (hot wire) to input (function generator) signals. Also shown is the frequency response of an eight-state linear model representing the actuator dynamics mathematically (solid lines). The compensator frequency response (thin dashed-dotted lines) represents an inversion of the model which is applied in the experiment to create a more uniform frequency response.

response. This mathematical model becomes an integral component in developing a control strategy for improving the jet response.

The mathematical model of the actuator dynamics consists here of an eight pole transfer function of the form

$$\frac{b_1 s^7 + b_2 s^6 + \cdots b_8}{s^8 + a_1 s^7 + a_2 s^6 + \cdots + a_8}.$$

The a and b parameters in this transfer function are obtained via an off-line fit to the experimental frequency response data using the Sanathanan & Koerner (1963) algorithm. The frequency response of the resulting model is shown by the solid line in figure 3. The model appears to replicate the actuator response very well in the frequency range of interest.

This actuator model is used to synthesize a discrete-time reference-signal-shaping filter. The objective here is to design this shaping filter such that the series connection of the filter and the actuator yields a *flat* frequency response magnitude with a *linear* phase lag response. This type of frequency response should improve reproduction of the input signal at the jet exit plane, especially for square wave excitation, with only a time delay between input and output. The shaping filter is designed by solving a weighted $\mathcal{H}_\infty$ model matching problem using the Matlab Mutools Toolbox (see Balas *et al.* (1995) for details). This type of model matching problem usually involves the mathematical inversion of the identified model. The inversion can lead to large filter gains, however, and hence an undesirable sensitivity to high frequency noise. A fourth order Bessel filter with an 800 Hz corner frequency is applied to the shaping filter output in order to attenuate the gain beyond 800 Hz yet to preserve the desired linear phase response.

The series connection of the shaping filter and the Bessel filter are collectively denoted the *compensator* for the jet actuation, shown schematically by the block in the right side of figure 1. The frequency response of the compensator itself is shown by the thin dash-dotted lines in figure 3, representing the applied excitation to the amplifier required to produce the desired frequency response (i.e., flat in terms of magnitude and linear in terms of phase). Note that the magnitude plot of the compensator in figure 3 is essentially an inverse of the actuator model magnitude. In these experiments, the shaping filter is implemented on a commercial digital signal processor (dSpace model 1103 DSP), while the Bessel filter remains analog.

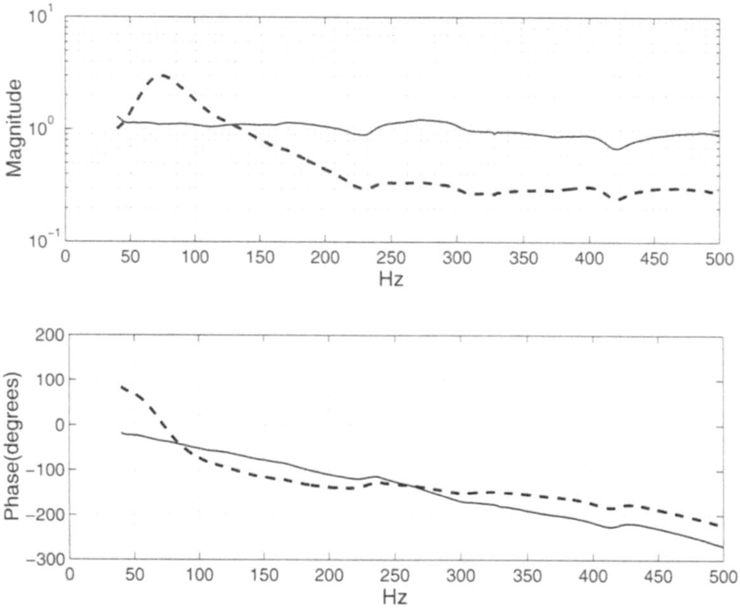

Figure 4. Bode plots representing the frequency response of the compensated jet on a linear frequency scale. The solid line shows the compensated actuator, while the uncompensated actuator is shown (for comparison) by the dashed line.

Application of the compensator to the jet data results in the compensated jet frequency response shown by the solid lines in figure 4. For reference, the uncompensated jet response is shown by the dashed lines; this is the same quantity as is represented by the thick dashed lines in figure 3. A linear frequency scale is used in figure 4 to further distinguish the difference between the compensated and uncompensated cases. The desired linear phase lag characteristic and almost constant magnitude in frequency response are both achieved in the compensated system. It should be noted that the RMS amplitude of the velocity perturbations, $U'_{j,RMS}$, may be matched for the compensated jet among different excitation conditions by adjusting the amplifier gain and applying King's Law to convert voltage to velocity. When $U'_{j,RMS}$ is matched among different excitation conditions, the real effect of applied frequency and duty cycle on jet behavior may be ascertained.

The effect of the compensator on the actual waveform output at the jet exit is shown in the examples in figure 2. Results are shown for the uncompensated jet response (solid line in m/s) and for a compensated jet in which the RMS amplitude of the velocity perturbations, $U'_{j,RMS}$, is set to 1.7 m/s (dashed line). The compensator has a significant effect on the jet exit velocity by forcing it to more closely resemble a square wave with essentially the desired duty cycle. Further, fixing $U'_{j,RMS}$ among different excitation cases allows a more appropriate comparison to be made among jets with different frequencies of excitation. The relatively low level undershoot observed in the compensated square wave signal is due to nonlinearities in the actuator dynamics which are not represented by the model. As will be seen, the overall improvement in square waveform shape by the compensator has a significant effect on jet behavior.

4 Results

Smoke visualization now reveals the effects of compensation and of different excitation conditions on jet behavior. In the present set of experiments, the gain to the amplifier is adjusted so that for all excitation frequencies and duty cycles explored, including both uncompensated and compensated jets, $U'_{j,RMS}$ is fixed at approximately 1.7 m/s.

As a baseline reference, smoke visualization for the unforced jet in crossflow with a jet-to-crossflow velocity ratio of $R = 2.58$ is shown in figure 5(a). Effects of the actuator compensation during random noise excitation show that both the uncompensated and compensated jet look nearly identical to the unforced jet. Since acoustic energy is supplied to the jet at a wide range of random frequencies, the jet is unable to respond to any one frequency or group of frequencies in a coherent way. These observations suggest that random excitation is inefficient for mixing enhancement.

When sine wave excitation is applied to the jet, compensation has the primary benefit of equalizing the magnitude of jet response at different frequencies, as discussed in M'Closkey *et al.* (2002). Sine wave excitation is seen to have relatively little impact on jet penetration and spread as compared with the unforced jet.

Jet compensation makes a significant difference when square wave excitation of the jet is desired, especially at specific frequencies and specific values of the pulse width τ. Figures 5(b) and (c) show smoke images for jet evolution without and with compensation, respectively, for square wave excitation at 110 Hz and a duty cycle of 31%. As indicated in figure 2, the primary difference between uncompensated and compensated jet response at the nozzle exit for this frequency is in the waveform shape. Figures 5(b) and (c) demonstrate that, with compensation of the square wave, deeply-penetrating and rather distinct vortex rings are introduced into the flowfield at the frequency of excitation, resulting in significant increases in jet penetration and spread. It is clear from hot wire measurements that these results are largely achieved because of the ability of the compensated jet to more accurately replicate the distinct pulsations of fluid introduced by a square wave.

Jet compensation permits a comparison of the effects of excitation frequency and duty cycle on jet behavior, since U'_j has been made independent of frequency. The improved jet penetration and spread during square wave excitation is actually seen to occur only for select frequencies and pulse widths τ. Similar to the jet behavior in figure 5(c), when square wave excitation is applied at 55 Hz and 15% duty cycle (figure 5(d)) and at 73.5 Hz and 22% duty cycle (figure 5(e)), individual vortex rings are formed which penetrate well into the crossflow. These three cases all have approximately the same characteristic pulse width τ, within the range 2.7-3.0 msec. When the duty cycle (and hence τ) is increased so that $\tau > 3$ msec for a given frequency, the penetration and spread of the jet are significantly reduced. When the duty cycle is reduced so that $\tau < 2.7$ msec, reasonably good penetration and spread are still achieved, consistent with the observations of Johari *et al.* (1999), yet they are not as great nor as visibly evident as in figures 5(c), (d), and (e).

The observation of an optimal temporal pulse width τ to achieve deeply-penetrating pulses of vorticity also appears to be consistent with the suggestion by Gharib *et al.* (1998) of the existence of a universal time scale (or "time of pinchoff") required for coherent vortex ring formation. These researchers find that a piston-generated vortex ring attains a maximum circulation and is disconnected from its trailing jet at critical values of the ratio of piston stroke to diameter, corresponding to a "universal time scale of vortex formation". While the present experiments involve the acoustically pulsed, continuous jet, rather than piston-driven starting vortex rings, the observation of a critical time scale for distinct ring formation suggests a similar physical mechanism for enhanced penetration.

The frequencies of square wave excitation in figures 5(c), (d), and (e) (110 Hz, 55 Hz, and 73.5 Hz, respectively) generate the highest observed level of jet penetration. These cases represent frequencies which are, respectively, one-half, one-quarter, and one-third the preferred frequency for the unforced jet in crossflow under these flow conditions (approximately 220 Hz). Square wave excitation at 44 Hz (one-fifth of 220 Hz) and a duty cycle of 13% (corresponding to $\tau = 2.95$ msec) yields similar results to those in figures 5(c)-(e). Hence when distinct pulses of fluid are introduced at periods which are two, three, four, or five times that associated with natural vortex shedding for the jet, and with a specific pulse width in the range $\tau \approx 2.7$-3.0 msec, significant jet penetration can be achieved. When compensated jet excitation is applied at non-subharmonic frequencies but with duty cycles corresponding to $\tau \approx 2.7$-3.0 msec, jet penetration and spread are not as extensive as at the subharmonic frequencies, shown in M'Closkey *et al.* (2002). This optimal jet response for subharmonic forcing appears to be analogous to the enhancement of planar shear layer spread during subharmonic acoustic forcing as studied, for example, by Ho & Huang (1982). Compensated square wave excitation at the preferred frequency for the unforced jet, 220 Hz, with a duty cycle corresponding to $\tau = 2.82$ msec, produces jet penetration which is only slightly improved over the unforced jet, as indicated in figure 5(f). In fact, in no instance did square wave excitation at frequencies higher than about 140 Hz yield good jet penetration. It appears that when the duty cycle α must be close to or greater than 50% in order to achieve pulse widths of $\tau \approx 3$ msec for a specified frequency, distinct pulses of vorticity and hence strong jet penetration are difficult to achieve.

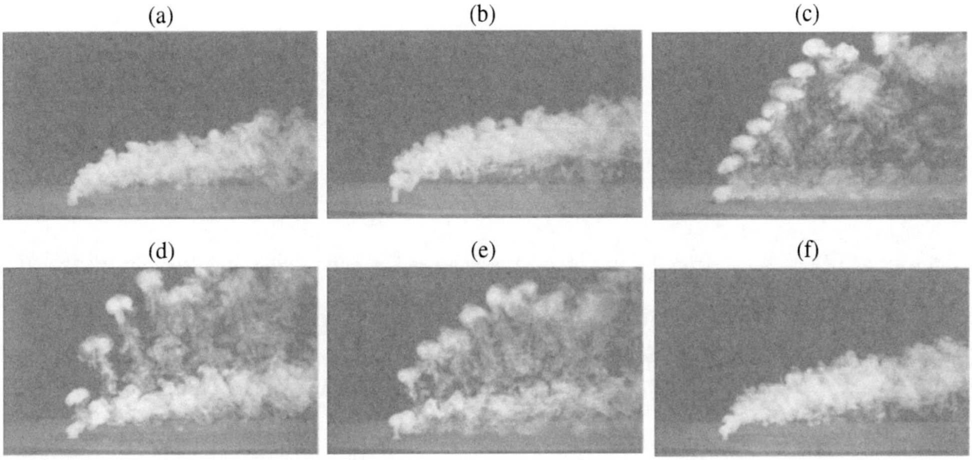

Figure 5. Smoke visualization of the jet in crossflow at an average jet-to-crossflow velocity ratio of R = 2.58, where $\bar{U}_j = 3.1$ m/s and $\bar{U}_\infty = 1.2$ m/s, and for an RMS amplitude of the velocity perturbations, U'_j, matched at 1.7 m/s. Images shown are for photographs taken with an exposure time of 1/2000 sec, yielding an "instantaneous" image. Results shown are for: (a) the "unforced" jet (in the absence of any acoustic excitation), (b) "uncompensated" square wave excitation at 110 Hz and a duty cycle of 31%, (c) "compensated" square wave excitation at 110 Hz and a duty cycle of 31%, (d) "compensated" square wave excitation at 55 Hz and a duty cycle of 15%, (e) "compensated" square wave excitation at 73.5 Hz and a duty cycle of 22%, and (f) "compensated" square wave excitation at 220 Hz and a duty cycle of 62%.

5 Conclusions

Characterization of the dynamics of the actuation system for the forced jet in crossflow is described here, and a robust methodology is evolved whereby these dynamics may be used to develop a jet compensation system for open loop jet control. Application of the compensator allows significantly improved temporal waveforms to be attained at the jet exit, so that accurate comparisons among different conditions of jet excitation can be made. Optimal jet penetration and spread are observed to occur for square wave excitation at subharmonics of the "natural" vortex shedding frequency for the jet (at one-fifth, one-quarter, one-third, and one-half of this shedding frequency). In all optimal cases, however, maximum jet penetration and spread correspond to square wave excitation at duty cycles corresponding to a specific value of excitation pulse width τ of approximately 2.7-3.0 msec. At other duty cycles (and at other frequencies), jet penetration is not as great. This characteristic value of τ appears to depend on specific experimental conditions (e.g., jet velocity, crossflow velocity, etc.), and merits further exploration. The present approach defines a methodology by which this dependence, and other aspects of optimizing the forced jet in crossflow, may be studied with confidence.

References

Balas, G. J., Doyle, J. C., Glover, K., Packard, A., and Smith, R. (1995). *μ-Analysis and Synthesis Toolbox*, The MathWorks, Natick, MA.

Broadwell, J. E. and Breidenthal, R. E. (1984). Structure and mixing of a transverse jet in incompressible flow. *J. Fluid Mech.*, 148:405–412.

Fric, T. F. and Roshko, A. (1994). Vortical structure in the wake of a transverse jet. *J. Fluid Mech.*, 279:1–47.

Eroglu, A. and Breidenthal, R.E. (2001). Structure, penetration, and mixing of pulsed jets in crossflow. *AIAA J.*, 39(3), 417–423.

Gharib, M., Rambod, E., and Shariff, K. (1998). A universal time scale for vortex ring formation. *J. Fluid Mech.*, 360, 121–140.

Ho, C.-M. and Huang, L.-S. (1982). Subharmonics and vortex merging in mixing layers. *J. Fluid Mech.*, 119: 443-473.

Johari, H., Pacheco-Tougas, M., and Hermanson, J.C. (1999). Penetration and mixing of fully modulated turbulent jets in crossflow. *AIAA J.*, 37(7), 842–850.

Kamotani, Y. and Greber, I. (1972) Experiments on a turbulent jet in a cross flow *AIAA J.*, 10:1425–1429.

Kelso, R. M., Lim, T. T., and Perry, A. E. (1996). An experimental study of round jets in cross-flow. *J. Fluid Mech.*, **306**, 111–144.

M'Closkey, R. T., King, J., Cortelezzi, L., and Karagozian, A. R. (2002). The actively controlled jet in crossflow. *J. Fluid Mech.*, Vol. 452, pp. 325–335.

Sanathanan, C. K. and Koerner, J. (1963). Transfer function synthesis as a ratio of two complex polynomials. *IEEE Trans. Automatic Control*, 8, 56–58.

Schuller, T., King, J., Majamaki, A., Karagozian, A. R., and Cortelezzi, L. (1999). An experimental study of acoustically controlled gas jets in crossflow. *Bull. Amer. Phys. Soc.*, 44(8), 111.

Smith, S. H. and Mungal, M. G. (1998). Mixing, structure and scaling of the jet in crossflow. *J. Fluid Mech.*, 357:83–122.

Vermeulen, P. J., Grabinski, P., and Ramesh, V. (1992). Mixing of an acoustically excited air jet with a confined hot crossflow. *J. Engr. for Gas Turbines and Power, ASME Transactions*, 114:46–54.

Low Velocity Ratio Transverse Jets Influenced by Concentric Synthetic Jets *

John Diep and Lorenz Sigurdson

Vortex Fluid Dynamics Lab, Department of Mechanical Engineering, University of Alberta, Canada

Abstract. The objective was to discover and evaluate methods of smokestack downwash reduction using unsteady forcing. The flow is essentially a low momentum jet in a crossflow. After preliminary investigations of several concepts, the final wind-tunnel model applies an annular synthetic jet coaxially to a turbulent or laminar pipe flow at the stack exit. The crosswind Reynolds number based on stack diameter is on the order of 1000 to 7000. Primary diagnostics are hot-wire measurements and image processing of smoke flow visualization photographs of the plume gas. For a laminar pipe flow and increasing forcing amplitude, the initially turbulent plume can be made to completely relaminarize and then become turbulent again at higher amplitudes. Therefore an increased forcing does not always increase mixing. At higher amplitudes vortices appear at the forcing frequency. An abrupt transition of the vortices' arrangement occurs at a particular amplitude that coincides with an apparent jump in mixing. For turbulent pipe flow, plume height increases by as much as 1.8 stack diameters. This data is collapsed using a simple model based on the idea that synthetic jet momentum dominates the plume at high forcing amplitudes.

1 Introduction

This paper concerns an attempt at flow control of a jet in a cross-flow. An annular synthetic jet (zero net mass flux) is produced coaxially to the jet plume emerging from a cylinder representing a smoke or flare stack. Results indicate that this type of forcing can reduce the smoke-stack downwash of harmful plume gases to ground level which can occur for high wind velocities. This technique is expected to also be relevant to oil-field flaring where enhanced mixing of the plume with the ambient air could improve combustion; and to un-premixed cross-jets emerging from the walls of combustion chambers. This flow geometry and forcing techniqur have similarities to an experiment designed to control the separation bubble shape[1] on an axisymmetric forward facing step (Sigurdson, 1995). That apparatus at times may have been producing a synthetic jet which was transverse to the crossflow set up by the global flow. Experiments in other laboratories have looked at the behaviour of two dimensional conventional jets in response to the application of a single synthetic jet on both sides (Smith and Glezer, 1998). To the authors' knowledge, the current work is the first time that both an annular synthetic jet has been created and used in this manner to force a cross-jet.

---

* A version of these results is presented in Diep and Sigurdson (2002).

[1] This might now be called a form of "virtual surfacing" as discussed by Glezer in Chapter IV Section G.

2 Experimental Details

The experimental rig consists of a one inch diameter (D) cylinder in a wind tunnel. It is designed with an outer annulus through which a sinusoidal velocity fluctuation is introduced near the jet exit. This annulus extends the length of the cylinder and through the wind tunnel wall to a piston-driven plenum outside. This particular plenum/annulus design concept was proposed by Coward (2000) to minimize the length of the high viscous loss annulus. It has a resonance chamber connected to the annulus as is shown in Figure 1. The chamber is based on an 8" diameter speaker. The piston is connected to a B&K Vibration Exciter Type 4809, and is excited by an in-house built power oscillator. This shaker table-piston arrangement is used instead of a speaker because of the larger forces it is capable of applying. Sinusoidal movement of the piston creates a similarly varying pressure in the chamber which produces an oscillating flow within the annulus, and resultantly, jetting at the stack tip. The stack tip is constructed of two concentric thin-walled brass tubes. The wall thickness of both tubes is 0.74 mm (0.029"), with 25.4 mm (1") and 20.6 mm (13/16") outer diameters for the inner and outer tube, respectively. This sizing results in a plume area of 57% of the total stack cross section, with a 1.64 mm gap (d_a) between the tubes.

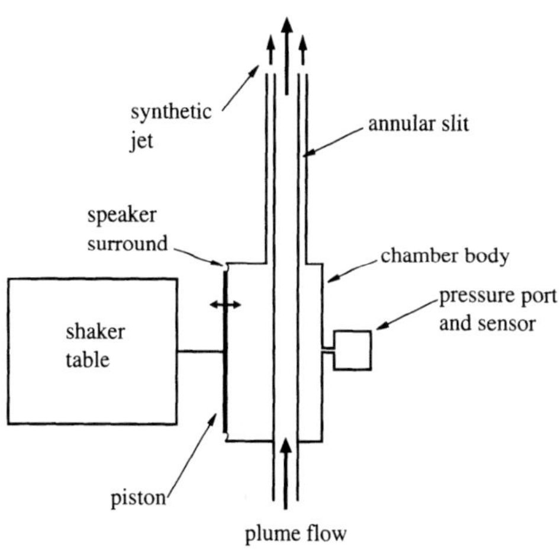

Figure 1. Schematic of the Forcing Apparatus (the wind tunnel is not shown)

The apparatus is designed to operate with an axisymmetric laminar or fully developed turbulent plume flow which is seeded with an aerosol for flow visualization. This aerosol is provided by a Radio Shack chemical fog machine. The plume is driven by a small DC centrifugal fan powered by a variable DC power supply. The fog machine fills a plenum with fog that is then allowed to mix with ambient air and cool, as the fog is slightly greater than room temperature upon leaving the fog machine. This aerosol is then drawn out of the plenum past a valve before entering the fan inlet. To ensure that the plume is fully developed there are approximately 50 straight pipe diameters between a contraction and the exit plane of the plume. Plume flow rate was calculated by using a calibration of the plume centerline velocity, W_{CL}, against the measured pressure drop across this contraction and approximately 30 diameters of straight pipe, P_c.

A suction-type open-circuit wind tunnel is used to create the cross-flow. The stack extends halfway across the test section which has a cross section of 30.5 cm by 30.5 cm (12" by 12").

The acquisition and processing of all pressure and hot-wire sensor data is done using an 800 MHz Pentium equipped with a 12-bit AT-MIO-16-E National Instruments data acquisition board and running LabVIEW® 4.1.1. Synthetic jet velocity W_{SJ} is characterized by a centerline mean hot-wire reading taken a vertical distance z=3 mm from the orifice exit plane (z/d_a=2). This location is used as a compromise between being too close to the jet orifice where flow reversal takes place, and too far downstream of the jet to measure the jet speed that is affecting the emerging plume close to the exit plane.

Photographs for all experiments were taken with a Nikon F3 35mm camera using Kodak 3600ASA TMAX black and white film, and a General Radio Type 1540 Strobolume for a flash.

3 Laminar Pipe Flow Results

A version of these results were presented in Diep and Sigurdson (2001), Sigurdson and Diep (2001a), and Sigurdson and Diep (2001b). This section describes observations made of the special case of a plume from a laminar pipe flow forced with a varying amplitude 60 Hz synthetic jet. During the course of these observations no effort was made to keep the plume flow constant while forcing was applied, but P_c only deviated a maximum of 15% from the average of readings taken over the course of observations. The cross-stream velocity (U_∞) was approximately 0.5 m/s and the unforced centerline velocity of the pipe flow was measured to be 0.9 m/s using the hot-wire. Assuming a parabolic laminar velocity profile within the pipe, the average plume velocity (W_p) can be approximated to be 0.45 m/s, giving a plume flow Reynolds number of 570. The velocity ratio of the plume jet velocity to the cross-stream velocity is $R = 0.9$. The nondimensional forcing frequency based on the stack external diameter and W_p is 3.4. This is 3.4 times the estimated frequency of the naturally occurring structures apparent in the unforced case shown in Figure 2a. From this view, the vortex structures within the plume are clearly visible. Vorticity in these structures originate from the boundary layer of the plume flow while inside the pipe. This vorticity is shed into the plume jet as vortex rings, which are then tilted and distorted by the velocity gradient between the cross-flow and the slower fluid in the wake of the plume and stack (Moussa et al., 1977). Some downwashing is evident in this Figure. This would be expected as that is known to be the case for a plume from a turbulent pipe flow with a similar velocity ratio.

The ratio of W_{SJ} to W_p was varied from 0 to 1.3. As low-amplitude synthetic jetting is applied ($W_{SJ} \approx 0.1$ m/s), the three-dimensional vortex structure on the upstream side of the plume disappears, resulting in the image shown in Figure 2b. Large vortices remaining in this forced plume are reminiscent of the inner vortices appearing on the lower surface of the unforced plume. These large vortices in the forced plume become gradually smaller in size with increasing forcing amplitude. Eventually the plume becomes completely relaminarized when $W_{SJ} \approx 0.3$ m/s, as shown in Figure 2c. This represents the minimum observed mixing of the plume with the cross-flow. This value of W_{SJ} is nearing the cross-stream velocity.

With increased forcing amplitude, an instability begins to grow from the relaminarized plume. This instability is shown in Figure 2d when $W_{SJ} \approx 0.9$ m/s. This value is close to the centerline

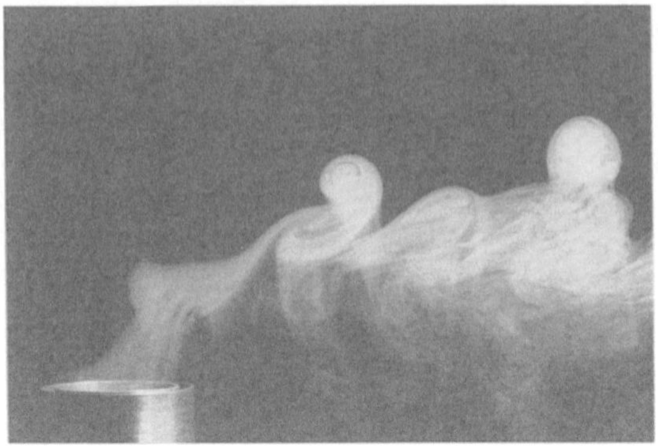

(a) Reference Photograph of Laminar Plume: No forcing, $U_\infty \approx 0.5$ m/s, $\mathrm{W_p} \approx 0.45$ m/s.

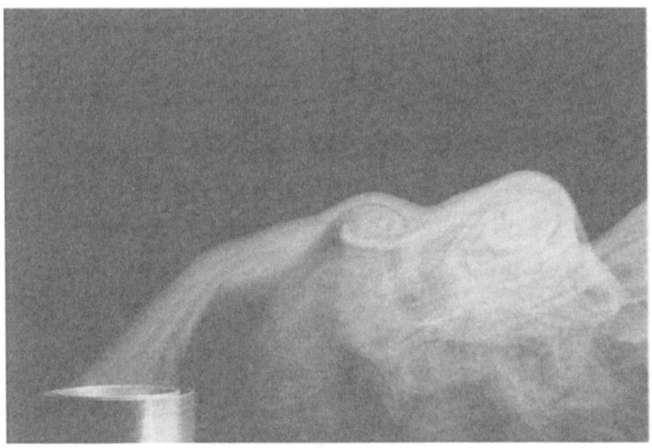

(b) The outer structure disappears, and the inner structure appears to remain, $U_\infty \approx 0.5$ m/s, $\mathrm{W_p} \approx 0.45$ m/s, $\mathrm{W_{SJ}} \approx 0.1$ m/s.

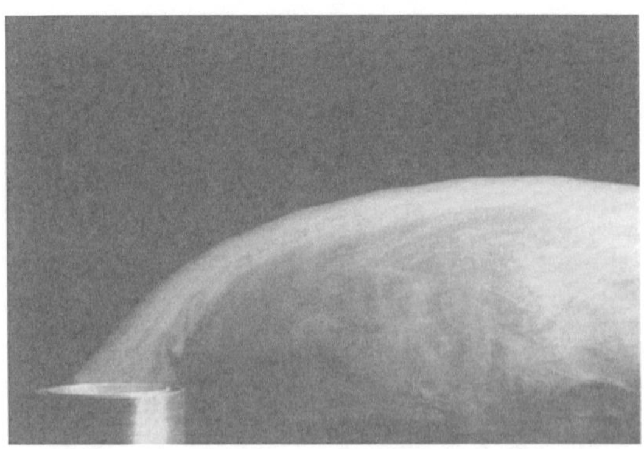

(c) Plume relaminarizes, $U_\infty \approx 0.5$ m/s, $\mathrm{W_p} \approx 0.45$ m/s, $\mathrm{W_{SJ}} \approx 0.3$ m/s.

Figure 2. Change in Structure of Laminar Plume With 60 Hz Synthetic Jet Forcing

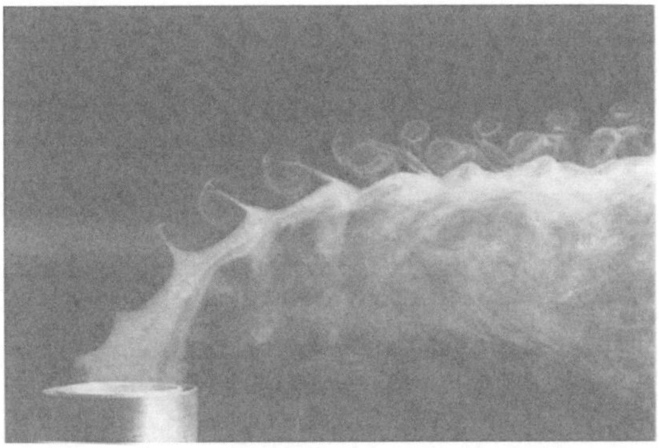

(d) Upstream-pointing mushroom vortices emerge from upper surface, $U_\infty \approx 0.5$ m/s, $W_p \approx 0.45$ m/s, $W_{SJ} \approx 0.9$ m/s.

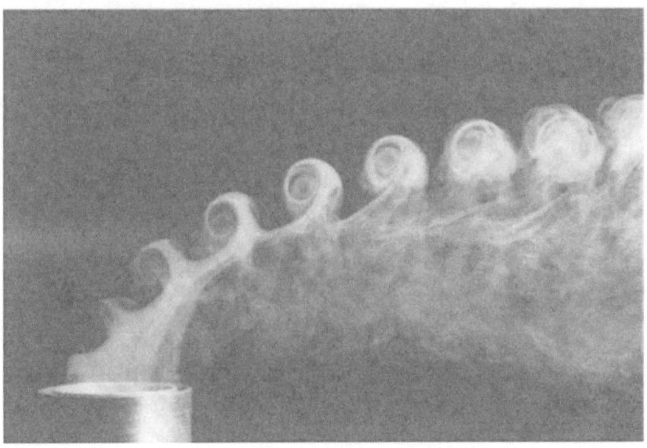

(e) After the abrupt transition, vortices on the upper surface contain mainly one sign of vorticity and point downstream, $U_\infty \approx 0.5$ m/s, $W_p \approx 0.45$ m/s, $W_{SJ} \approx 0.9$ m/s.

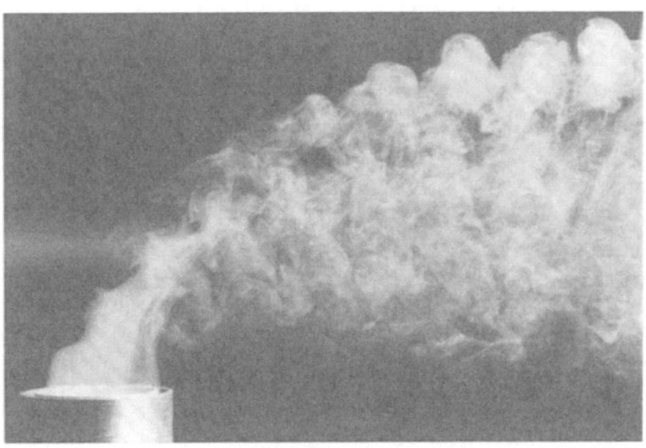

(f) Lifting off and more turbulent, $U_\infty \approx 0.5$ m/s, $W_p \approx 0.45$ m/s, $W_{SJ} \approx 1.2$ m/s.

Figure 2: Change in Structure of Laminar Plume With 60 Hz Synthetic Jet Forcing value of the pipe flow, measured at 0.9 m/s using the hot-wire. From this side-view there are structures resembling mushroom vortices pointing upstream on the upper surface of the plume. "Pointing" refers to the direction the pairs would self-propagate. The smoke intensity in these structures is rather small, indicating not a lot of plume fluid enters these. Mixing has presumably increased from the laminar case however. With a very small increase in synthetic jetting amplitude, there is an abrupt transition in the vortex structures' arrangement to that seen on the upper surface of the plume in Figure 2e. Between these photographs forcing amplitude was increased by 6% (calculated from chamber pressure measurements) to make sure the transition did not occur while photographs were being taken. This transition point is so sensitive that the apparatus can be set for transition to occur intermittently with no controls being touched. After this transition, the vortices appear to point downstream, and appear to more strongly consist of vorticity out of the page by the right-hand-rule. It also appears that after the transition these vortices contain more plume fluid than did the upstream-pointing vortices before transition. This is indicated by the brighter smoke contained in the vortices. A leap in mixing is implied, of the plume with the cross-stream fluid. As forcing amplitude is further increased, the vortices on the upper surface of the plume continue to lift off and form pinched-off lobes, as seen in Figure 2f where $W_{SJ} \approx 1.2$ m/s. The plume is clearly higher than in the unforced case, as well as showing an increased plume width, which suggests a significantly improved mixing between the plume and surrounding air. With even greater forcing amplitude the plume continues to rise, but becomes too turbulent to discern any structure within it. In any case, downwash is significantly reduced if not eliminated.

4 Turbulent Pipe Flow Results

This section describes results that occur when the pipe flow in the stack is turbulent (Diep and Sigurdson, 2002).

4.1 Introduction to Plume Measurements

For these experiments the camera is positioned to include the entire stack and illuminated portion of the plume, which are then used to measure plume height. At least several other researchers have used photographs to make experimental plume height measurements, including Vadot (1967), Fan (1967), Overcamp and Hoult (1971), and Hoult and Weil (1972). Plume measurements in this case are performed on digital scans of contact sheets printed with images of photographic negatives, resulting in a pixel resolution of around 0.02 stack diameters. Measurements of these scanned images are made with Matrox Inspector®, an image analysis program.

Plume measurements centered at 7.5 stack diameters downstream of the stack centerline is made for each photo. This location is the furthest downstream where the plume smoke is clearly visible in all of the photographs, and also corresponds to the locations of plume height data points published by other authors (Snyder and Lawson, 1991; Johnston and Wilson, 1997). Figure 3 shows an example of a typical photograph and how these measurements are made. The upper mean edge of the plume is determined by a horizontal line[2] placed by inspection so that there are approximately equal portions of white smoke above the line as there are portions of black background visible below the line within the vicinity of $x = 7.5D$. The opposite is done for the

[2] In Figure 3 the line thicknesses have been enhanced for clarity, and are normally 1 pixel in width.

lower mean edge of the plume. The plume centerline at $x = 7.5D$ is approximated as the average of the upper and lower mean plume edges, Z_p. For the remainder of this paper, the term "plume height" will refer to this value.

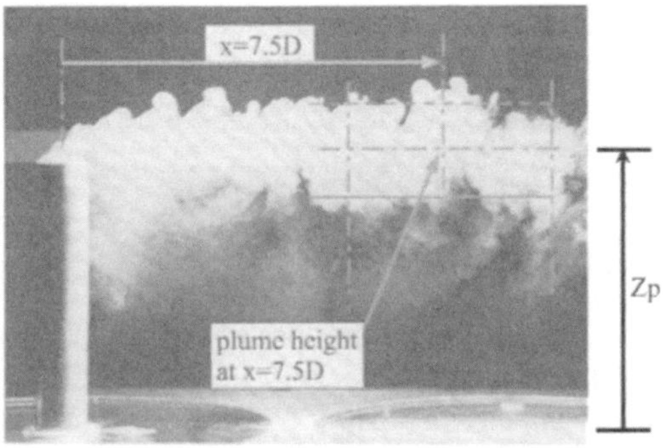

Figure 3. Typical plume photograph with turbulent pipe flow.

In these photographs with turbulent pipe flow, the plume is illuminated from above by a 1.0 cm to 1.5 cm wide light sheet along its centerline. Data collected from these photographs is meant to be representative of the whole plume. The basis for doing this is provided by Johnston (1994). He used laser-induced fluorescence to show that concentration centroids taken from an imaged vertical streamwise plane matched centroids taken from perpendicular cross sections.

4.2 Experimental Results

Figure 4 shows selected results of these measurements. In this Figure, closed symbols represent reference plume heights of unforced plumes, and are plotted against the plume velocity ratio (W_p/U_∞). Open symbols represent plume heights of plumes forced with a 40 Hz synthetic jet, and are plotted against the synthetic jet velocity ratio (W_{SJ}/U_∞). The value R in the legend is the velocity ratio of the original plume without forcing. For clarity, only plumes forced with 40 Hz synthetic jets are shown in this graph. It was found that these measurements do not depend on synthetic jetting frequency in the tested range (40 Hz–100 Hz), and the 40 Hz synthetic jet gives the widest possible amplitude range.

In the forced cases of Figure 4, it can be seen that except for Case 13, increased synthetic jetting amplitude results in a monotonic increase in plume height. Data points lying on the vertical axis, for the forced cases at zero forcing amplitude, show an expected amount of scatter in plume heights (±1.0D), owing to their differing plume velocity ratios. Of course, this range would be larger with a greater range of tested plume velocity ratios. As forcing amplitude increases, the plume heights of all the forced cases appear to converge in Z_p until beyond W_{SJ}/U_∞ of around 2, the data has a maximum spread of approximately ±0.3D.

The plume height curves of these forced cases appear to converge at high forcing amplitude regardless of their plume to crosswind velocity ratios. This convergence suggests that at high amplitudes of forcing, the synthetic jet velocity and not the plume velocity is controlling the plume height. Heuristically, the synthetic jet momentum can be seen as overwhelming the plume momentum as the main source of vertical momentum flux into the combined plume. In essence, the synthetic jetting is becoming the entire plume. If this was truly the case, a greatly idealized outlook would predict that at higher forcing amplitudes the forced plume heights should approach unforced plume heights. This can be seen happening to some extent in Figure 4 as data points corresponding to strongly forced plumes begin to lie within the region of the reference unforced plume data. This may or may not be coincidental because of course there are other factors that also have an effect on plume rise.

4.3 Applying Momentum Ratio

A corrected momentum ratio (Briggs, 1975; Johnston and Wilson, 1997) may be used in an attempt to combine the momenta of the originally present plume flow as well as the forcing being applied upon it. Such an expression is shown in Equation 1:

$$M_t = {\alpha_t}^{1/2}\left(\frac{\rho_s}{\rho_a}\right)^{1/2}\frac{W_t}{U_\infty} \tag{1}$$

where α_t is the correction factor. Note that the subscript t is used to denote application over the total stack exit area of the plume and synthetic jetting, A_t. Hence, W_t represents the average vertical velocity of the plume and synthetic jet over A_t. The correction factor α_t is meant to account for the difference in momentum between an assumed top-hat velocity profile of the plume as it leaves the stack, and the actual viscous velocity profile containing a boundary layer. Values of α_t range from 2.0 for a laminar parabolic profile to 1.0 for a top-hat profile. For the turbulent cases of the present study α_t ranges from 1.09 to 1.11 due to the relatively flat turbulent plume profile.

A more intuitive form of Equation 1 is shown in Equation 2 below:

$$M_t = \left(\frac{F_t}{\rho_a A_t {U_\infty}^2}\right)^{1/2} \tag{2}$$

where F_t is the total additive momentum flux of the original plume (F_p) and the synthetic jet (F_{SJ}). This is defined formally by:

$$F_t = F_p + F_{SJ} \tag{3}$$

Figure 5 shows the results of applying this approximation. Plume height is plotted against M_t, the total momentum velocity ratio calculated over the total stack exit area. Several data points from the unforced cases are included for reference, and these are plotted against M_p. The main difference to notice in Figure 5 compared to Figure 4 is that the large initial data spread in Z_p for small or no forcing is not shown. Instead, when looking at the data points corresponding to zero forcing amplitude, indicated by the over-sized open symbols, they can be joined with a smooth curve. Scatter in the vertical data is approximately $\pm 0.25D$, which is close to the lowest scatter of the converging data series in Figure 4.

Although the data points of the forced plumes in Figure 5 are generally grouped together, this group tends to depart from the trend of the reference data points of the unforced cases. While this is more noticeable for the larger forcing amplitudes, this departure is also present to a lesser

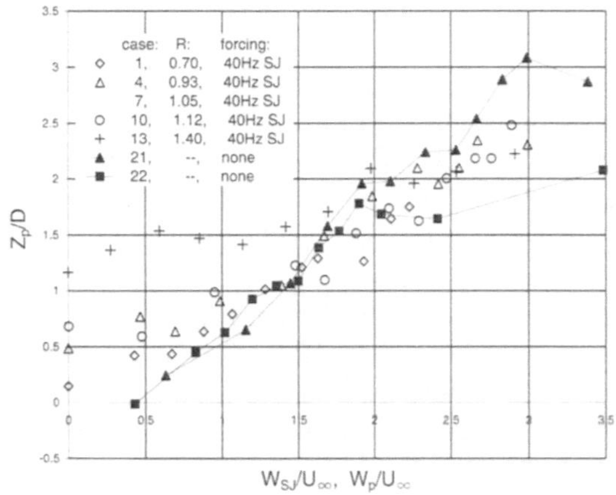

Figure 4. Plume Heights of Synthetic Jet-Forced and Reference Plumes: Forced data is plotted against synthetic jet velocity normalized by cross-wind velocity and unforced plume data is plotted against average plume velocity normalized by cross-wind velocity using the same axis.

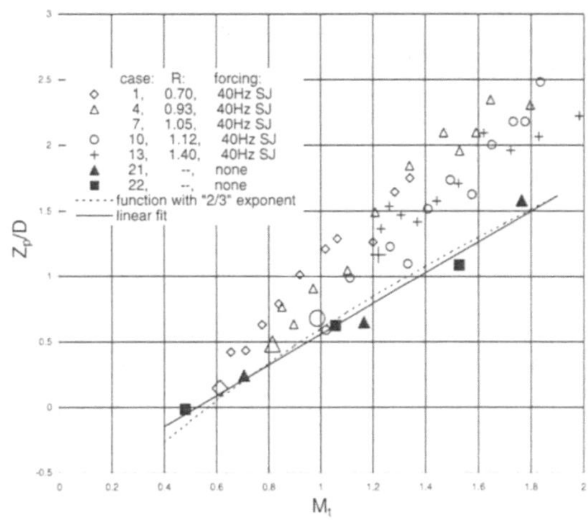

Figure 5. Plume Height vs. Momentum Ratio Calculated Using Total Area of Plume and Forcing, M_t: Enlarged symbols represent data points of zero forcing amplitude for each case. Shown for reference are solid symbols representing unforced plumes. The solid line is a linear regression curve fit to the reference unforced data points. The dashed line represent the function $Z_p/D = C_1 M_t^{2/3} + C_2$ which is suggested by plume rise equations in Briggs (1975). The constants have been adjusted by inspection to fit the reference unforced data points.

extent for lower and zero forcing amplitudes, and depends on plume and forcing geometry. This happens because the total stack exit area, A_t, was used to calculate the momentum velocity ratio M_t in Equation 2. Considering the case of an unforced stack plume, A_t in the numerator of this equation should be substituted with plume area A_p, because this is the only surface having momentum flux. So in the case of using Equation 2 to evaluate M_t for the unforced stack plume, the area term A_t is too large by A_{SJ}/A_p, about 40% for this particular geometry, resulting in a momentum velocity ratio M_t that is 15% too small. This relative error results in larger absolute errors with higher momentum velocity ratios, thus increasing the departure from the reference cases. A similar argument can be made for the limiting case as the synthetic jet momentum divided by the plume momentum approaches infinity. This is equivalent to having no plume flow, with all vertical momentum provided by the synthetic jet, and the correct area to use in Equation 2 is A_{SJ}. Therefore, this limiting case results in an error in area of A_p/A_{SJ} of about 250%, giving a momentum velocity ratio that is about 36% too small.

A suggested solution to resolve the error in calculated area is to use a corrected area that is weighted linearly to the relative momentum fluxes of the plume and synthetic jetting, F_p and F_{SJ}, respectively. A first guess for the empirical weighting is shown in Equation 4:

$$A_M = \frac{F_p A_p + F_{SJ} A_{SJ}}{F_t} \tag{4}$$

where F_t is defined by Equation 3. The area calculated using Equation 4 satisfies both the limits of no forcing as well as forcing approaching infinite amplitude, in that it is equal to A_p when F_{SJ} equals zero, and approaches A_{SJ} when F_p/F_{SJ} approaches zero. Figure 6 shows the results of using this corrected area. The same data is used as in Figure 5 except it is plotted against M_M, the momentum velocity ratio using the momentum-weighted area, as defined by Equation 5:

$$M_M = \left(\frac{F_t}{\rho_a A_M {U_\infty}^2} \right)^{1/2} \tag{5}$$

where F_t is given by Equation 3.

The results of applying the momentum-weighted area are shown in Figure 6, where Z_p is plotted against M_M. A dotted line showing Briggs'(1975) model of plume height with a variable entrainment coefficient is also shown in this Figure, and shows a rough agreement in slope with the bulk of the data points although it is shifted vertically. A linear regression fit of all the forced data points shown in Figure 6 has a slope that is 19% lower than a similar fit for the reference unforced data points. As expected, the points corresponding to no forcing now lie on the same curve as the reference unforced cases also shown in the plot. Also expected was a shift of the strongly-forced data points towards larger values of the momentum velocity ratio. The plots for the forced series are still grouped closely, with a spread of $\pm 0.25D$ throughout most of their range. Aside from the grouping of the forced data points, there are two other important features of the data that support the model used to achieve this. First is the contiguity of the beginnings and ends of different data series. Secondly is the fact that even for high synthetic jetting amplitudes, the trend of plume height for the forced cases is close to the plume heights of the unforced reference cases. What this suggests is that the simplified equivalent plume model using a momentum weighted area is appropriate for applying to a plume under this type of synthetic jet forcing.

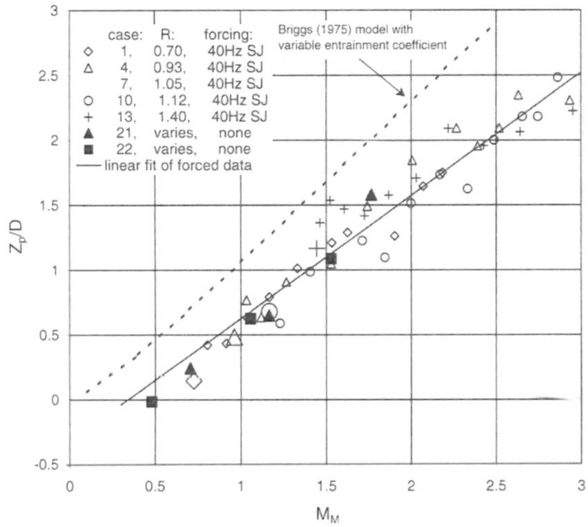

Figure 6. Plume height under Synthetic Jet Forcing vs. Momentum Ratio Calculated Using Momentum-Weighted Area, M_M: Enlarged symbols represent data points of zero forcing amplitude for each case. A linear regression of the forced data points is indicated by a solid line. Shown for reference are solid symbols representing unforced plumes.

5 Conclusions

5.1 Laminar Pipe Flow

For the single velocity ratio tested here for a plume resulting from a laminar pipe flow, increasing the synthetic jet velocity causes:

1. relaminarization when it is near the cross-stream velocity,
2. an increasingly complex periodic 3-D vortex structure,
3. an abrupt transition in the vortex structure when the forcing velocity is near the peak in the cross-jet velocity profile,
4. an increased mixing at the highest amplitudes.

The conclusions regarding under what conditions the relaminarization and abrupt transition occur can only be confirmed by testing other velocity ratios of jet to cross-stream.

One important conclusion from the observed relaminarization is that increased forcing does not always increase mixing. There would appear to be an island of calm in the turbulence otherwise observed as the forcing amplitude is increased. It should be noted that the pipe velocity would have to be reduced an order of magnitude or more to get the jet to be laminar on its own accord.

The abrupt transition noted in 3. coincides with a significantly increased mixing of the plume fluid with the ambient fluid. This may not be coincidental but rather be an indication of the reason why the transition occurs. As the forcing becomes more powerful, the vortices created by it are expected to become stronger and entrain plume fluid more quickly. This may be why the sign of the plume fluid now appears to dominate the structure resulting on the upstream side of the cross-jet.

These vortex structures are seen in the photographs as primarily 2-D but are actually quite strongly 3-D. While the two transformations discovered in this plume vortex structure with laminar pipe flow are clear and distinct, no corresponding behaviour in the case of turbulent pipe flow has been observed, either through visual interpretation or measurements of the photographs.

5.2 Turbulent Pipe Flow

When forcing a plume resulting from a turbulent pipe flow, two major results are achieved:

1. Downwash is reduced or eliminated in every case studied.
2. Average plume height increases with increased forcing.

It is observed that with very high forcing, plume height data appears to depend only on forcing amplitude, and not the original plume velocity. In an attempt to consider both the effect of the original plume as well as its forcing, a simple model can be used which essentially adds the momenta of the plume and forcing. When applying this model to evaluate a corrected momentum ratio that takes both these momenta into account, the data collapses somewhat when plotted against this ratio. This collapse can be improved by reasoning that the equivalent plume should consist of largely synthetic jet momentum when forced at high amplitude, and only plume momentum when forcing is turned off. Incorporating this idea into an empirical momentum-weighted area term in the equations results in a significantly improved collapse of the data.

Acknowledgments

Thanks to Dave Pape for building the rig, Terry Nord for his help with the electronics, Albert Yuen, Don Fuhr, Dallas Chapple and Bernie Faulkner for their advice and assistance, and Andrew Coward and Chris Apps for their ideas and help whenever it was needed. Financial support is appreciated from the Natural Sciences and Engineering Research Council of Canada, Luscar Ltd., and the University of Alberta.

References

Briggs, G. (1975). Plume rise predictions. In *Lectures on Air Pollution and Environmental Impact Analyses*. American Meteorological Society. 59–111.

Coward, A. (2000). Personal communications.

Diep, J. and Sigurdson, L. (2001). Cross-jet influenced by a concentric synthetic jet. *Physics of Fluids* 13(9):S16.

Diep, J. and Sigurdson, L. (2002). Reduction of smokestack downwash using a concentric synthetic jet. Manuscript submitted for publication.

Fan, L. (1967). Turbulent buoyant jets into stratified or flowing ambient fluids. Technical Report KH-R-15, W.M. Keck Laboratory of Hydraulics and Water Resources.

Hoult, D. and Weil, J. (1972). Turbulent plume in a laminar cross flow. *Atmospheric Environment* 6:513–531.

Johnston, C. (1994). Downwash of stack gas plumes. Master's thesis, University of Alberta.

Johnston, C. and Wilson, D. (1997). A vortex pair model for plume downwash into stack wakes. *Atmospheric Environment* 31(1):13–20.

Moussa, Z., Trischka, J., and Eskinazi, S. (1977). The near field in the mixing of a round jet with a cross-stream. *Journal of Fluid Mechanics* 80(1):49–80.

Overcamp, T. and Hoult, D. (1971). Precipitation in the wake of cooling towers. *Atmospheric Environment* 5:751–765.

Sigurdson, L. (1995). The structure and control of a turbulent reattaching flow. *Journal of Fluid Mechanics* 298:139–165.

Sigurdson, L. and Diep, J. (2001a). Mixing in a cross-jet enhanced by a coaxial annular synthetic jet. Manuscript submitted for publication.

Sigurdson, L. and Diep, J. (2001b). Control of smokestack downwash using a concentric annular synthetic jet. In *Bulletin of the American Physical Society 54th Annual Meeting of the Division of Fluid Dynamics, November 18–20, 2001, San Diego, California* 46(10):46.

Smith, B. and Glezer, A. (1998). The formation and evolution of synthetic jets. *Physics of Fluids* 10(9):2281–2297.

Snyder, W., and Lawson, R. (1991). Fluid modeling simulation of stack-tip downwash for neutrally buoyant plumes. *Atmospheric Environment* 25A(12):2837–2850.

Vadot, L. (1967). *Étude de la Diffusion des Panaches de Fumée dans l'Atmosphère.* Centre Interprofessionnel Technique d'Études de la Pollution Atmosphérique, Paris.

Separation Control using Synthetic Jet Actuators

Michael Amitay[1] and Ari Glezer[2]

[1] Aerospace, Transportation and Advanced Systems Laboratory, Georgia Tech Research Institute
[2] School of Mechanical Engineering, Georgia Institute of Technology

Abstract. The suppression of post-stall separation over an unconventional 2-D airfoil at moderate Reynolds numbers (up to 10^6) using transverse synthetic (zero net mass flux) jet actuators is discussed. As shown by the authors in earlier investigations, the apparent modification of the surface shape by the interaction domain between the actuator jets and the cross flow results in a local displacement of the cross flow streamlines. The concomitant modification of the streamwise pressure gradient upstream of where the flow nominally separates in the baseline configuration can lead to complete suppression of separation over a significant range of angles of attack in the post-stall domain. While in the absence of flow control the airfoil is stalled at angles of attack exceeding 5°, actuation leads to either completely or partially attached flow within the entire range of angles tested (up to 25°) that is accompanied by a dramatic increase in lift and a corresponding decrease in pressure drag. Actuation is typically effected at frequencies that are *an order of magnitude higher* than the characteristic (shedding) frequency of the airfoil [i.e., St ~ O(10) rather than St ~ O(1)]. When the actuation frequency St is O(1), the reattachment is characterized by a Coanda-like tilting of the separated shear layer and the formation of large vortical structures at the driving frequency that persist beyond the trailing edge of the airfoil and lead to unsteady attachment and consequently to a time-periodic variation in vorticity flux and in circulation. In contrast, the suppression of separation at high actuation frequencies [i.e., St = O(10)] is marked by the absence of organized vortical structures along the flow surface. Finally, the dynamics of the transient lift in controlled reattachment and separation are investigated using pulsed amplitude modulation of the actuation input.

1 Introduction

Optimum aerodynamic performance of aircraft wings that has been traditionally achieved by appropriate design of the airfoil section to avoid flow separation may be somewhat compromised when the wing design is driven by non-aerodynamic constrains (e.g., payload, or stealth). Hence, these designs may rely on either active or passive flow control techniques to prevent flow separation and to maintain acceptable aerodynamic performance throughout the normal flight envelope. Although the relative simplicity of passive flow control devices (e.g. vortex generators, oscillating cavities, etc.) makes them attractive means for delaying flow separation, they afford no proportional control and introduce a drag penalty in the absence of "natural" separation. In contrast, conventional active control approaches which are based on the manipulation of fundamental instabilities of the separated free shear layer and derive their effectiveness from its receptivity to relatively low-level actuation input can be largely innocuous except when activated and have the potential for delivering variable power.

Controlled reattachment of separated flows over lifting surfaces at moderate and high angles of attack with the objective of improving aerodynamic performance and extending the flight envelope has been the focus of a number of investigations since the early eighties. Active control techniques that have achieved varying degrees of separation control by manipulation of the unstable separated free shear layer have included external and internal acoustic excitation (e.g., Ahuja and Burrin, 1994 and Zaman et al., 1987), vibrating ribbons or flaps (e.g., Bar-Sever, 1989), and steady and unsteady blowing/bleed (e.g., Williams, 1991 and Seifert et al., 1993).

The reattachment of a separated shear layer on an stalled airfoil by means of internally-driven acoustic excitation of the boundary layer upstream of separation (through a small rectangular orifice near the leading edge) was first investigated by Huang et al. (1987) and Hsiao et al. (1990, 1994), on NACA 63_3-018 airfoil). While the actuation frequency used by Huang et al. (1987) was limited to the shedding frequency of the airfoil (i.e., dimensionless frequency $St \sim 1$, scaled with the separated flow region), Hsiao et al. (1990) reported similar lift recovery over a much broader range of excitation frequencies (up to $St = 20$) that *far exceed the unstable frequency of the separating shear layer*. In the follow-on work of Chang et al. (1992), the actuation level was quantified using the orifice momentum coefficient (based on the amplitude of the velocity oscillations) and the authors reported that actuation at $St = 2$ and $C_\mu < 10^{-4}$ led to a post-stall increase in lift of up to a 50%. However, as noted in the earlier work of Hsiao et al. (1990) separation could also be controlled up to $St \sim 20$ at higher actuation levels.

Flow control work in the 50s showed that separation on lifting surfaces could be avoided or at least delayed by the streamwise injection of fluid near the surface using a tangential wall jet. The work of Seifert et al., (1993) demonstrated that separation over a flap (at 25% chord of a NACA 0015 airfoil) can be effectively controlled with substantially less mass flux by replacing the continuous wall jet at the flap hinge with oscillatory injection at the unstable frequency of the separating shear layer. These authors reported that oscillatory blowing at reduced frequencies within the range $1 < St < 3$ combined with a low-level steady blowing yielded effective actuation for maximum aerodynamic performance augmentation. Similar approach was used in a later investigation (Seifert et al., 1996) to control leading edge separation on the same airfoil where the dimensionless frequency for correlating C_μ was scaled with the length over which reattached flow was maintained. In a related numerical simulation of the post-stall ($\alpha = 22^\circ$) flow about an NACA0012 airfoil actuated at $St \cong 1$ by a zero net mass flux jet, Donovan et al. (1998) demonstrated that the increased lift had substantial oscillations (up to 20% peak-to-peak) at the actuation frequency.

The work described here is part of an ongoing investigation of the flow mechanisms of a novel control approach of the performance of lifting surfaces through fluidic modification of their apparent aerodynamic shape. Control is effected by surface-mounted fluidic actuators based on synthetic jet technology having the distinct feature that they do not require a fluid source and are formed locally from the fluid adjacent to the flow surface in which they are embedded (Smith and Glezer, 1997). Thus, the interaction between these jets and the cross flow over the surface leads to the formation of a domain that displaces the local streamlines of the cross flow thereby inducing an 'apparent' or 'virtual' change in the shape of the surface and an *apparent* modification of the flow boundary and the local streamwise pressure gradient.

While conventional flow control schemes have primarily focused on the mitigation of flow separation by coupling to the instability of the separating shear layer on the time scale of the flow about the airfoil, fluidic modification of the apparent aerodynamic shape of aero-surfaces does not necessarily rely on coupling to global flow instability and thus can be applied over a broader range of flow conditions. Furthermore, this approach to flow control can accommodate broader band control algorithms because the actuation frequency is typically at least an order of magnitude higher than the characteristic frequency of the flow. Smith et al. (1998) and Amitay et al. (1998, 1999) demonstrated the utility of transverse synthetic (zero mass flux) jet actuators operated at $St = O(10)$ for the suppression of separation over an unconventional airfoil at moderate Reynolds numbers (up to 10^6) and consequently a substantial increase in lift and decrease in pressure drag. Recent experiments by Erk (1997) demonstrated suppression of separation on an FX61-184 airfoil at Reynolds numbers up to $3x10^6$ using synthetic jet actuation at frequencies up to $St \sim O(100)$. As shown by Amitay et al. (1999), the quasi-steady lift coefficient of the attached flow can be further augmented by exploiting the flow transients associated with the onset of separation using a prescribed unsteadiness of the locally separated flow domain that is induced by a temporally-modulated control input.

The present paper reviews some of the elements of the work on the modification of the aerodynamic performance of the unconventional airfoil of Amitay et al. (1998, 1999a, 1999b) using fluidic control that is effected by transverse synthetic jet actuators near the airfoil's leading edge (§3.1). The control effectiveness at actuation frequencies that are either on of the order of or well above the natural shedding frequency of the airfoil [i.e., $St \sim O(1)$ and $O(10)$] are discussed in §3.2. Finally, The mechanisms of flow transients that are associated with the reattachment and separation processes are discussed in §3.3.

2 Experimental Apparatus and Procedure

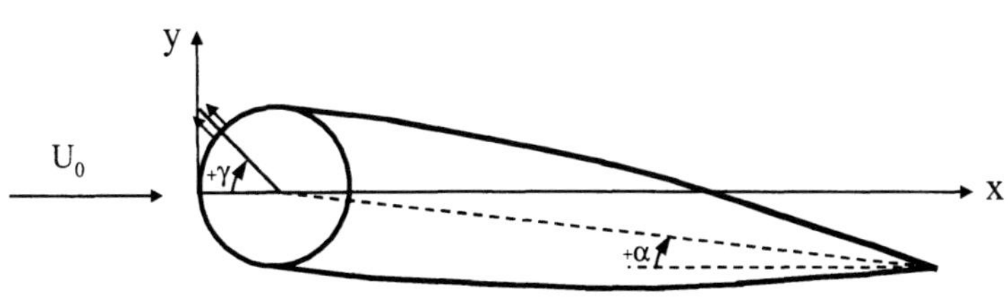

Figure 1. Unconventional airfoil model.

The application of transverse synthetic jets for separation control over an unconventional airfoil are demonstrated using a thick airfoil model (Figure 1). The airfoil's leading edge is formed by a 62 mm circular cylinder mounted within an aerodynamic fairing that is based on a uniformly stretched NACA four-digit series symmetric airfoil for which the thickness to chord ratio is $t/c = 0.24$ ($c = 254$ mm). The surfaces of the cylinder and of the fairing are well polished and the transition between the surface of the cylinder and the edge of the fairing is designed to have a close fit such that the surface discontinuity at the edge of the overlap domain is less than 0.1 mm (and is too small to trigger local separation).

The airfoil is mounted in the test section of an open-return, low-speed wind tunnel having a square test section measuring 91 cm on a side (the maximum air speed is 32 m/s with a free-stream turbulence level less than 0.25%). The upper and lower walls of the wind tunnel are adjusted to compensate for blockage created by the airfoil. Distributions of surface pressure are measured using 47 pressure taps that are located in the spanwise mid-plane and are equally spaced circumferentially around the cylinder and additional 45 pressure taps at mid span along the top and bottom surfaces of the fairing.

The center section of the leading edge cylinder houses a pair of parallel synthetic jet actuators each having a flush-mounted 140 mm x 0.5 mm orifice that are 2.5 mm apart azimuthally and co-linear with the axis of the cylinder. The jet actuators can be positioned at an azimuthal angles γ between -90° to +90° (at $\alpha = 0°$) relative to the incoming flow direction by rotating the cylinder independently of the fairing. Two fences are placed at the edges of the actuator orifice to maintain a nominally two-dimensional flow in the controlled section of the airfoil. High-frequency synthetic jets are produced at St = 10, 14.7 and 20 (f_{act} = 740Hz, 1088Hz and 1480Hz, respectively) using piezo-electrically driven diaphragms mounted in compact shallow cavities underneath the surface of the cylinder. Low-frequency synthetic jets are produced through the same orifices (in the absence of the piezoelectric drivers) at St = 0.95, 2.05, and 3.3 (f_{act} = 71Hz, 148Hz, and 246Hz, respectively) by conventional pressure speakers mounted on opposite ends of the cylinder cavity outside of the wind tunnel's test section.

In the present work, the performance of the actuators is measured using the momentum coefficient,

$$C_\mu = \frac{\bar{I}_j}{\frac{1}{2}\rho_o U_o^2 c}$$

where $\bar{I}_j$ is the time-averaged momentum flux per unit length during the outstroke and is given by

$$\bar{I}_j = \frac{1}{\tau}\rho_j b \int_0^\tau \langle u_j^2(\phi) \rangle \, d\phi \,,$$

τ = T/2 (T is the period of the diaphragm motion), ρ_j and ρ_o are the jet and free-stream fluid densities, respectively, b is the jet orifice width, c is the chord, U_o is the free-stream velocity, and $\langle u_j(\phi) \rangle$ is the phase-averaged velocity at the jet exit plane. The velocity within the jet orifice is measured using a miniature hot-wire sensor and the velocity traces are rectified when the velocity reverses its direction at mid-cycle. Cross stream distributions of the streamwise and cross-stream velocity components are measured in the wake of the airfoil using X-configuration hot wire miniature sensors that are mounted on a computer-controlled traversing mechanism.

3 Results and Discussion

3.1 Modification of the Time-Averaged Aerodynamic Performance

This section reviews briefly some of the effects of jet actuation on the aerodynamic performance (as measured by pressure distributions, lift and pressure drag) of the unconventional airfoil described in §2. The distributions of the static pressure coefficient about the circumference of the

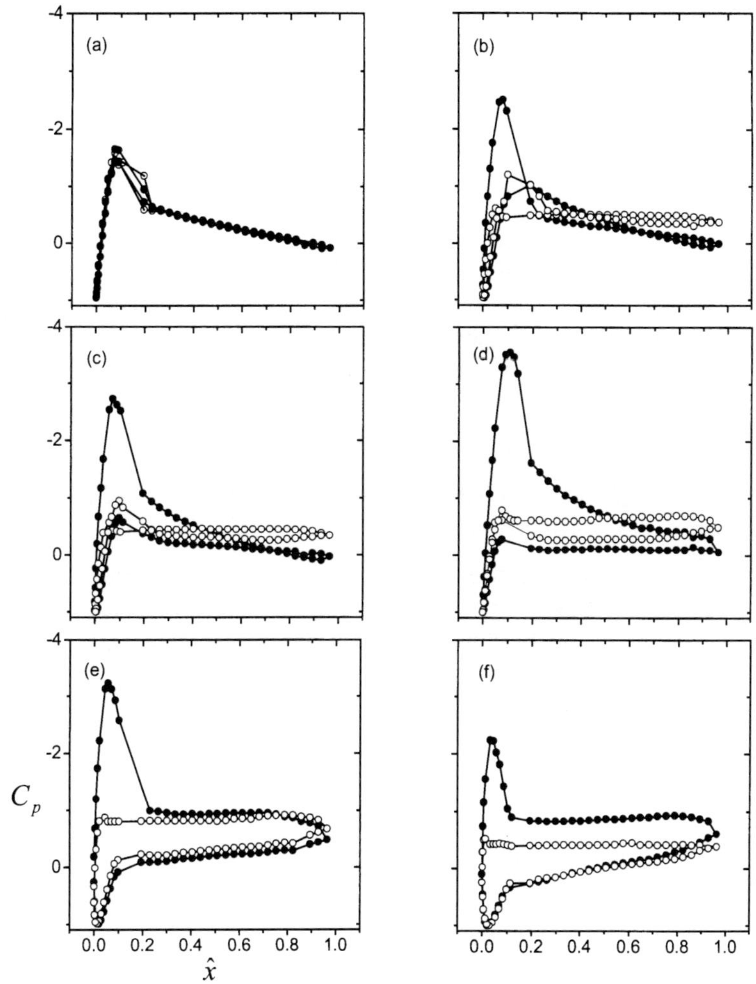

Figure 2. Pressure coefficient distributions around the airfoil at different angles of attack. Actuated (•), and baseline (o). $\alpha = 0^{\circ}$ (a), $\alpha = 5^{\circ}$ (b), $\alpha = 10^{\circ}$ (c), $\alpha = 15^{\circ}$ (d), $\alpha = 20^{\circ}$ (e) and $\alpha = 25^{\circ}$ (f).

airfoil in the presence and absence of actuation (closed and open symbols) are shown in Figures 2a-f ($\alpha = 0^{\circ}$, 5°, 10°, 15°, 20° and 25°, respectively) for $Re_c = 310{,}000$. The jet angle for the controlled flow is $\gamma = 60^{\circ}$ (relative to the oncoming free stream), the jet momentum coefficient is $C_{\mu} = 3.5 \cdot 10^{-3}$, and $\hat{x} = x/c$.

In the absence of flow control, the airfoil stalls at angles of attack exceeding 5° (Figure 2b). At $\alpha = 0^{\circ}$ (Figure 2a), the baseline flow is attached over the entire airfoil surface. These data show that while the suction peak is almost the same on both sides of the airfoil, there are slight differences in the domains where the fairing blends into the cylinder. Although for the present momentum coefficient the effect of the jets on the attached flow is small, it appears that despite the (top-to-bottom) asymmetry of the actuation the pressure distributions on the top (where the jets are located) and bottom surfaces become more similar when the jet are activated. For

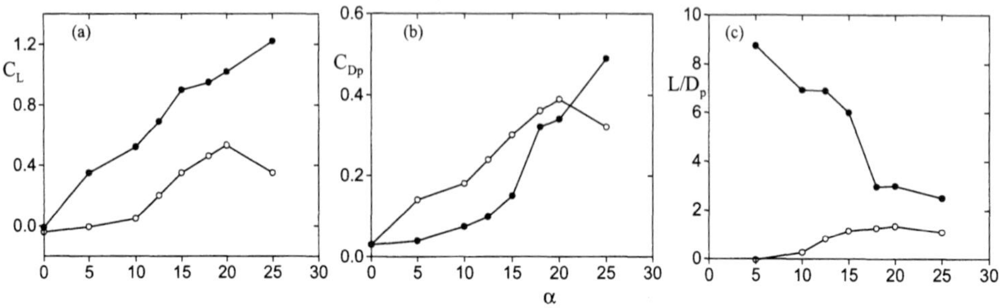

Figure 3. Lift coefficient (a), pressure drag coefficient (b) and lift-to-pressure drag ratio (c) vs. angle of attack, α for $\gamma = 60^{\circ}$. Actuated (•), and baseline (○).

$\alpha \le 15^{\circ}$ (Figures 2b-d), the flow becomes fully reattached when control is applied, and the resulting pressure distributions exhibit a large suction peak in the vicinity of the maximum thickness (near $\hat{x} = 0.1$). A rapid pressure recovery occurs for $0.1 < \hat{x} < 0.2$ followed by a more gradual pressure recovery towards the trailing edge.

It is interesting to note that at moderate angles of attack the pressure within some streamwise domains in the absence of actuation is higher on the upper (suction) surface than on the lower (pressure) surface of the airfoil resulting in a reduction in lift (e.g., $0.075 < \hat{x} < 0.30$ and $0.06 < \hat{x} < 0.22$ in Figures 2b and 2c, respectively). However, when the jets are activated and the flow becomes fully attached, the pressure on the upper surface is continuously lower than on the lower surface from the leading edge and along most of the airfoil (through $\hat{x} = 0.62$ for $\alpha = 5^{\circ}$ and $\hat{x} = 0.72$ for $\alpha = 10^{\circ}$). The existence of small domains near the trailing edge in which the pressure difference between the two surfaces is reversed indicates that the streamwise rate of pressure recovery resulting from the reattached flow on the upper surface exceeds the rate of recovery associated with the curvature of the bottom surface. This effect decreases with increasing angle of attack where the pressure distribution on the lower surface becomes almost invariant with $\hat{x}$ (Figures 2d and 2e) and finally decreases with $\hat{x}$ (Figure 2f).

At $\alpha = 20^{\circ}$ (Figure 2e), the jet actuation produces only partial reattachment. The flow remains attached through $\hat{x} = 0.22$ but subsequently separates and the pressure distribution within the separated flow domain is almost identical to the corresponding distribution in the absence of control. When $\alpha = 25^{\circ}$ (Figure 2f), the baseline flow separates farther upstream ($\hat{x} = 0.02$) and as a result the pressure within the separated flow domain is actually higher than for $\alpha = 20^{\circ}$. However, even though the attained suction peak in the presence of actuation is lower than at $\alpha = 20^{\circ}$, separation still occurs at the same streamwise location ($\hat{x} = 0.2$) and as a result the pressure in the separated region of the actuated flow is considerably lower than in the baseline flow. While the increase in the pressure difference (compared to the baseline flow) results in increased lift, at these angles of attack there is also a significant increase in pressure drag. Nevertheless, as shown in Figure 3 below, the lift-to-pressure drag ratio increases compared to the baseline case.

The pressure distribution on the airfoil is integrated to yield the lift coefficient, the pressure drag coefficient and the lift-to-pressure drag ratio over a range of angles of attack (Figures 3a-c, respectively). The corresponding distributions of the baseline flow are also shown for comparison (open symbols). Without control, the aerodynamic performance of this airfoil is rather poor (e.g., the lift-to-pressure drag ratio at $\alpha = 5^\circ$ is 0.115), and the airfoil is stalled even at low angles of attack. However, with actuation ($\gamma = 60^\circ$, $C_\mu = 3.5 \cdot 10^{-3}$), the lift coefficient increases almost linearly with angle of attack as might be expected for a conventional airfoil. Furthermore, the attached flow in the presence of actuation results in a substantial reduction in the pressure drag coefficient. For example at $\alpha = 5^\circ$, the pressure drag decreases by 45% with respect to the baseline case. Even at relatively high angles of attack ($\alpha = 25^\circ$), where the flow is only partially attached, the lift curve has yet to exhibit the downturn characteristic of stall. However, the monotonically increasing pressure drag (which exceeds the drag coefficient of the baseline airfoil at $\alpha = 25^\circ$) and the pressure distribution (Figure 2f) suggest that stall is imminent. Figure 3c shows that transverse synthetic jet actuation significantly improves the lift-to-pressure drag ratio through the entire range of angles of attack in the present investigation. At $\alpha = 5^\circ$, this ratio is increased by almost *two orders of magnitude*. However, the improvement in the L/D_p diminishes as the angle of attack increases.

3.2 The Role of the Actuation Frequency

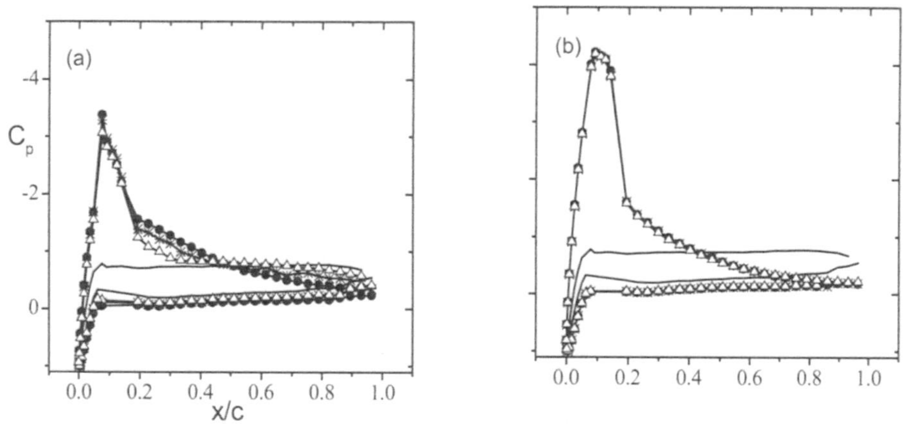

Figure 4. Variation of the pressure coefficient with the dimensionless forcing frequency; St ~ O(1) (a) and St ~ O(10) (b). (The distribution for the baseline flow is shown in a solid line).

As noted in §1, much of the earlier work that focused on extension of the pre-stall envelope of various conventional airfoils by utilized active (time-periodic) actuation upstream of separation. The characteristic period of the actuation typically has been comparable to the time of flight over the separated flow domain corresponding to a dimensionless frequency St, which is the ratio between the latter and former time scales, of order 1 (i.e., the actuation frequency is comparable to the shedding frequency of the airfoil). The natural receptivity of the separating shear layer to this frequency band results in a Coanda-like tilting of the shear layer towards the surface of the airfoil and therefore in partial restoration of the lift. In contrast to this approach, in the present work the actuation frequency is selected to be at least an order of magnitude higher than the "natural" (or shedding) frequency of the airfoil in order to bypass flow separation altogether.

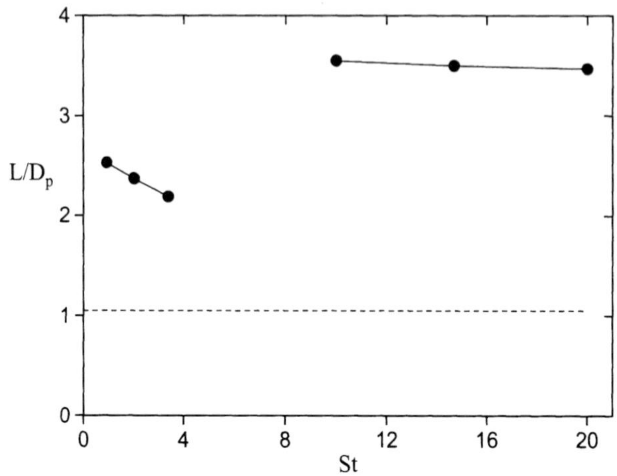

Figure 5. Variation of L/D$_p$ with dimensionless actuation frequency, $\gamma = 60°$. (---) baseline.

To demonstrate the effect of the actuation frequency on the control effectiveness, the synthetic jet actuators are activated at six frequencies corresponding to St = 0.95, 2.05, 3.4, 10, 14.7 and 20 (71Hz, 148Hz, 246Hz, 740Hz, 1088Hz and 1480Hz, respectively). At St = 0.95, 2.05 and 3.4, the jets are driven by audio speakers while at high frequencies the jets are formed by compact, piezo-electrically-driven actuators (cf. §2). The actuators are calibrated at each operating frequency outside of the wind tunnel, and then once again in the presence of a cross flow in the wind tunnel. It is found that when the operating frequencies are below St < 4, the jets driven by the speaker actuators are spanwise-uniform (along the long dimension of the orifice). However, at higher frequencies the speaker-driven jets develop spanwise non-uniformities due to three-dimensional acoustic effects within the actuator cavity. Velocity measurements (not shown here) within the actuator orifice using a miniature hot-wire sensor reveal that for St > 4, the performance of the speaker-driven actuator jets is markedly different in the presence and absence of a cross flow. In fact, in the presence of a cross flow, the velocity of the speaker driven jets at a given spanwise position along the orifice decreases by more than 80% and thus their ability to affect the separating shear layer is substantially reduced. In contrast, the velocity of the high-frequency jets produced by the piezoelectric actuators is spanwise-uniform, virtually invariant with tunnel speed and leads to a complete reattachment of the separated flow.

Distributions of the pressure coefficient around the airfoil for St = 0.95, 2.05 and 3.4 and St = 10, 14.7 and 20 are shown in Figure 4a and Figure 4b, respectively (the shedding frequency of the baseline flow is St = 0.7). Here, $Re_c = 310{,}000$, $\gamma = 60°$, $C_\mu = 3.5 \cdot 10^{-3}$ and $\alpha = 17.5°$. The pressure distribution for the stalled flow (solid line) is also shown for reference. Actuation at low frequencies (Figure 4a) results in a very sharp suction peak near $\hat{x} = 0.075$, which corresponds to the location of the separation in the baseline flow. Concomitantly, the degree of pressure recovery towards the trailing edge is reduced with increasing control frequency, leading to an increase in pressure-drag. The pressure distributions for St $\geq$ 10 (Figure 4b), which is over an order of magnitude higher than the shedding frequency (St = 0.7), are significantly different and exhibit larger and wider suction peaks with a corresponding larger increase in the lift coefficient. Following the suction peak the pressure difference between the suction and pressure sides is smaller than at corresponding stations at the low actuation frequencies resulting in lower pressure drag. Moreover, while at low actuation frequencies (Figure 4a) the pressure distribution varies with St, at high actuation frequencies (Figure 4b) the pressure distribution appears to be independent of St, suggesting that when the actuation frequency is high enough, the details of the flow reattachment become independent of the frequency.

The variation of the ratio of lift to (pressure) drag, L/D_p, with actuation frequency is shown in Figure 5. Two distinct domains are immediately apparent. In the first domain (I), where the actuation frequencies are of the same order as the shedding frequency (St < 4), L/D_p decreases with increasing St (from 2.65 for St = 0.95 to 2.35 for St = 3.3), which may be the result of reduced receptivity of the separated shear layer. In the second domain (marked as II on the figure), the actuation frequency is at least an order of magnitude higher than the shedding frequency (i.e., St ≥ 10) and the lift to pressure drag ratio is higher ($L/D_p \approx 3.2$). It appears that L/D_p is almost independent of the actuation frequency, suggesting that the mechanism that leads to the suppression of separation is not associated with the stability of the separated shear layer. Note that bandwidth limitations of the low- and high-frequency actuators prevented overlap within the domain 4 < St < 10.

The power spectra of the streamwise velocity measured at $\hat{x} = 2$ on the lower side of the airfoil's wake at a cross-stream elevation where the streamwise velocity deficit is half the maximum deficit are shown in Figures 6a-d (the spectrum of the baseline flow is reproduced, using a dashed line, in each of the figures for reference), where the dimensionless frequency is $v^+ = (f \cdot c)/U_o$. As evident from the cross-stream distributions of the velocity fields (not shown) when the flow is actuated (and nominally attached to the upper surface of the airfoil) the cross-stream extent of the wake is reduced substantially and is accompanied by a reduction in the magnitudes of rms velocity fluctuations within the wake. At St = 0.95 (Figure 6a) there is a strong spectral component at the actuation frequency and the entire spectrum appears to be attenuated by approximately 3.5. As St is increased to 2 and 3.3, the spectral peak at the actuation frequency shifts towards the decaying part of the spectrum (it is virtually indistinguishable from the background at St = 3.3) and the attenuation imposed by the collapse of the separated flow domain increases to 4.5 and 7, respectively. The spectrum of the actuated flow at St = 10 (Figure 6d) is remarkably different from the spectral distributions at the lower actuation frequencies. There is a stronger attenuation (well over an order of magnitude) at both the low and high spectral ends, and the spectrum includes a distinct inertial sub-range for $v^+ > 3$ over almost two decades (which includes the actuation frequency). The reduced power at all spectral components (compared to actuation at lower St) indicates that the attached flow removes less energy from the uniform stream resulting in a lower drag.

3.3 The Dynamics of Flow Reattachment and Separation

In what follows, the dynamics of the reattachment and separation processes following an onset and termination of pulsed amplitude modulation of the actuator (control) input are presented for which the dimensionless frequency of the actuation signal is either St ~ O(1) or O(10). The

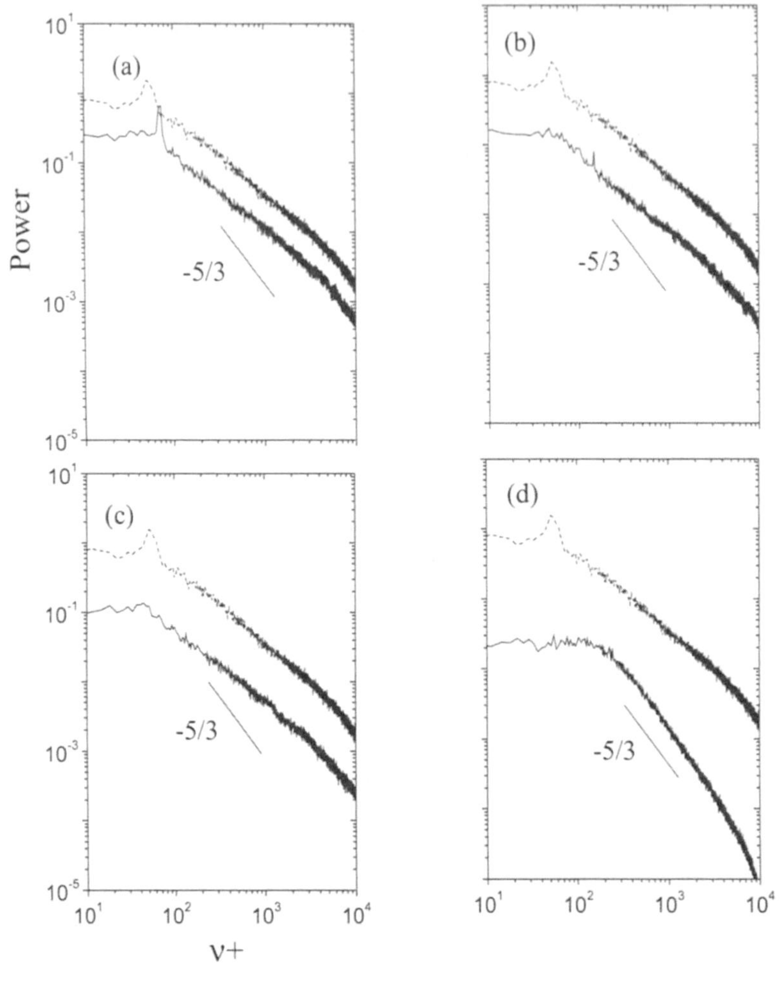

Figure 6. Power spectra for St = 0.95 (a), 2.05 (b), 3.3 (c), and 10 (d). (---) baseline.

modulation is synchronized with the actuator's driving signal such that the leading edge of the modulating waveform coincides with a zero crossing of the actuator signal and continues for 0.5 sec. Using x-wire anemometry, the transients of the flow resulting from the pulsed excitation is measured in detail across the near wake at $\hat{x} = x/c = 2$. The airfoil is placed at an angle of attack of 17.5^0, the jets issue from the suction side of the airfoil at $\gamma = 60^0$ where the corresponding downstream distance from the leading edge is $\hat{x} = 0.062$ (γ is the angle between the jets' centerline and the free stream), the momentum coefficient is $3.5 \cdot 10^{-3}$ and the free stream velocity is 18.5 m/s ($Re_c = 310{,}000$).

The phase-averaged cross-stream distributions of the dimensionless spanwise vorticity is computed from the streamwise and cross-stream velocity distributions (not shown) and are

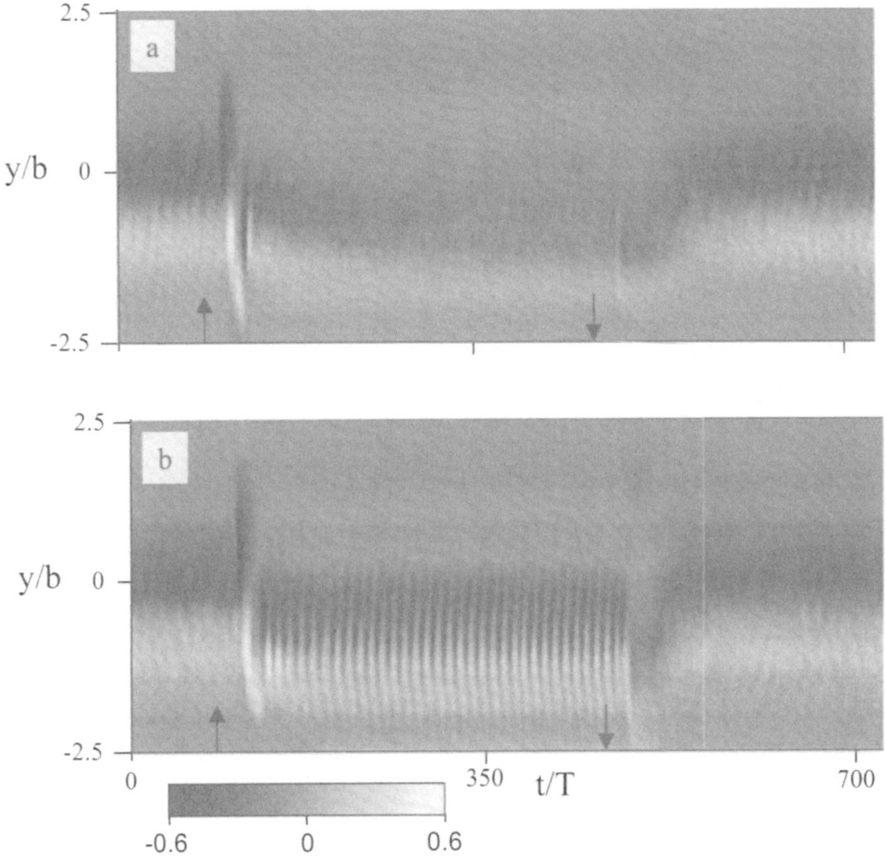

Figure 7. Phase-averaged color raster plot of the cross-stream distribution of the spanwise vorticity; St = 10 (a) and 0.95 (b).

shown in Figures 7a and b for St = 10 and 0.95, respectively (the up and down arrows denote the times where the control is turned on and off, respectively). Note that $\hat{y}$ corresponds to the leading edge of the airfoil (cf. Amitay and Glezer, 2001) and U_∞, T (the actuation period), and c are used to render all variables dimensionless. When the flow is separated (i.e., before and after the pulse modulated excitation is applied), the vorticity distribution in the wake is comprised of a train of vortical structures of alternating sign (clockwise vorticity is taken to be negative) having a nominal passage frequency of 50 Hz. Nevertheless, the total vorticity flux across the wake during one period of the (baseline) shedding frequency is approximately zero.

As was shown by Smith et al. (1998), the actuation leads to flow reattachment and the establishment of a higher (positive) lift force on the airfoil, which must be accompanied by a change in the vorticity flux and a net increase in circulation associated with positive (counter-clockwise) vorticity. However, following the reattachment, a strong clockwise vortex indicating *a reduction in lift* is initially advected past the measurement station. This vortex is inherent of a separated flow and it is associated with the negative "trapped" vorticity that must

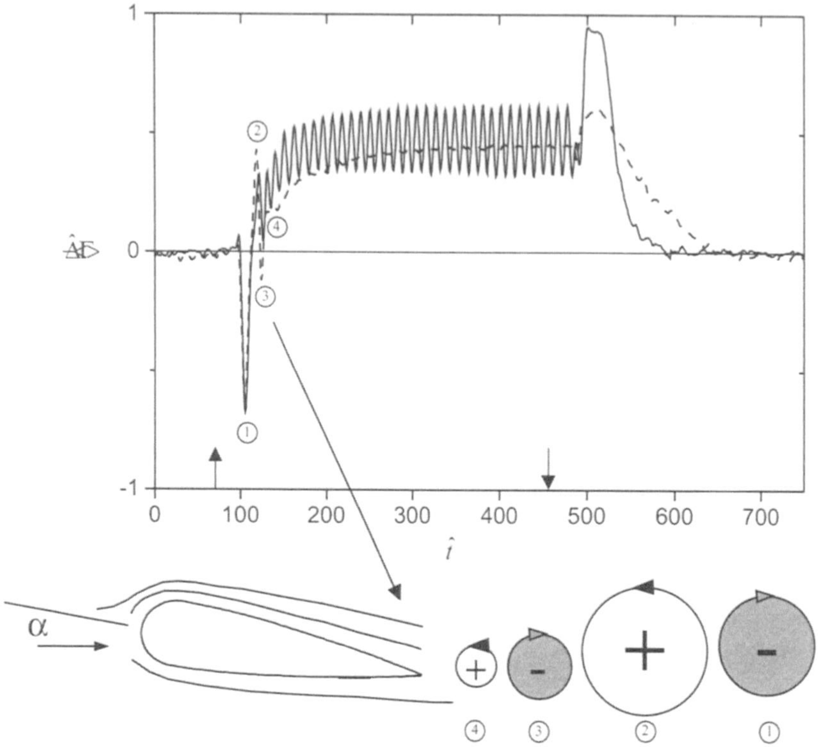

Figure 8. Phase-averaged circulation increment for $\alpha = 17.5^{\circ}$ and $\gamma = 60^{\circ}$. St = 0.95 (solid), and 10 (dash).

shed off the airfoil. The shedding of the negative vorticity is followed closely by a stronger counter-clockwise vortex indicating the re-establishment of lift ("starting vortex"). These two large vortices are followed by a series of smaller vortices of alternating signs and diminishing strength. It is remarkable that the initial transient following the application of the high-frequency pulse-modulated control (Figure 7a) are very similar to that of the low-frequency actuation (Figure 7b). However, in contrast to the reattachment at the higher reduced frequency in which the shedding of organized vortical structures appears to subside following the transient (for $\hat{t} > 300$, Figure 7a), the reattachment at a reduced frequency of order one appears to be accompanied by the coherent shedding of a train of strong vortices at the actuation frequency.

When the control is turned off, the flow separates again and the airfoil loses its lift. This reduction in lift is accompanied by a decrease in circulation and the shedding of negative (clockwise) vorticity. However, immediately following the termination of the control, a counter-clockwise vortex indicating *a momentary increase in lift* is advected past the measurement station before the separated vorticity field is established. These flow transients are similar at both actuation frequencies suggesting that the separation process is similar.

These data suggest that when the flow is actuated at a low reduced frequency, which is of the same order of the natural shedding frequency to which the separated flow is inherently receptive, the control input is amplified and the reattachment is manifested by the passage of shear layer vortices along the surface of the airfoil. However, when the actuation frequency is high enough, the interaction of the jets with the flow occurs at a smaller length scale, which leads to local modification of the apparent shape of the airfoil, and suppresses the shedding of large-scale vortices (for more details of the interaction of a synthetic jet with a cross flow see Honohan et al., 2000).

As noted by Amitay et al. (1999a), the time rate of change of the circulation is given by the phase-averaged vorticity flux

$$\frac{d\langle\hat{\Gamma}\rangle}{d\hat{t}} = \int_{-\infty}^{\infty} \langle\hat{U}\rangle \langle\hat{\Omega}_z\rangle \, d\hat{y}$$

Here, $\langle\hat{\Gamma}\rangle$, $\langle\hat{U}\rangle$, $\langle\hat{\Omega}_z\rangle$ and $\langle\hat{t}\rangle$ are the normalized phase-averaged circulation, streamwise velocity, spanwise vorticity and time, respectively. By integrating the vorticity flux, $d\langle\hat{\Gamma}\rangle/d\hat{t}$, the phase-averaged increment (relative to the baseline flow) of the circulation, $\Delta\langle\hat{\Gamma}\rangle$, is estimated and shown in Figure 8 (St = 0.95 and 10 are shown using solid and dashed curves, respectively). Note that the integration does not account for contributions of the fluctuating components, and it is assumed that because the measurement station is located only one chord-length downstream of the trailing edge of the airfoil, which is shorter than the wavelength of the shedding frequency, the interaction between successive vortices within this domain is minimal.

When the flow reattachment begins, $\Delta\langle\hat{\Gamma}\rangle$ exhibits a similar transient at both control frequencies. The incremental change in the circulation with respect to the baseline flow - $\Delta\langle\hat{\Gamma}\rangle$ initially diminishes to a value of -0.6 (resulting in a momentary decrease in the lift coefficient) and then recovers to a value of 0.33 and 0.43 (for St = 0.95 and 10, respectively) with the shedding of the second counter-clockwise vortex. It appears that the shedding of the "starting vortex" causes partial trailing edge separation, which is manifested by the shedding of another (weaker) clockwise vortex followed by a train of vortices of alternating signs (see cartoon). The circulation (and lift coefficient) ultimately converges to its attached value, which is in good agreement with the lift coefficient obtained from previous pressure measurements (Smith et al., 1998). However, while for high frequency actuation the circulation ultimately reaches a steady level, low frequency actuation results in oscillations of $\Delta\langle\hat{\Gamma}\rangle$ at the actuation frequency with peak-to-peak fluctuations of up to 55% of the mean level for the attached flow. When the control is turned off the circulation initially increases before settling to the baseline-stalled level, which is similar to the transient variation of lift on a pitching airfoil during dynamic stall.

4 Conclusions

Earlier approaches to partial restoration of lift for various configurations of stalled airfoils sought to affect the fundamental instability of the separating shear layer and thereby exploit a Coanda-like deflection of that shear layer towards the airfoil surface. This has been typically achieved by using time-periodic (mostly time-harmonic) actuation input having a dimensionless frequency (that nominally scales with the length of the separated flow domain) of order one, which is also within the receptivity of the separating shear layer. The present work explores the potential of bypassing flow separation by direct alteration of the streamwise pressure distribution along the surface that is induced by a local displacement of the cross flow as a result of its interaction with surface-mounted actuator jets. The quasi-steady interaction domain is formed using transverse synthetic (zero net mass flux) jets that are typically operated at frequencies that are at least an order of magnitude higher than the "natural" (or shedding) frequency of the airfoil.

The utility of this actuation approach for the prevention of flow separation over unconventional airfoils where aerodynamic design may be constrained and compromised by other design considerations is investigated on an unconventional symmetric airfoil for which the baseline configuration is normally stalled for angles of attack exceeding 5°. Actuation is effected in the vicinity of the leading edge (on both sides of the stagnation point) at a dimensionless frequency that is at least an order of magnitude larger than the characteristic shedding frequency of the airfoil. The actuation results in fully or partially attached flow with substantial augmentation of lift over a broad range of angles of attack (up to the maximum angle tested, $\alpha = 25^{\circ}$). Over a smaller range of angles of attack (up to 17.5°), the increases in lift (up to 100% compared to the corresponding baseline) is also accompanied by a decrease in pressure drag (up to 45%).

In the present experiments, the synthetic jet actuators are operated over a range of frequencies up to $St = 20$ (which is about 30 times higher than the natural shedding frequency of the separated flow). While at low actuation frequencies (i.e., $St < 4$) the recovered lift to (pressure) drag ratio L/D_p decreases with increasing St, actuation at $St \geq 10$ results in larger L/D_p that appears to be invariant with the actuation frequency. Spectral measurements in the wake of the airfoil indicate that there is a substantial difference in the nature of flow attachment at low and high actuation frequencies. While at low actuation frequencies the spectra are typically dominated by a large spectral component at the actuation frequency, actuation at higher frequencies result in a *featureless spectrum* and the emergence of a spectral band having a –5/3 slope indicating enhanced dissipation.

The flow transients that are associated with flow attachment and separation are investigated using time-modulated control input using phase-locked two-component hot-wire anemometry in the wake of the airfoil. Flow attachment begins with the shedding of a strong clockwise vortex indicating *a momentary reduction in lift* that is followed by a stronger counter-clockwise vortex indicating the re-establishment of lift. The flow transients associated with the application of the actuation are similar at low and high actuation frequencies. However, while at high actuation frequencies the reduced wake of the attached flow ultimately reaches a quasi steady symmetric vorticity distribution, at $St \sim O(1)$ there is coherent shedding of a train of vortices of alternating sign at the actuation frequency and therefore oscillations of the circulation (with peak-to-peak fluctuations of up to 45% of the mean level for the attached flow).

Acknowledgements

This work was supported in part by AFOSR (monitored by Drs. T. Beutner, M. Glauser and J. M. McMichael), DARPA, and by The Boeing Company, St. Louis. The authors gratefully acknowledge contributions by Drs. D.R. Smith, V. Kibens, and D.E. Parekh.

References

Ahuja, K. K. and Burrin, R. H. (1984). Control of Flow Separation by Sound. AIAA Paper 84-2298.

Amitay, M., Honohan, A., Trautman, M. and Glezer, A. (1997). Modification of the Aerodynamic Characteristics of Bluff Bodies using Fluidic Actuators. AIAA Paper 97-2004.

Amitay, M., Smith, B. L. and Glezer, A. (1998). Aerodynamic Flow Control using Synthetic Jet Technology. AIAA Paper 98-0208.

Amitay, M. and Glezer, A. (1999). Aerodynamic Flow Control of a Thick Airfoil Using Synthetic Jet Actuators. FEDSM99-6922, Proceedings of the 3rd ASME/JSME Joint Fluids Engineering Conference, San Francisco, California.

Amitay, M., Kibens, V., Parekh, D. E. and Glezer, A. (1999). Flow Reattachment Dynamics over a Thick Airfoil Controlled by Synthetic Jet Actuators. AIAA Paper 99-1001.

Bar-Sever, A. (1989). Separation control on an airfoil by periodic forcing., *AIAA J. 27*(6), 820-823.

Chang, R.-C., Hsiao, F.-B., and Shyu, R.-N. (1992). Forcing Level Effects of Internal Acoustic Excitation on the Improvement of Airfoil Performance. *Journal of Aircraft 29*(5), 823-829.

Donovan, J. F., Kral, L. D. and Cary, A. W. (1998). Active Flow Control Applied to an Airfoil". AIAA Paper 98-0210.

Erk, P. (1997). Separation Control on a Post-Stall Airfoil Using Acoustically Generated Perturbations. Ph.D. Dissertation, Technische Universitat Berlin, Germany.

Hsiao, F.-B., Liu, C.-F., and Shyu, J-Y. (1990). Control of Wall-Separated Flow by Internal Acoustic Excitation. *AIAA Journal 28*(8), 1440-1446.

Hsiao, F.-G., Shyu, R.--N., and Chang, R.-C. (1994). High Angle-of-Attack Airfoil Performance Improvement by Internal Acoustic Excitation. *AIAA Journal 32*(3), 655-657.

Huang, L. S., Maestrello, L. and Bryant, T. D. (1987). Separation Control over an Airfoil at High Angles of Attack by Sound Emanating from the Surface. AIAA Paper 87-1261.

Seifert, A., Bachar, T., Koss, D., Shepshelovich, M. and Wygnanski, I. (1993). Oscillatory Blowing: A Tool to Delay Boundary-Layer Separation. *AIAA Journal 31*(11), 2052-2060.

Seifert, A., Darabi, A., and Wygnanski, I. (1996). Delay of Airfoil Stall by Periodic Excitation. *Journal of Aircraft 33*(4), 691-698.

Smith, B. L. and Glezer, A. (1998). The Formation and Evolution of Synthetic Jets. *Physics of Fluids*, Vol. 10, No. 9.

Smith, D. R., Amitay, M., Kibens, V., Parekh, D. E. and Glezer, A. (1998). Modification of Lifting Body Aerodynamics using Synthetic Jet Actuators. AIAA Paper 98-0209.

Williams, D., Acharya, M., Bernhardt, J., and Yang, P. (1991). The Mechanism of Flow Control on a Cylinder with the Unsteady Bleed Technique. AIAA Paper 91-0039.

Zaman, K. B. M. Q., Bar-sever, A., and Mangalam, S. M. (1987). Effect of Acoustic Excitation on the Flow over a Low-Re Airfoil. *Journal of Fluid Mechanics,* Vol. 182, 127-148.

Adaptive Neural Network Based Approach for Active Flow Control

Moshe Idan[1], Anthony J. Calise[2], Ali T. Kutai[3], and David E. Parekh[4*]

Georgia Institute of Technology, Atlanta, GA, U. S.

Abstract. This paper presents an intelligent control system architecture for micro-adaptive flow control applications. The proposed adaptive non-model based control approach is motivated by the difficulty in constructing an accurate physical model of the micro-adaptive flow control system. The control architecture employs a Neural Network based adaptive control process, and a novel method for protecting the adaptive system from poorly modeled actuator nonlinearities, such as saturation and hyteresis. We give a simple illustration of these features by considering a ducted rotor model.

1 Introduction

Active flow control is a multi-disciplined approach to modifying flow field characteristics to achieve a desired aerodynamic performance. It combines sensing, actuation, and flow physics. In contrast to passive techniques, such as wall shaping or coatings, *active* control involves introducing a time-dependent forcing of the system. This is often achieved through acoustics, electromagnetic forces, mechanical actuation, thermal inputs, or fluxes of mass and/or momentum. Active flow control is most effective when one leverages a natural instability of the flow to amplify the control authority of the actuator.

By imposing this type of control one is able to achieve a state of flow that would not naturally occur. A classic example, for which many feedback control approaches have been considered, is the distributed and direct manipulation of turbulent structures in a wall-bounded flow to achieve significant viscous drag reduction. In other cases, one can concentrate actuation to a compact region where the flow is highly receptive to properly configured actuation. Examples of this include keeping wings from stalling at high angles of attack, dramatic increases in mixing rates in plumes, and direct control of vortex formation and asymmetry on forebodies and highly swept wings. The term "micro-adaptive" is often used to refer to the leveraging of natural flow instabilities to amplify the control authority of micro actuators in combination with adaptive feedback control.

Open-loop approaches are adequate when the state of the flow can be adequately measured, the flow phenomenon is sufficiently deterministic, and the actuation approach is robust. Consider

[1] Visiting Scholar from the Faculty of Aerospace Engineering, Technion - IIT, Haifa 32000, Israel.

[2] Professor, School of "Aerospace Engineering.

[3] Graduate Research Assistant.

[4] Director, Aerospace, Transportation, & Advanced Systems Laboratory, Georgia Tech Research Institute

[*] Research supported by AFOSR under grant F49620-01-1-0024.

mixing enhancement in jet flows. Knowledge of the flow velocity and nozzle geometry is often sufficient to determine an appropriate shear layer forcing frequency. Forcing amplitudes based on previous characterization of the flow can be readily scheduled. However, there are many more cases where some form of feedback control is required to achieve the desired performance gains, such as turbulent viscous drag reduction.

Feedback control of dynamical systems, addresses three main aspects: (a) improved tracking in response to commands, (b) robustness to model uncertainty and (c) disturbance rejection, or in other words, maintaining tracking performance in the presence of external disturbances. These aspects of automatic control have been addressed in the literature by a multitude of synthesis and analysis approaches, ranging from classical frequency domain techniques, time domain (state-space) methods, optimal control, and more recently, multivariable robust and intelligent (adaptive) control solutions. These approaches require a dynamic model of the controlled system. The robustness and disturbance rejection issues are addressed by such concepts as stability margins, uncertainty modeling, loop shaping, performance weighting functions, H_∞ synthesis and more. Many of these techniques incorporate at least partial system state estimation, or on-line model parameter identification (indirect adaptive) algorithms.

The direct adaptive control approach provides robustness to parametric uncertainty, which is in particular attractive for the micro adaptive flow control (MAFC) application, where accurate modeling is challenging. In this paper, we propose the use a nonlinear adaptive control technique, which incorporates dynamic inversion and feedback linearization, augmented by a neural network (NN) element to account for modeling errors and uncertainty. This approach was successfully utilized in a multitude of aerospace applications (McFarland and Calise, 1997, Rysdyk et al., 2000, Johnson 2000, and Idan et al., 2001), including some actual flight test experiments (Calise et al., 2000 and Calise et al., 2001). In the following, we present an outline of this control design methodology, following a sample design case and simulation results.

2 Control System Architecture

The adaptive NN based control system is constructed as a "model reference adaptive control" (MRAC) scheme presented in Figure 1. This conceptual layout incorporates an approximate dynamic inversion block, a linear compensator and an on-line adaptive NN. The system is driven by the outputs of a reference model block. NN adaptation can be strongly affected by the commonly encountered saturation or quantization of the actuator inputs, as is the case in MAFC. This is addressed by the recently developed pseudo-control hedging (PCH) technique, incorporated in Figure 1, and which will be detailed later.

The main advantage of the proposed control setup is in its ability to adapt to parametric uncertainty in actuator and plant model, which may be highly nonlinear. The NN can compensate for a wide range of modeling (inversion) errors. The compensator design is straightforward and relies mainly on linear control theory. The NN adaptation rule results from nonlinear stability analysis, which ensures that the error signals and network weights are bounded.

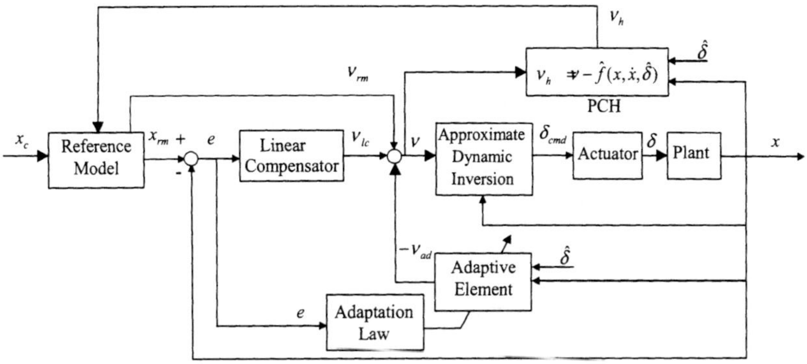

Figure 1. Model Reference Adaptive Control Setup, Including Approximate Dynamic Inversion and Pseudo-Control Hedging Compensation.

3 Approximate System Linearization

One of the common methods for controlling nonlinear dynamical systems is based on approximate feedback linearization (Isodori, 1989). We illustrate this approach for the special case in which the system has full relative degree, and each controlled variable (elements of x) has a relative degree of two

$$\ddot{x} = f(x, \dot{x}, \delta) \tag{1}$$

where $x, \delta \in \Re^n$. We introduce a pseudo-control input ν such that the dynamic relation between it and the system state is linear

$$\ddot{x} = \nu \tag{2}$$

where

$$\nu = f(x, \dot{x}, \delta) \tag{3}$$

Ideally, the actual system controls (δ) are obtained by inverting Eq. (3). Since the function $f(x,\dot{x},\delta)$ is usually not known exactly, an approximation is introduced

$$\nu = \hat{f}(x, \dot{x}, \delta) \tag{4}$$

which results in

$$\ddot{x} - \nu = \Delta(x, \dot{x}, \delta) \tag{5}$$

where the modeling error is given by

$$\Delta(x, \dot{x}, \delta) = f(x, \dot{x}, \nu) - \hat{f}(x, \dot{x}, \nu) \tag{6}$$

where the modeling error is given by

$$\Delta(x,\dot{x},\delta) = f(x,\dot{x},v) - \hat{f}(x,\dot{x},v) \tag{6}$$

The approximation function $\hat{f}$ is chosen such that its inverse with respect to δ is readily computable. Thus, given the pseudo-control signal (v), the actuator command is computed as

$$\delta_{cmd} = \hat{f}^{-1}(x,\dot{x},v) \tag{7}$$

Approximate dynamic inversion produces a model inversion error that will be adaptively compensated using an on-line neural network.

As shown in Figure 1, the total pseudo-control signal is constructed of three components

$$v = v_{rm} + v_{lc} - v_{ad} \tag{8}$$

where v_{rm} is the pseudo-control component generated by the reference model, v_{lc} is the output of the linear compensator, and v_{ad} is generated by the adaptive element introduced to compensate for the model inversion error. In the case of perfect actuation $(\delta = \delta_{cmd})$, the commanded pseudo-control signal generated by the reference model equals $\ddot{x}_{rm}$, the acceleration of the reference model state.

4 Linear Compensator Design

A linear compensator is designed for each degree of freedom assuming perfect inversion $(\hat{f} = f)$. If the controlled variable has relative degree two (as illustrated in the preceding section), the state tracking error dynamics associated with the linearized plant have two poles at the origin. The linear compensator is designed so that the error dynamics are stabilized. This is most often achieved using standard proportional-derivative (PD) controllers, although additional integral action can be incorporated to improve steady state performance. In general, the linear compensator can be designed using any technique as long as the linearized closed loop system is stable.

For the second order system, PD compensation is expressed by

$$v_{lc} = \begin{bmatrix} K_P & K_D \end{bmatrix} e \tag{9}$$

where the state tracking error is defined by

$$e = \begin{bmatrix} x_{rm} - x \\ \dot{x}_{rm} - \dot{x} \end{bmatrix} \tag{10}$$

The compensator gain matrices $K_P, K_D \in \Re^{n\times n}$ are chosen so that the tracking error dynamics given by

$$\dot{e} = Ae + B(v_{ad} - \Delta) \tag{11}$$

$$A = \begin{bmatrix} 0 & I \\ -K_P & -K_D \end{bmatrix}, \qquad B = \begin{bmatrix} 0 \\ I \end{bmatrix} \tag{12}$$

are stable, i.e., the eigenvalues of A are prescribed. It is evident from Eq. (11) that the role of the adaptive component, ν_{ad}, is to cancel Δ.

Eqs. (6-8) imply that the model inversion error Δ is a function of the pseudo-control ν and consequently of the NN output ν_{ad}. To guarantee existence and uniqueness of a solution for ν_{ad} that produces any (unknown) Δ, it is assumed that the map $\nu_{ad} \mapsto \Delta$ is a contraction. It can be shown that this is equivalent to the following two requirements on $\hat{f}$:

$$\text{sign}\left(\frac{\partial \hat{f}}{\partial \delta}\right) = \text{sign}\left(\frac{\partial f}{\partial \delta}\right) \tag{13}$$

$$\left|\frac{\partial \hat{f}}{\partial \delta}\right| > \left|\frac{\partial f}{\partial \delta}\right| / 2 > 0 \tag{14}$$

The linearized closed loop system is driven by the output of an at least 2nd order reference model. The reference model is hedged in the presence of saturation or failure using the pseudo-control hedging methodology, presented next.

5 Pseudo-Control Hedging

PCH introduces a modification to previous work on NN based model reference adaptive flight control (Johnson et al., 2000, Johnson and Calise, 2001). It is used to address NN adaptation difficulties arising from various actuation anomalies, including actuator position and/or rate saturation, discrete (magnitude quantized) control, actuator dynamics, and partial or complete actuator failures. NN training difficulties occur when unmodeled actuator characteristics are encountered. For example, unless an adaptive process is specifically protected against it, saturations that result from failed operation quickly lead to NN wind-up. The main idea behind the PCH methodology is to limit or hedge the reference model of a MRAC architecture to prevent the adaptive element from attempting to adapt to these characteristics, when they are present, while not affecting the NN adaptation to other sources of inversion error, for which compensation is possible.

To briefly review the PCH concept, consider the case of full model inversion, in which the plant dynamics is as in Eq. (1). The pseudo-control signal defined in Eq. (4) represents the desired acceleration, while the actuator commands are given by Eq. (7). The dynamic inversion element is designed while neglecting the actuator model. Hence, this actuator command (δ_{cmd}) will not equal the actual actuator position (δ) due to its dynamics, saturation and/or failure. The pseudo-control hedge signal ν_h is defined as the difference between the commanded pseudo-control input and the actually achieved pseudo-control, which is non-zero only when the commanded actuator position is different from its actual value. To compute this difference, a measurement or an estimate of the actuator position $(\hat{\delta})$ is required. This estimate is then used to compute the pseudo-control hedge as

$$\nu_h = \hat{f}(x,\dot{x},\delta_{cmd}) - \hat{f}(x,\dot{x},\hat{\delta}) = \nu - \hat{f}(x,\dot{x},\hat{\delta}) \tag{15}$$

The PCH signal is next introduced as an additional input into the reference model, forcing it to "move back". If the reference model update *without PCH* was of the form

$$\ddot{x}_{rm} = f_{rm}(x_{rm},\dot{x}_{rm},x_c) \tag{16}$$

where x_c is the external command signal, then the reference model update *with PCH* is set to

$$\ddot{x}_{rm} = f_{rm}(x_{rm},\dot{x}_{rm},x_c) - \nu_h \tag{17}$$

The instantaneous pseudo-control output of the reference model that is used as an input to the linearized plant model is not changed by the use of PCH and remains

$$\nu_{rm} = f_{rm}(x_{rm},\dot{x}_{rm},x_c) \tag{18}$$

Hence, the effect of the PCH signal on the pseudo-control is introduced only through the reference model dynamics. PCH is a central component in the controller architecture for the MAFC application, to accommodate the possible position and rate saturation, inherent time delays, unknown dynamics and quantization of active flow control devices.

6 Neural Network for Inversion Error Compensation

In this paper, a nonlinear single hidden layer (SHL) NN is suggested to compensate for the inversion error of the complex MAFC system. The SHL NN was chosen because of its universal approximation property (Funahashi, 1989 and Hornik et al., 1989). For an input vector $\bar{x}$, which is constructed of the measured states, the reference model outputs and the pseudo-control signal, the output of the SHL NN is given by

$$\nu_{ad} = W^T \sigma(V^T \bar{x}) \tag{19}$$

where V and W are the input and output weighting matrices, respectively, and σ is a sigmoid activation function. Although ideal weighting matrices are unknown and usually cannot be computed, they can be adapted in real time using the following NN weights training rules:

$$\dot{W} = -\left[(\sigma - \sigma' V^T \bar{x})\eta + \kappa\|e\|W\right]\Gamma_W \tag{20}$$

$$\dot{V} = -\Gamma_V\left[\bar{x}\eta W^T \sigma' + \kappa\|e\|V\right] \tag{21}$$

where Γ_W and Γ_V are the positive definite learning rate matrices, σ' is the partial derivative of the sigmoids σ with respect to the NN inputs $\bar{x}$, and κ is the e-modification parameter. η is defined by for any positive definite $Q \succ 0$. A and B in the above equations are the tracking error dynamics matrices defined in Eq. (12).

7 Output Feedback Adaptive Control

The adaptive NN based approach presented above relies heavily on the full state information (full state feedback control), as is evident in the linear controller Eq. (9), the input vector $\bar{x}$ to the NN and the adaptation rules Eqs. (20-22). It also lacks robustness to unaccounted for dynamics in the process. That is, the dimension of the state space is assumed known. This assumptions are overly restrictive for the MAFC. Applications.

To overcome these restriction, output feedback solutions have been proposed, many of which incorporate NN based state estimation (Kim and Lewis, 1998, and Seshagiri and Khalil, 2000). The drawback of these techniques is the requirement for a system model. Recently, an output feedback adaptive approach has been developed, which avoids state estimation (Calise et al., 2001). Instead, a sequence of previously measured data is used to estimate the model inversion error. In addition, a modification and augmentation of the linear *dynamic* compensator lead to a NN adaptation rule that depends on output tracking error only. This output feedback approach is most attractive for the MAFC application due to the fact that a system model is not required. The only knowledge needed is the relative degree of the transmission between the control effectors and the measurement, which can be evaluated from the physical understanding of the system.

Figure 2: The MAFC Controller Ducted

8 Design Example and Simulation Results

The closed loop adaptive control architecture of Figure 1 was applied to a sample ducted rotor system developed by GTRI. This new vehicle concept is enabled by active flow control. It incorporates a single-piece propeller within a simple torroidal body. A picture of the prototype mounted on a dynamic measurement stand is shown in Figure 2.

A set of four fixed stators below the main rotor provides a nominal yaw force to counter the torque generated by the motor. Piezo-electric synthetic jet actuators embedded in these stators are used to implement active control of separation on the stators. By switching between attached and separated flow, a large change in both lift and drag can be generated. The flow control actuators on the stators can be individually controlled. By varying lift on these stators, one changes the yawing moment produced by the vanes. By varying drag differentially, one can produce pitch/roll moments. This approach mitigates the need for a counter-rotating rotor or a tail rotor and also enables the use of a simple propeller rather than a rotor system with cyclic and collective control. However, unlike conventional control approaches, the control of the forces and moments are strongly coupled since one cannot independently control lift and drag on each stator.

We illustrate below several simulated results in which parametric uncertainty has been introduced to illustrate NN adaptation. It is assumed that each actuator is capable of providing 25% of the control force needed to balance the propeller-induced torque. The propeller torque acts as a disturbance to the yaw degree of freedom of the torroid. The control objective is to maintain zero degrees of heading (yaw) angle in the presence of this disturbance. In general the disturbance is not constant, but can vary significantly since the voltage to the propeller is varied to control the vertical degree of freedom.

The action of the synthetic jets is mainly that of an on-off device, with a narrow region over which the force varies with the applied voltage. The difficulty from a control perspective lies in the fact that the force/voltage ratio within this region is nonlinear, poorly modeled, and can change significantly with operating conditions. So, mainly we know there is a significant control effect from the actuators, and that it suddenly varies from no effect to full effect, but the precise variation is only approximately known. There is also a hysteretic effect. Control is allocated to the stators based on a linear model, with a specified gain and maximum amplitude. The commands to the actuators are 'daisy chained' so that i^{th} actuator is not commanded until all the preceding i-1 actuators in the chain have been estimated to be fully actuated. This approach minimizes the effect of actuator modeling error in controlling the heading angle of the torroid. The goal is the achieve control without chattering any of the actuators between their full on and full off positions.

We first demonstrate the effect of hedging (PCH) on the adaptive process for the situation in which the actuators are modeled as behaving precisely as they are represented by the allocation scheme, except for an unaccounted constant force that the stators provide without actuation. Thus, when not saturated, there is a constant inversion error in the controller design process, plus a varying error due to the effect of the propeller disturbance. The propeller disturbance is modeled as a constant plus the sum of three sinusoids with frequencies of 1.5, 3 and 4.5 rad/s.

Figure 3 shows the result for this case. The upper portion of this figure compares the commanded and achieved heading responses. The raw heading command in this case is simply to maintain constant, zero degree heading. However, as explained in Eq. (17), this command is modified by the hedging signal whenever the actuator is saturated. As can be seen, it is nearly impossible to distinguish the difference between the actual heading and the 'hedged' commanded position in heading. The middle portion of Figure 1 shows the difference between the commanded force (normalized by the maximum force), and the achieved force. The actuation process in this case is linear except for saturation, which is rather significant over portions of the time history. The bottom portion of Figure 1 displays the fact that the NN output perfectly tracks the inversion error, which is mainly due to the disturbance in this case. Thus the network acts as a nearly perfect estimator for the disturbance (except for a constant effect), even when the actuators are saturated. Without PCH, the adaptive process diverges quickly once the actuators are saturated, causing instability in the entire process.

We next illustrate what happens when the actuator is nonlinear. In this case we have underestimated the gain of each actuator by 50%, and inserted a 20% hysteretic effect as well (the width of the hysteresis loop is 20% of the width of the linear range for voltage). Underestimating the gain translates into saying that the range of voltage over which there is a proportional effect is 50% narrower than expected. This causes a dead zone in the daisy-chain allocation, within which there is no increase in control torque for an increase in voltage. The overall nonlinear effect of the actuation process is quite complex when the dead zone is combined with the hysteretic effect. A hysteretic model was added to the PCH path (see Figure 1) as well, but the width of the hysteresis loop is also underestimated by 50% in this path.

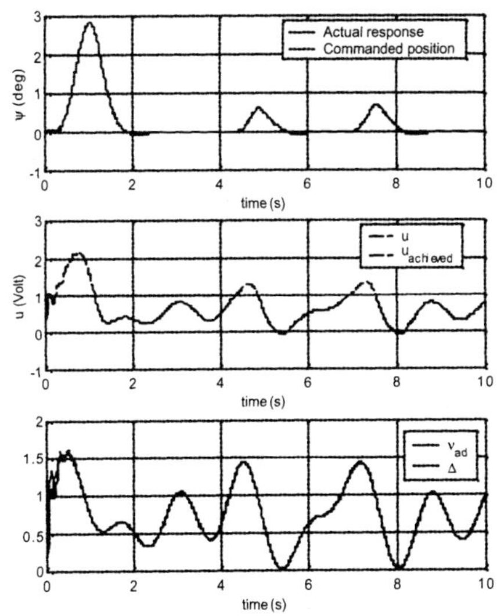

Figure 3. Responses with linear actuator model.

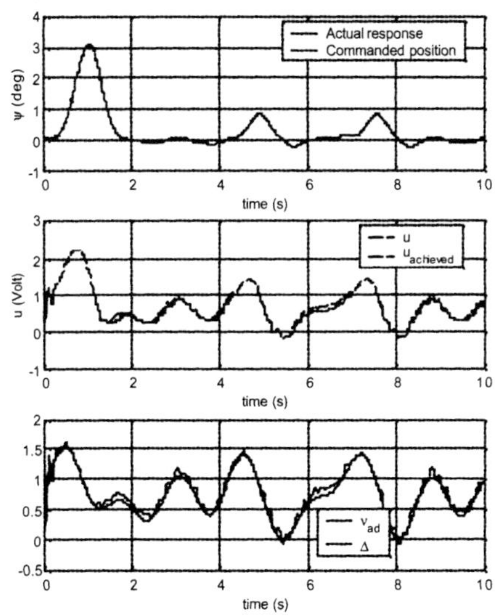

Figure 4. Responses with nonlinear actuator model.

Figure 4 shows the result for this case. From the upper portion of this figure it can be seen that hedging occurs throughout the time history, even when the actuator is not saturated. This is due to the addition of hysteresis. Note that essentially perfect tracking of the hedged command is achieved, even though both the gain and the hysteresis model used in computing the hedging signal are in error. The middle portion compares the commanded and achieved normalized control forces, which also differ throughout the time history, due to the fact that inversion is based on a linear model for the actuation process. The lower portion of Figure 4 shows that the NN output tracks the inversion error, although not quite as accurately as in Figure 3.

If the hysteresis model is not included in the PCH loop, then the heading angle is still well regulated, but there are undesirable high frequency oscillations in the control. What happens is that the adaptive process is not shielded from seeing this nonlinear effect, and attempts unsuccessfully to adapt to it.

9 Summary

This paper has presented an intelligent control system architecture for micro-adaptive flow control applications. The architecture is motivated by the difficulty in constructing a reasonably accurate physical model of the micro-adaptive flow control system. It employs several novel components which include a neural network based adaptive control process, and a novel method for protecting the adaptive system from poorly modeled actuator nonlinearities, including saturation and hyteresis. Future research will examine the output feedback approach of Calise et al. (2001) that eliminates the need for state information, and that also provides robustness to unmodelled flow dynamics.

References

Calise, A.J., Sharma, M. and Corban, J.E. (September 2000) "Adaptive Autopilot Design for Guided Munitions," *AIAA Journal of Guidance, Control, and Dynamics*, Vol. 23, No. 5.

Calise, A., Hovakimyan, N. and Idan, M. (August, 2001). "Adaptive Output Feedback Control of Nonlinear Systems using Neural Networks," *Automatica*, Vol. 37, No. 8.

Calise, A. J., Lee, S. and Sharma, M. (September 2001). "Development of Reconfigurable Flight Control Law for Tailless Fighter Aircraft", *AIAA Journal of Guidance, Control, and Dynamics*, Vol. 24, No. 5.

Funahashi, K. (1989). "On the Approximate Realization of Continuous Mappings by Neural Networks", *Neural Networks,* Vol. 2.

Hornik, K., Stinchcombe, M. and White, H. (1989). "Multilayer Feedforward Networks are Universal Approximators", *Neural Networks,* Vol. 2.

Kim, Y. and Lewis, F. (1998). "High Level Feedback Control with Neural Networks", World Scientific, NJ.

Idan, M., Johnson, M. and Calise, A. J (August 2001). "A Hierarchical Approach to Adaptive Control for Improved Flight Safety," *American Institute of Aeronautics and Astronautics Guidance, Navigation and Control Conference.*

Isodori, A. (1989). "Nonlinear Control Systems". Springer Verlag, Berlin.

Johnson, E. (2000). *Limited Authority Adaptive Flight Control.* Ph.D. Thesis, Georgia Institute of Technology.

Johnson, E. and A. Calise, A. J. (June, 2001). "Neural Network Adaptive Control of Systems with Input Saturation", *American Control Conference.*

McFarland, M. B. and Calise, A. J. (1997). "Multilayer Neural Networks and Adaptive Nonlinear Control of Agile Anti-Air Missiles". *American Institute of Aeronautics and Astronautics (AIAA-97-3540), Guidance Navigation and Control Conference,* August 1997.

Rysdyk, R., Leonhardt, B. and Calise, A. J. (August, 2000). "Development of an Intelligent Flight Propulsion Control System: Nonlinear Adaptive Control", *American Institute of Aeronautics and Astronautics Guidance Navigation and Control Conference.*

Seshagiri, S. and Khalil, H. (2000). "Output Feedback Control of Nonlinear Systems Using RBF Neural Networks", *Institute of Electrical and Electronics Engineers Transactions on Neural Networks,* Vol. 11(1).

Real-Time Optimization of Jet Mixing via Evolution Strategies

Paul Wickersham[1] and David Parekh[2*]

[1] Woodruff School of Mechanical Engineering, Georgia Institute of Technology, Atlanta GA, U. S.
[2] Georgia Tech Research Institute, Georgia Institute of Technology, Atlanta, GA, U. S.

Abstract. This paper presents initial findings in the application of evolution strategies for real-time optimization of jet mixing enhancement. The experimental apparatus consists of a small-scale jet turbine and pulsed fluidic actuators operated under computer control. An optimization algorithm determines in real time the actuator frequency selected at each iteration based on minimizing the plume centerline temperature. This approach successfully determined an optimal actuator frequency that corresponds to the global minimum for the system without any *a priori* information about the flow or actuator characteristics. Additional results are presented that address robustness to noise and changes in plant conditions and convergence characteristics.

1 Introduction

In this work we employ an optimization strategy for the control of flows based on a class of algorithms from the discipline of "soft computing." Specifically, we consider the application of evolution strategies for optimization in real-time of a physical experiment. The broad class of evolution strategies, of which genetic algorithms are a part, differ from traditional gradient search techniques in that they do not assume any knowledge of the derivatives of the function being optimized. Instead, the optimal (based on a set of preferred metrics) value of a function is obtained by a random natural selection from the available sample set of independent variables. Major contributions to the development of evolution strategies have been made by (Rechenberg, 1971), (Schwefel, 1974), and (Hoffmeister and Back, 1991). General information regarding these techniques can be found in books by (Coley, 1999) and by (Haupt and Haupt, 1998). The specific approach followed in this work is similar to that implemented by Koumoutsakos et al. (1998) in optimizing numerically modeled jet flows. The jet mixing enhancement approach follows that previously described by Parekh et al. (1998).

The objective of the search in this case is to select the set of actuator parameters that maximize the jet mixing as indicated by plume downstream temperature or velocity. In this application, the independent variables include actuator number, placement, frequency, phasing, and amplitude. The initial results presented in this paper focus on actuator frequency. From

* This work was sponsored in part by AFOSR. The idea of applying evolution strategies to jet mixing control resulted from collaborations with Drs. Petros Koumoutsakos and Jon Freund.

[1] Graduate research assistant.

[2] Director, Aerospace, Transportation, and Advanced Systems Laboratory.

iteration to iteration (or generation to generation) a new set of parameters are selected based on a random selection process, and the value of the function for each of those parameters is compared against the function values corresponding to a previous set of parameters. The parameters that provide more desirable function values survive and are the basis for selection of a new set of parameters. Algorithms for defining the population of independent variables that form each generation can be as simple as one member per generation to numerous members that mutate and propagate a new generation of members.

The specific search strategy that we employ is a simple 1 + 1 strategy in which only two function evaluations at a time are compared to determine the better parameter set. The initial evaluation of each set is referred to as the 'parent' and the successive evaluation as the 'child'. This is akin to simulated annealing except that instead of using a Boltzmann distribution as the basis for the selection of successive generations, we employ a restart process after a finite number of iterations to assure that the optimization is not trapped in a local minimum.

2 Experimental Apparatus and Approach

The laboratory apparatus consists of a small-scale, 10-cm-diameter, jet turbine with pulsed fluidic actuators at the engine nozzle exit plane for plume mixing enhancement. The actuators function much like a siren. Compressed air enters an actuator from the top and moves into a cylindrical plenum. Inside the plenum is a cylindrical drum with 12 slots, which is connected to a high power servomotor directly below it. Control of the motor speed sets the frequency of the excitation; the 'amplitude' of the excitation is set by controlling the flow rate through the actuator.

The miniature engine used in the experiments is a RAM 750 manufactured by R.A. Microjets. It has a single-stage centrifugal compressor and a single-stage axial turbine. It is capable of producing approximately 20 lbs. of thrust and high subsonic exhaust flows at temperatures over 700° C. Figure 1 shows the engine installed on the test stand with the compact actuators on opposite sides of the nozzle.

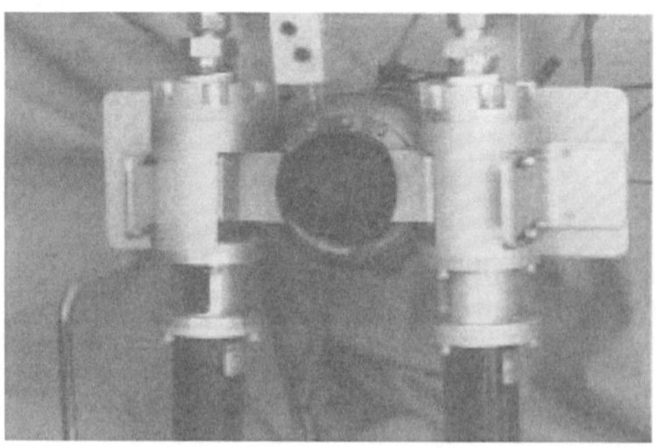

Figure 1. GTRI Laboratory configuration using the RAM 750 miniature engine for demonstration of turbulent jet mixing control. The centerline temperature probe is located 4 diameters downstream.

Initial experiments with the jet engine demonstrated that the Mach number of the exhaust did not remain constant throughout a run, but was slightly dependent on motor speed. This was most likely the result of electromagnetic interference from the actuator motors with the electronic control module that came with the jet engine. To remedy this 'problem' the electronic control unit that came with the jet engine was bypassed, and the engine was placed under computer control using a closed loop PID control algorithm to hold the Mach number steady. However, in some runs the computer control of the jet engine was purposely not implemented in order to test the ability of the evolutionary algorithm to adapt to noise. The actuators were operated under computer control with the evolution algorithm controlling in real time the pulsing frequency based on jet centerline temperature measured 4 diameters downstream of nozzle exit.

Three factors contributed to the jet characteristics differing significantly from those observed in generic jet flows from axisymmetric laboratory nozzles with top-hat exit velocity profiles. First, the presence of the centerbody results in a jet exhaust that decays much more rapidly than one would expect from simply scaling generic jet data based on the engine nozzle diameter. Second, being a hobby class engine, the engine is not designed to produce uniform non-swirling flow from the exit, and the engine speed control tolerance is only good to a few percent due to it susceptibility to EMI. Third, in these flow control experiments, only one actuator is turned on. This provides information regarding the control authority of a single actuator as well as simulates failure of the second unit. While these factors are undesirable in a canonical flow experiment, they are ideal for evaluating the general robustness of the optimization technique in the presence of realistic complexities and noise sources.

3 Characterization of Jet and Actuators

3.1 Characterization of Baseline Jet

The centerline downstream temperature profiles for three different Mach numbers on the baseline jet are shown in Figure 2. The variable T* is the nondimensional temperature, and is defined in the following way:

$$T^* = \frac{T_{Downstream} - T_{Ambient}}{T_{Exit} - T_{Ambient}}$$

The x-axis is the downstream distance of the temperature probe in jet exit diameters. This figure shows the effect of the centerbody on downstream temperature. Near the exit, T* is less than unity due to obstruction by the centerbody, and the decay in centerline temperature starts within the first two diameters in contrast to the longer potential cores typical of simple, axisymmetric jet flows. A downstream distance of 4 diameters was chosen as the location to measure temperature for all optimization runs.

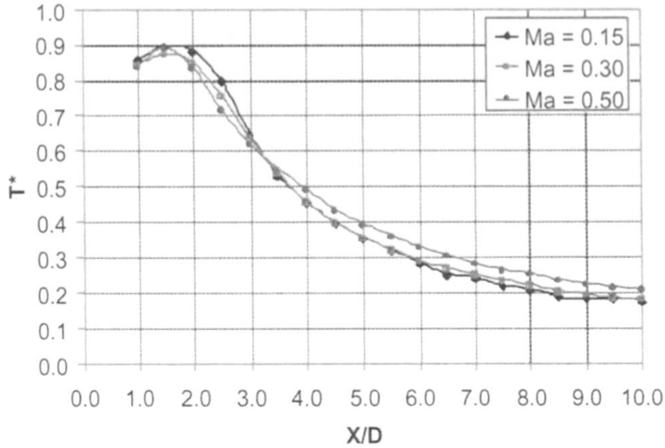

Figure 2. Centerline temperature profile of unpulsed engine exhaust for three different Mach numbers. The centerbody extends one diameter downstream.

3.2 Characterization of Pulsed Actuators

A velocity profile was obtained for each of the actuators and a representative result is shown in Figure 3a. When the actuator motor is not spinning (i.e., the pulsing frequency is zero) and air is allowed to flow through the actuator, the resulting profile is not symmetric, as seen in the figure. However, for all pulsing frequencies, the actuator velocity profile essentially takes on the nominally symmetric shape labeled "spinning" in the figure. A typical hot-wire measurement of the actuator centerline velocity is presented in Figure 3b.

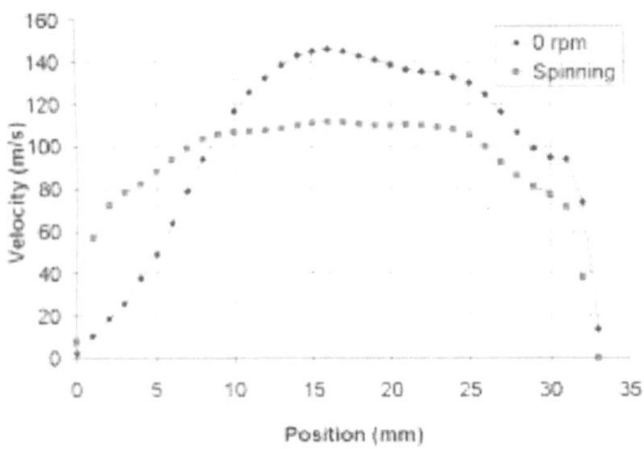

Figure 3a. Average velocity profile along the actuator slot for a stationary and spinning actuator. All angular speeds and actuator pressures resulted in similar profiles.

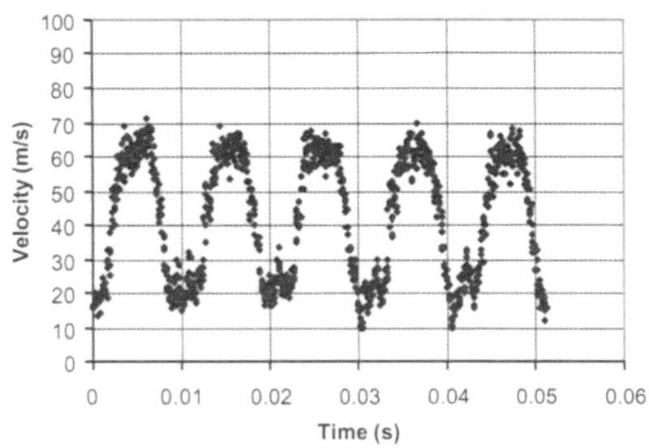

Figure 3b. Actuator centerline velocity variation over time for 100 Hz pulsing. For this experiment, the mean velocity is 42.0 m/s.

3.3 Characterization of Pulsed Jet Mixing

In Figure 4, the dependence of T* on pulsing amplitude was characterized for two different pulsing frequencies. The pulsing frequency is nondimensionalized as the Strouhal number

$$ST = \frac{PulsingFrequency \times NozzleDiameter}{JetExitVelocity}$$

where:

The pulsing amplitude is specified as a percent mass flow rate (%MFR) relative to the mass flow rate of the jet engine. Both curves show an initial increase in T* with increased pulsing amplitude followed by a steady decrease. The two curves are fit with 5th order polynomials for the range of amplitudes measured. From the figure it is clear that choice of amplitude and frequency both affect the resulting nondimensional downstream temperature T*.

The next three figures show the dependence of T* on pulsing frequency for several different Mach numbers. Figure 5a is a representative result of an 'exhaustive frequency sweep' in which the frequency is slowly increased (or decreased) across a given range which is limited only by the top speed of the actuator motor. The baseline T* indicated in the figure is the nondimensional temperature with no pulsing.

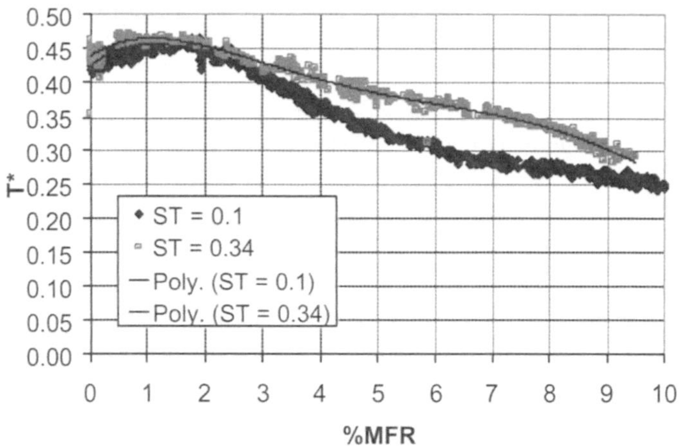

Figure 4. T* dependence on pulsing amplitude (measured in percent mass flow rate) at Mach 0.15 for two different Strouhal numbers.

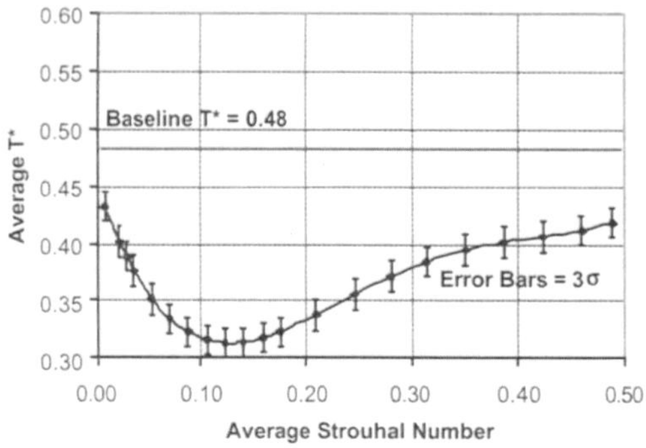

Figure 5a. T* dependence on Strouhal number at Mach 0.30 and 5% MFR. The curve is a 4th order polynomial curve fit to the average T* and Strouhal number values.

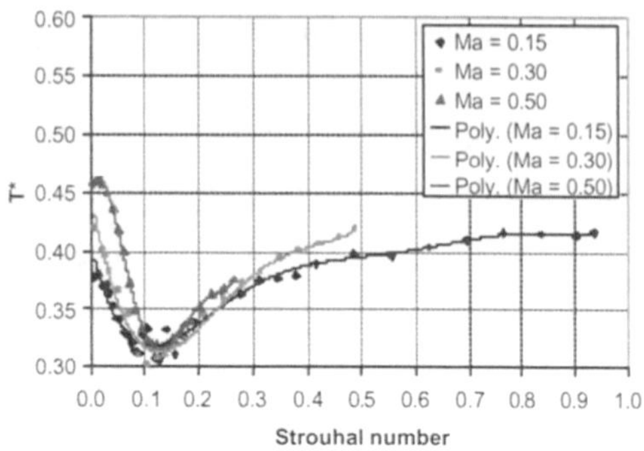

Figure 5b. Comparison of Strouhal effects for three different Mach numbers at 5% MFR.

Similar results were obtained for other Mach numbers. Figure 5b compares the results obtained for three different Mach numbers. Increasing the Mach number also has the effect of decreasing the Strouhal number domain because of the definition of the nondimensional Strouhal number.

4 Evolution Algorithm Results

A typical experimental result is shown in Figure 6. Starting with a random initial guess for excitation frequency, the algorithm determines the next guess by random selection. The random distribution governing the "rolling of the dice" is a uniform distribution of fixed width (referred to as the window width) centered about the previous guess. The function evaluation consists of measuring a short-time (about 1 second) average centerline temperature at a streamwise location 4 diameters downstream of the jet exit. Whichever frequency ('parent' or 'child') corresponds to a lower temperature is selected as the 'parent' for the next iteration or 'generation'. Every ten iterations, the window width is either increased or decreased based on the 1/5-rule noted above. There is also a 5% probability of 'mutation' which means that 5% of the rolls are not limited to the current window size, but may fall anywhere in the entire search space. The history of variations in terms of specified excitation frequency and the corresponding measured temperature for a single experiment is shown in Figure 6 as a function of iteration number. The algorithm is able within a few iterations to get very close to the optimal value. As the optimization converges toward an optimal value, the width of the random distribution is automatically scaled down as is evident from the reduced variations in the 'child' Strouhal numbers at the latter iterations. The expected convergence value of 0.12 is sometimes slightly different from the value to which the algorithm converged. This is because of: 1) the normal variability of the engine plume characteristics from run to run, and 2) measurement error which can slow convergence of the algorithm as is approaches the global minimum (see Figure 5a for order of error on a typical curve). This illustrates the naturally adaptive nature of the technique.

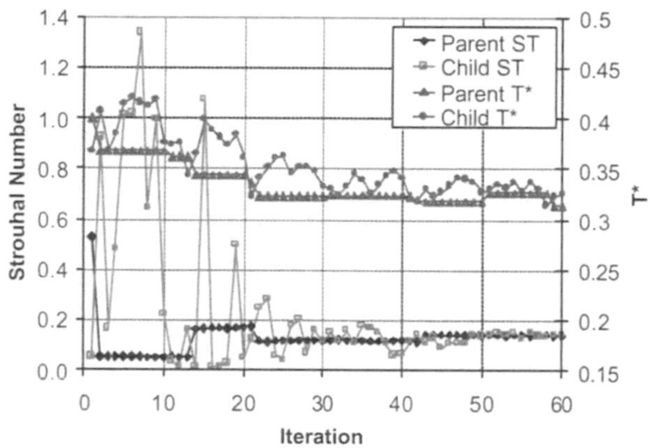

Figure 6. Example of evolutionary algorithm convergence to the global minimum. In this run, the Mach number was 0.15 and pulsing amplitude was 5% MFR. Each iteration lasted one second.

4.1 Effect of Noise

Figure 7 illustrates the robust nature of the evolutionary algorithm to noise and uncertainty. Here, the jet engine Mach number is allowed to vary with motor speed (due to electromagnetic interference) as previously mentioned. This has the additional effect of increasing the uncertainty in the measurement of the centerline temperature. Despite allowing these additional 'errors' into the system, the algorithm is still able to find the global minimum. This graph is a typical result obtained under these conditions. There was no mutation used in the algorithm at the time this was run.

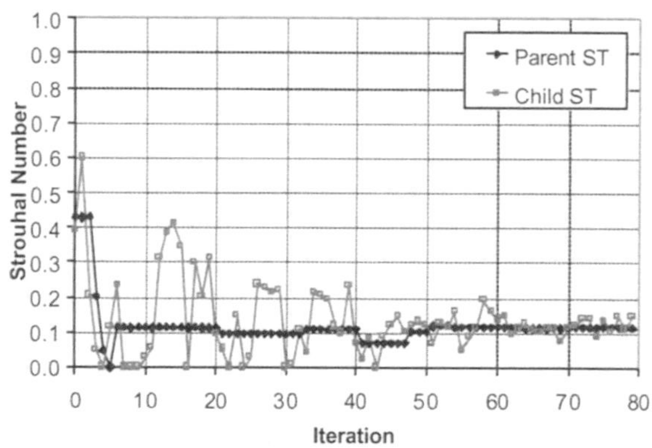

Figure 7. Example of convergence to the global minimum. In this run, the Mach number was not held constant by computer control. The Mach number is nominally 0.15 and the actuator pulsing amplitude is 5% MFR. For this run, the measurement time was increased to 2 seconds per iteration.

4.2 Convergence Characteristics

However, despite the fact that the algorithm usually converged to the global minimum, sometimes it did not. Figure 8 demonstrates one of the pitfalls that all optimization algorithms are susceptible to, convergence to a local minimum. In this case the local minimum only exists when the jet engine Mach number is allowed to vary with motor speed. Note that the 'convergence' is much less stable than the convergence to the global minimum above.

In order to gain a better understanding of how quickly and repeatedly the evolutionary algorithm converges to the global minimum, ten runs were performed at different times using the same conditions (Mach number, %MFR). Figures 9a and 9b show all of the Strouhal and T* convergence data (respectively) for the runs with the average convergence curves superimposed on top. These figures show the envelope of variations for this set of cases.

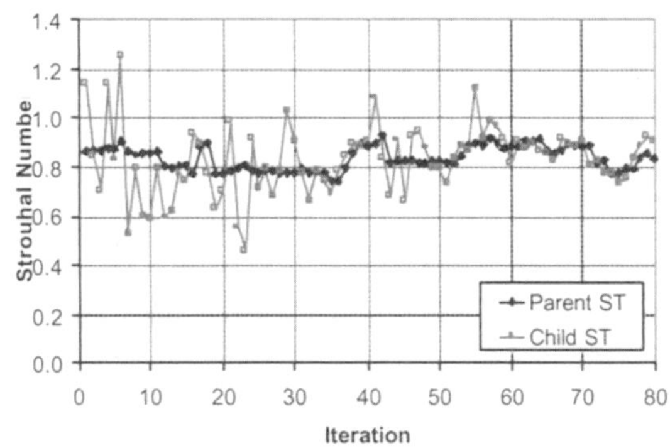

Figure 8. Example of convergence to a local minimum. In this run, the Mach number was not held constant by computer control. The Mach number is nominally 0.15 and the actuator pulsing amplitude is 5% MFR. Measurement time is 2 seconds per iteration.

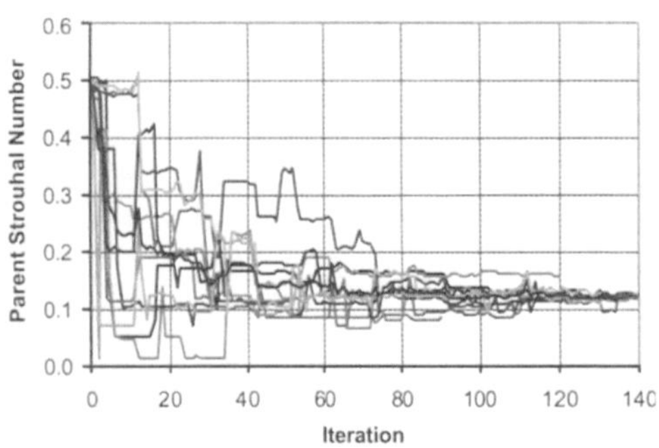

Figure 9a. Strouhal convergence data for 10 runs. The Mach number is 0.15 and pulsing amplitude is 5% MFR. Iteration time is one second per iteration. Bold curve represents average

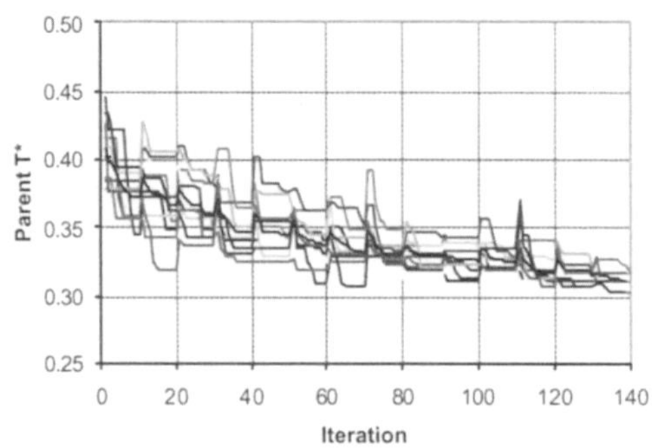

Figure 9b. T* convergence data for 10 runs. The Mach number is 0.15 and pulsing amplitude is 5% MFR. Iteration time is one second per iteration. Bold curve represents average.

Figure 10 shows how quickly the algorithm converges to the global minimum in a different way. The fraction of T*'s and Strouhal numbers within a specified range is calculated for every five iterations. For example, for the first bin (iterations 1 through 5), there are no T*'s within the specified T* range (0.3 to 0.33) and 6% of the Strouhal numbers are within the specified Strouhal range (0.1 to 0.14).

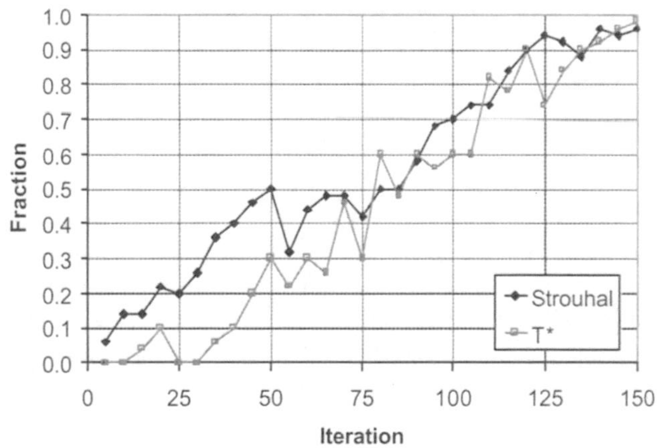

Figure 10. Fraction of parent Strouhal numbers between 0.1 and 0.14 and T*'s between 0.3 and 0.33 for ten separate runs. Iteration time is one iteration per second.

4.3 Adaptation to System Changes

Additional runs were performed to test the ability of the evolutionary algorithm to adapt to changing throttle conditions during the run (representative run presented in Figure 11). For this

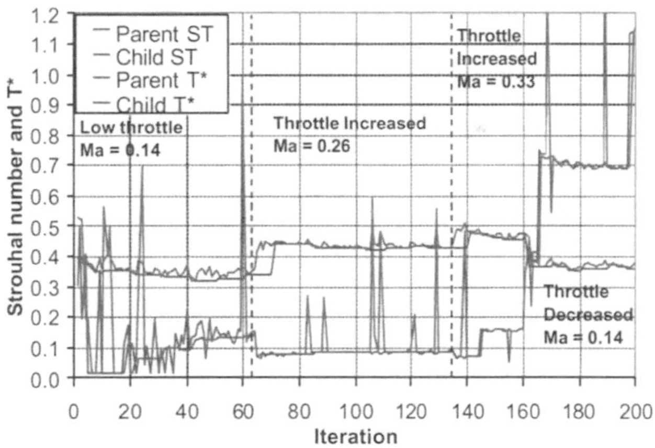

Figure 11. Convergence of the evolutionary algorithm under changing throttle conditions.

experiment , the jet engine Mach number was not held constant under computer control, allowing local minimums to be present. For the representative run shown in Figure 11, the algorithm is able to stay near the global minimum for two throttle increases, but then moves to the local minimum after the throttle is decreased. If allowed to run much longer, the algorithm would eventually find its way out of the local minimum and back to the global minimum.

5 Summary

This paper has presented an approach and successful implementation of evolution strategies for real-time optimization of jet mixing enhancement. The experimental apparatus consists of a small-scale jet tubine engine and pulsed fluidic actuators with variable frequency and amplitude control. The initial results presented here focused on optimization of a single parameter, actuator frequency, but the approach is easily generalizable to the multi-parameter case. The utility of this approach is expected to be especially powerful in the complex, multi-parameter cases.

In the results presented here, the optimization technique is consistently able to select the forcing frequency that produces the global minimum in jet centerline temperature regardless of the engine operating speed with no prior knowledge of the characteristics of the jet or of the actuators. This optimization is possible even in the presence of large measurement uncertainty and engine speed unsteadiness. Additionally, this approach is able to adjust to changes in the state of the system, for example, abrupt changes in engine operating point. The approach routinely gets close to the optimal value within 10 iterations. However, since no independent metric, such as a desired T* range, is specified, statistical convergence often takes over 50 iterations. This is the most generic case, where the algorithm objective is simply to find the global minimum in temperature.

References

Rechenberg, I. (1971). Evolutionsstrategie – Optimierung technischer Systeme nach Prinzipien der biologischen Evolution, Fromman-Holzboog.

Schwefel, H. P. (1974). Numerische Optimierung von Computer-Modellen, Birkhauser, Basel.

Hoffmeister, F., and Back, T. (1991). "Genetic algorithms and evolution strategies: Similarities and differences," Proceedings, 1st International Conference on Parallel Problem Solving from Nature, Berlin, Springer.

Coley, D. A. (1999). *An introduction to Genetic Algorithms for Scientists and Engineers*, World Scientific Publishing Co. Pte. Ltd., Singapore.

Haupt R. L., and Haupt S. E. (1998). *Practical Genetic Algorithms*, John Wiley & Sons Inc., New York.

Koumoutsakos, P., Freund, J., and Parekh, D. E. (1998). "Evolution Strategies for Jet Flow Control," the 7th Proceedings of the Center for Turbulence Research.

Parekh, D. E., Kibens, V., Glezer, A., Wiltse, J. M., and Smith, D. M. (1996). "Innovative Jet Flow Control: Mixing Enhancement Experiments," 34th Aerospace Sciences Meeting and Exhibit, American Institute of Aeronautics and Astronautics Paper 96-0308.

MIX
Papier | Fördert gute Waldnutzung
FSC® C083411

Kolibri 360 GmbH
Ferdinand-Jühlke-Straße 7
99095 Erfurt, Deutschland
produktsicherheit@kolibri360.de